ADVANCES IN MODELING AND CONTROL OF VENTILATION

ADVANCES IN EXPERIMENTAL MEDICINE AND BIOLOGY

Recent Volumes in this Series

Volume 443
ADVANCES IN LACTOFERRIN RESEARCH
Edited by Geneviève Spik, Dominique Legrand, Joël Mazurier, Annick Pierce, and Jean-Paul Perraudin

Volume 444
REPRODUCTIVE TOXICOLOGY: *In Vitro* Germ Cell Developmental Toxicology, from Science to Social and Industrial Demand
Edited by Jesús del Mazo

Volume 445
MATHEMATICAL MODELING IN EXPERIMENTAL NUTRITION
Edited by Andrew J. Clifford and Hans-Georg Müller

Volume 446
MOLECULAR AND CELLULAR MECHANISMS OF NEURONAL PLASTICITY: Basic and Clinical Implications
Edited by Yigal H. Ehrlich

Volume 447
LIPOXYGENASES AND THEIR METABOLITES: Biological Functions
Edited by Santosh Nigam and Cecil R. Pace-Asciak

Volume 448
COPPER TRANSPORT AND ITS DISORDERS: Molecular and Cellular Aspects
Edited by Arturo Leone and Julian F. B. Mercer

Volume 449
VASOPRESSIN AND OXYTOCIN: Molecular, Cellular, and Clinical Advances
Edited by Hans H. Zingg, Charles W. Bourque, and Daniel G. Bichet

Volume 450
ADVANCES IN MODELING AND CONTROL OF VENTILATION
Edited by Richard L. Hughson, David A. Cunningham, and James Duffin

Volume 451
GENE THERAPY OF CANCER
Edited by Peter Walden, Uwe Trefzer, Wolfram Sterry, and Farzin Farzaneh

Volume 452
MECHANISMS OF LYMPHOCYTE ACTIVATION AND IMMUNE REGULATION VII: Molecular Determinants of Microbial Immunity
Edited by Sudhir Gupta, Alan Sher, and Rafi Ahmed

ADVANCES IN MODELING AND CONTROL OF VENTILATION

Edited by

Richard L. Hughson

University of Waterloo
Waterloo, Ontario, Canada

David A. Cunningham

University of Western Ontario
London, Ontario, Canada

and

James Duffin

University of Toronto
Toronto, Ontario, Canada

Springer Science+Business Media, LLC

Library of Congress Cataloging-in-Publication Data

Advances in modeling and control of ventilation
p. cm. -- (Advances in experimental medicine and biology ; 450)
Includes bibliographical references and index.

1. Respiration--Regulation--Congresses. 2. Respiration--Regulation--Mathematical models--Congresses. I. Hughson, Richard L. II. Cunningham, David (David A.) III. Duffin, James. IV. Series.
QP123.A29 1998
612.2--dc21 98-40483
CIP

Proceedings of Advances in Modeling and Control of Ventilation, held September 17 – 21, 1997, in Huntsville, Ontario, Canada

ISBN 978-1-4757-9079-5 ISBN 978-1-4757-9077-1 (eBook)
DOI 10.1007/978-1-4757-9077-1

Originally published by Plenum Press, New York in 1998.
Softcover reprint of the hardcover 1st edition 1998

http://www.plenum.com

10 9 8 7 6 5 4 3 2 1

PREFACE

The seventh "Oxford Conference" on Modeling and Control of Ventilation was held in the beautiful setting of Northern Ontario at the Grandview Inn in Hunstville. This meeting was called the Canadian Conference on Modeling and Control of Ventilation (CCMCV) to follow on LCMCV held in London, England, three years ago. The beautiful view over Fairy Lake greeted everyone in the morning and provided an ideal setting for many discussions about respiratory physiology and modeling.

The Oxford Conferences began in 1971 when Dr. Richard Hercynski (a mathematical modeler with an interest in respiratory physiology) and Dr. Dan Cunningham (a respiratory physiologist with an interest in modeling) decided to organize a meeting "Modelling of a Biological Control System: The Regulation of Breathing" in Oxford, England, in 1978. The meeting was a success, and it spawned a series of meetings that have continued to today. A second conference was organized at Lake Arrowhead, California, in 1982. After this, conferences were repeated at three-year intervals. My first Oxford Conference was at the abbey in Solignac, France, in 1985. Next, we met in the cabins overlooking Grand Lake, Colorado, in 1988. In 1991, we traveled to the training institute at the base of Mt. Fuji (or at least they tell us Mt. Fuji was out there—we never saw it because of a typhoon rolling through). Our last meeting was at Royal Holloway College (University of London) where we got to dine in a castle among artwork that required guards and an electronic security system.

Sadly, we had to note at this conference the death of Dr. Dan Cunningham. His contributions to the Oxford Conferences and to respiratory physiology were noted in a dinner presentation by Peter Robbins, who was one of Dan's students.

The Oxford Conferences have attracted scientists who have an interest in developing an integrative view of respiratory control in health and disease. There are no concurrent sessions at the meeting so all scientists can participate fully in the discussions. The meetings have a truly international flavor. A large contingent from Japan, the United Kingdom, and Europe joined the North American scientists. The many graduate students had a great opportunity to interact with the senior investigators and to canoe through Algonquin Park with Professors Honda, Severinghaus, and others.

Graduate students also took part in a competition at this meeting for the best presentation in the general areas of "Control of Breathing" and "Modeling of Breathing." The winners are shown in the photos on the next page.

No conference can be a success without the help of many individuals and organizations. The primary person who deserves credit for the flawless meeting is Ms. Betty Bax of the Faculty of Applied Health Sciences at the University of Waterloo. My co-organizers

Figure 1. Dr. Richard Hughson looks on as Judith Thornton of Oxford University receives her award in the "Control of Breathing" category from conference co-organizer, Dr. Jim Duffin, for her paper "Cardiorespiratory responses to the imagination of exercise and altered perception of exercise load."

Figure 2. Ravi Mohan of the University of Toronto receives his award in the "Modeling of Breathing" category from conference co-organizer, Dr. David Cunningham, for his paper "Measurement of chemoreflex model parameters."

and co-editors Jim Duffin and David Cunningham provided great support in all areas. Dr. Martin Holroyde at Glaxo Wellcome Inc. and Dr. Bert Taylor at the University of Western Ontario kindly arranged financial support.

The next meeting will be held in the Boston area in the year 2000. We hope those who have not had the opportunity to experience the stimulating atmosphere of the Oxford Conferences will be able to join us.

Richard L. Hughson
Waterloo, Ontario
January, 1998

Ms. Thornton and Mr. Mohan were judged to have presented the best papers from a total of 19 student papers presented at the recent CCMCV meetings. Each student received a CCMCV canoe paddle and a cheque for $250. Honorable mentions for excellent presentations were given to Ms. X. Ren of Oxford University and Ms. E. Sarton of Leiden University in the Control of Breathing category, and to Mr. Z. Topor of the University of Calgary and Mr. D. Young of Harvard-MIT in the Modeling of Breathing category. Further details about the conference can be found at the conference web site http://www.ahs.uwaterloo.ca/cmcv. This site will be maintained until the next conference.

ACKNOWLEDGMENTS

Sponsorship was provided by:

- Glaxo Wellcome Inc. (Mississauga, Ontario)
- School of Kinesiology and Faculty of Health Sciences, University of Western Ontario (London, Ontario)

CONTENTS

ADVANCES IN MODELING AND CONTROL OF VENTILATION

EFFECT OF PRIOR O_2 BREATHING ON HYPOXIC HYPERCAPNIC VENTILATORY RESPONSES IN HUMANS

A. Masuda,[1] T. Kobayashi,[2] Y. Ohyabu,[3] T. Nishino,[4] S. Masuyama,[5] H. Kimura,[5] T. Kuriyama,[5] H. Tani,[6] T. Komatsu,[7] and Y. Honda[8,*]

[1]Department of Physiology and Biochemistry
School of Nursing, Chiba University
[2]Health Science Center
Tokyo University of Mercantile Marine
[3]Department of Physical Education
Kogakuin University
[4]Department of Anestheology
[5]Department of Chest Medicine
School of Medicine, Chiba University
[6]Department of Physical Therapy
International University of Health and Welfare
Ohdawara, Japan
[7]School of Allied Medical Sciences
Chiba, Japan
[8]Department of Physiology
School of Medicine, Chiba University
Chiba 260, Japan

1. INTRODUCTION

In the previous communication[11], we reported that prior O_2 breathing lasted for 10 min effectively augmented the subsequent ventilatory level in isocapnic sustained hypoxia. In addition, we found that involvement of a humoral agent, excitatory amino acid glutamate, may be responsible for inducing this phenomenon.

In this report, we further examined the effect of prior O_2 breathing on progressive hypercapnia and compared with the previous post-hyperoxic hyperventilation during sustained hypoxia.

* Address for correspondence: Y. Honda, Ohmiyadai, 4-26-17, Wakaba-ku, Chiba, 264 Japan. Tel. +81-43-265-4857; Fax +81-43-266-6865

Advances in Modeling and Control of Ventilation, edited by Hughson *et al.*
Plenum Press, New York, 1998.

2. MATERIALS AND METHODS

Sixteen and 13 young college students participated in hypoxia and hypercapnia studies, respectively. Their age, height and body weight (mean ± SD) in the former group were 21.1 ± 2.2 yr, 161.5 ± 6.8 cm, and 54.9 ± 5.3 kg and those in the latter group were 22.5 ± 6.5 yr, 160.3 ± 6.5 cm and 53.9 ± 6.3 kg, respectively. The subject breathed in a respiratory circuit via a mouth-piece and inspiratory minute volume ($\dot{V}_I$), airway $P_{ET}O_2$ and $P_{ET}CO_2$ and arterial oxygen saturation by a pulse oximeter (SpO_2) were continuously monitored. A bypass circuit containing CO_2 absorber was also included in the respiratory circuit, thus allowing to control $P_{ET}CO_2$ at the desired level. For prior O_2 breathing, the subject first inspired 100% O_2 from a reservoir bag connected to the ventilatory circuit for a few minutes then switched to O_2 rebreathing from an another reservoir bag of 10 liter capacity for 10 min. $P_{ET}O_2$ was ascertained to reach the level higher than 600 mmHg. For hypoxia test, $P_{ET}O_2$ was rapidly decreased to about 50 mmHg, then rebreathed about 10 liter of 9% O_2 gas mixture while maintaining SpO_2 at 80% with normocapnic condition for 20 min. Ventilatory response to progressive hypercapnia was examined by Read's rebreathing method[15] which was modified to conduct under normocapnic condition so as to avoid an additional hyperoxic exposure. The hypoxic and hypercapnic tests conducted with and without prior O_2 breathing were termed $+O_2$ and $-O_2$ run, respectively. In the $-O_2$ run, room air breathing via the respiratory circuit was conducted instead of O_2 for 10 min. To keep a blindfold state to be noticed the experimental setup and its maneuver by the subject, a screen was placed in front of the subject. Two $+O_2$ and $-O_2$ runs were conducted in hypoxia and hypercapnia tests in random order for each subject.

3. RESULTS

Effect of prior O_2 breathing on sustained isocapnic hypoxia is illustrated in Fig. 1.

Although the amount of ventilation was augmented in the $+O_2$ run, the specific ventilatory profile in response to this hypoxic challenge, i.e., biphasic hypoxic ventilatory decline (Biph. HVD) was unchanged. Thus, the effect of prior O_2 breathing resulted in a parallel upward shift in the ventilatory response curve. Mean ventilatory response curve to progressive hypercapnia is represented in Fig. 2. The response curve was significantly shifted upward in the $+O_2$ run, but its slope was not significantly different from $-O_2$ run. Thus, both hypoxic and hypercapnic ventilatory responses were augmented in parallel in the $+O_2$ run from that in the $-O_2$ run.

We have previously found[11] that difference between $+O_2$ and $-O_2$ run in plasma glutamine level significantly correlated with that in amount of hypoxic hyperventilation. Additionally, the relationship between plasma glutamate level and hypoxic ventilatory response is shown in Fig. 3. Although no significant correlation was detected, tendency to substantially increase correlation coefficient, like in glutamine was seen, when plotted the parameters in difference between $+O_2$ and $-O_2$ run.

Increment of hypercapnic ventilatory response in $+O_2$ from that in $-O_2$ run was plotted against the slope of hypercapnic ventilatory response (Fig. 4). Significant correlation was found.

4. DISCUSSION

The major finding in this study is that prior O_2 breathing effectively augmented the hypoxic and hypercapnic ventilatory responses as a parallel upward-shift of both response

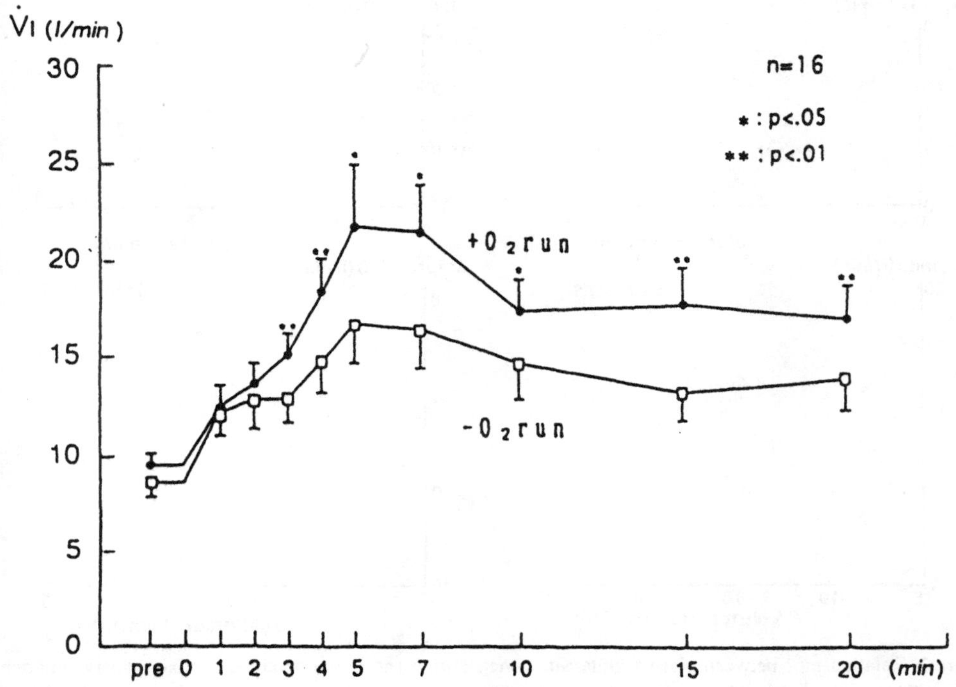

Figure 1. Mean ventilatory response to sustained isocapnic hypoxia in $+O_2$ (closed symbol) and $-O_2$ (open symbol). Ventilatory response in the $+O_2$ run was significantly shifted upward from $-O_2$ run, but specific profile of hypoxic ventilatory response, i.e., biphasic hypoxic ventilatory decline, was unchanged. The vertical bar express the magnitude of SE.

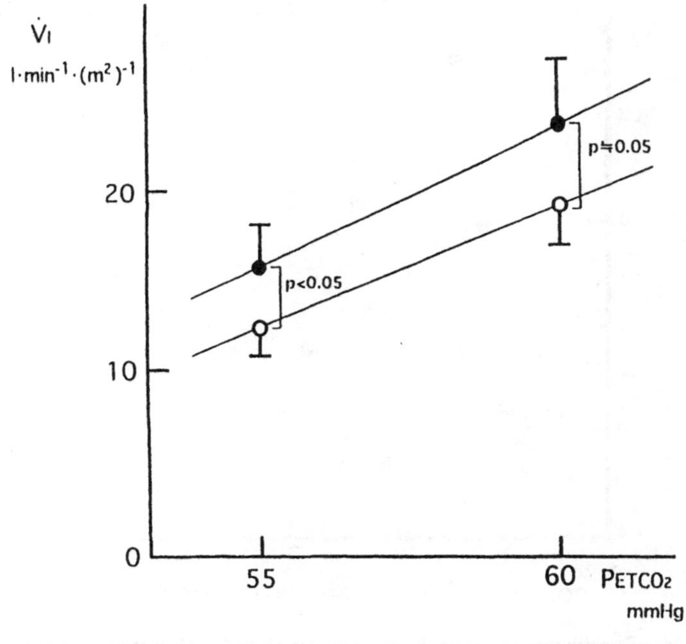

Figure 2. Mean ventilatory response to progressive hypercapnia. The response in the $+O_2$ run was significantly shifted upward in parallel with that of $-O_2$ run. The vertical bar represents the magnitude of SE.

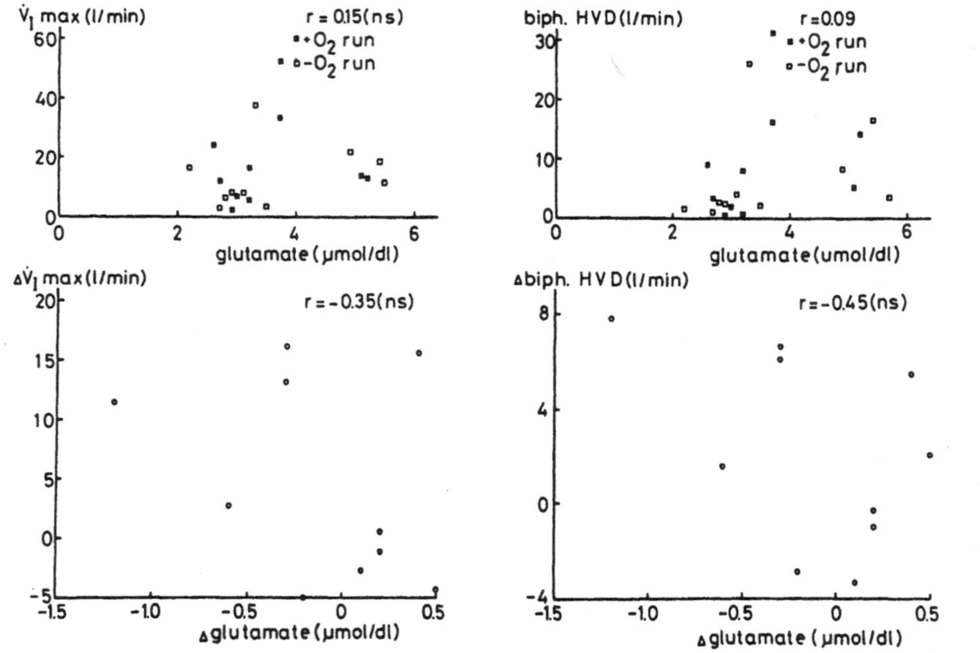

Figure 3. Relationships between plasma glutamate concentration and initial maximal hypoxic hyperventilation (two sections in the left hand side) and subsequent ventilatory decline (two sections in right hand side). When plotted their absolute magnitude, no definite trend was seen (two upper sections). However, when the difference between $+O_2$ and $-O_2$ run was plotted, the correlation coefficient improved substantially.

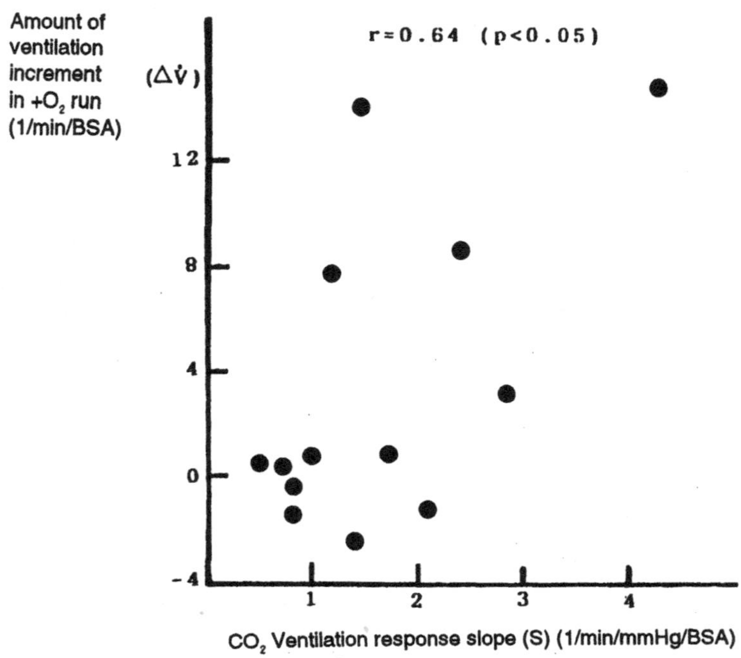

Figure 4. The amount of ventilatory increment in $+O_2$ run was plotted against the corresponding slope of CO_2-ventilation response curve. Significant correlation was found.

curves. Gradual development of hyperventilation during hyperoxia, particularly when maintained normocapnic condition, seems well established in humans[1,2,18] as well as other animal species[8,9,13,16] in awake condition. Number of studies claim that excess CO_2 release (Haldane effect) and decreased cerebral blood flow (CBF) are mainly accounted for this hyperpnea[2,6,9,16]. However, the present study demonstrated that even 30–40 min after termination of O_2 breathing both hypoxic and hypercapnic ventilations were significantly augmented. Because Haldane and CBF effects may no longer be effective in eliciting hyperpnea in this period, some other factors such as humoral agent(s) may be involved in this process. In fact, Becker et al.[1] found in awake humans inspiring 60% O_2 with normocapnic condition that ventilation still augmented 33% above the basal control level even 30 min after termination of hyperoxic exposure. Recently, this group[2] further made qauntitative analysis that Haldane and CBF effects can account for 85% of actual amount of hyperventilation during 75% O_2 breathing with normocapnia. However, remaining 15% ventilation left undetermined its origin. Accordingly, if we postulated that our finding, showing significant relationship between hypoxic hyperventilation and changes in plasma glutamine-glutamate system (Fig. 3), is also revealed in the brain-stem ventilatory control system, augmented ventilation in the $+O_2$ run can be explained. In supporting this hypothesis, Nattie and his collaborator[12,14] found long-lasting hyperpnea by stimulating the metabotrophic glutamate receptors in the retrotrapezoid nucleus (RTN) which they claimed the most CO_2 chemosensitive region in the ventral medulla. In addition, recent evidences[1,2] have demonstrated that augmented glutamate activated NO synthase from both O_2 plus arginine, thus resulting in increased guanosine 3′,5′-cyclic monophosphate (cGMP) formation and neuronal activity in the central nervous system. Since the amount of NO release is expected to be elevated by increasing its substrate, molecular O_2, during O_2 breathing[10], we suspect that this could be molecular mechanism to explain the hyperventilation observed in our experiment.

Becker et al.[2] found that the amount of increment of hyperoxic hyperventilation is significantly correlated with hypercapnic ventilatory chemosensitivity measured by Read's rebreathing method[15]. They reasoned that this finding can explain the main amount of hyperoxic hyperventilation on the basis of Haldane and CBF effect. As presented in Fig. 4, we also found significant correlation between amount of ventilatory increment in $+O_2$ run from that in $-O_2$ run and the slope of CO_2-ventilation curve in the study of CO_2 ventilatory response by the modified Read's method. Since our $+O_2$ run was conducted more than 30 min after O_2 breathing, the reason proposed by Becker et al.[2] may not be tenable to explain our finding. Perhaps, post-hyperventilation hyperpnea (PHH)[17,19] or ventilatory afterdischarge[5] could be a candidate. However, the possible mechanism to determine its magnitude by CO_2 chemosensitivity seems yet to be resolved.

Dillon and Waldrop[4] found CO_2 sensitive cells in in-vitro slice preparation obtained from caudal hypothalamus in rats. However, their ventilatory role is yet to be determined. Moreover, since Rosenstein et al.[16] demonstrated in the awake decerebrated cats that major findings by hyperoxic hyperpnea are not different from the intact ones, the role of CO_2 chemosensitivity in the hypothalamus may not be substantial in our results.

5. CONCLUSION

These consideration mentioned above led us to speculate that long-lasting augmented ventilation following hyperoxic exposure may be derived, at least in part, by glutamate receptor activation and subsequent NO release in the brain stem chemosensitive area.

ACKNOWLEDGMENTS

The authors greatly acknowledge the partial support to this work by the Research Institute for Science and Technology of Kogakuin University. We are also grateful for the kind help in discussing the data by Dr. D. Gozal.

REFERENCES

1. Becker, H., O. Polo, S. G. McNamara, M. Berthon-Jones, and C. E. Sullivan. Ventilatory response to isocapnic hyperoxia. J. Appl. Physiol. 78: 696–701, 1995.
2. Becker, H., O. Polo, S. G. McNamara, M. Berthon-Jones, and C. E. Sullivan. Effect of different levels of hyperoxia on breathing in healthy subjects. J. Appl. Physiol. 81: 1683–1690, 1996.
3. Daristotle, L., M. J. Engwall, W. Niu, and G. E. Bisgard. Ventilatory effects and interactions with change in PaO_2 in awake goats. J. Appl. Physiol. 71: 12544–1260, 1991.
4. Dillon, G. H., and T. G. Waldrop. In vitro responses of caudal hypothalamic neurons to hypoxia and hypercapnia. Neurosci. 5: 941–950, 1992.
5. Eldridge, F. L., and P. Gill-Kumar. Central neural respiratory drive and afterdischarge. Respir. Physiol. 40: 49–63, 1980.
6. Eldridge, F. L., and J. P. Kiley. Effects of hyperoxia on medullary ECF pH and respiration in chemodenervated cats. Respir. Physiol. 76: 37–49, 1987.
7. Garthwaite, J. Glutamate, nitric oxide and cell-cell signalling in the nervous system. Trends. Neurosci. 14: 60–67, 1991.
8. Gautier, H., and M. Bonora. Effects of carotid body denervation on respiratory pattern of awake cats. J. Appl. Physiol.: Respirat. Environ. Exercise Physiol. 48: 1127–1131, 1979.
9. Gautier, H., M. Bonora, and J. H. Gaudy. Ventilatory response of the conscious or anesthetized cat to oxygen breathing. Respir. Physiol. 65: 191–196, 1985.
10. Gozal, D., J. E. Torres, Y. M. Gozal, and S. M. Littwin Effect of nitric oxide synthase inhibition on cardiorespiratory responses in the conscious rats. J. Appl. Physiol. 61: 2068–2077, 1996.
11. Honda, Y., H. Tani, A. Masuda, T. Kobayashi, T. Nishino, H. Kimura, S. Masuyama, and T. Kuriyama. Effect of prior O_2 breathing on ventilatory response to sustained isocapnic hypoxia in adult humans. J. Appl. Physiol. 81: 1627–1632, 1996.
12. Li, A., and E. E. Nattie. Prolonged stimulation of respiration by brain stem metabotrophic glutamate receeptors. J. Appl. Physiol. 79: 1650–1656, 1995.
13. Miller, M. J., and S. M. Tenney. Hyperoxic hyperventilation in carotid-deafferented cats. Respir. Physiol. 23: 23–30, 1975.
14. Nattie, E. E., and A. Li. Retrotrapezoid nucleus glutamate injections: Long-term stimulation of phrenic activity. J. Appl. Physiol. 76: 760–772, 1994.
15. Read, D. J. C. Clinical method for assessing the ventilatory response to carbon dioxide. Aust. Ann. Med. 16: 20–32, 1967.
16. Rosenstein, R., L. E. McCarthy, and H. L. Borison. Slow respiratory stimulant effect of hypoxia in chemodenervated decerebrate cats. J. Appl. Physiol. 39: 767–772, 1975.
17. Swanson, G. D., D. S. Ward, and J. W. Belliville. Posthyperventilation isocapnic hyperpnea. J. Appl. Physiol. 40: 592–596, 1976.
18. Tansley, J. G., C. Clar, M. E. F. Pedersen, and P. A. Robbins. Human ventilatory response to acute hyperoxia during and after 8h of both isocapnic and poikolocapnic hypoxia. J. Appl. Physiol. 82: 513–519, 1997.
19. Tawadrous, F. O., and F. L. Eldridge. Posthyperventilation breathing patterns after active hyperventilation in man. J. Appl. Physiol. 37: 353–356, 1974.

INHIBITORY DOPAMINERGIC MECHANISMS ARE FUNCTIONAL IN PERIPHERALLY CHEMODENERVATED GOATS

Ken D. O'Halloran, Patrick L. Janssen, and Gerald E. Bisgard

Department of Comparative Biosciences
School of Veterinary Medicine
University of Wisconsin
Madison, Wisconsin 53706

1. INTRODUCTION

Dopamine (DA) is a prominent amine that is found in relatively large quantities in type I cells of the carotid body (CB) of different mammals[9]. Studies have indicated that DA in the CB acts as a neuromodulator of CB function. Much evidence indicates that DA has a modulatory role which is inhibitory to the chemosensory activity of the CB. Thus, exogenous administration of DA inhibits CB discharge[1,4,5,7,14,18,21–23] and depresses ventilation[2,5,6,10,18,22] in animals and human subjects[20], while peripheral DA receptor blockade stimulates ventilation and CB neural activity[4,10,12,23]. However, there have been a number of observations which suggest that CB inhibition may not be the only mechanism by which DA can inhibit ventilation. Ventilatory depressant effects of DA persist in hyperoxia[2,18,22] and in CB denervated (CBD) animals[1,2,6,22]. It is not clear, however, whether these effects are mediated through dopaminergic mechanisms. In the present study, we wished to: 1) examine if inhibitory dopaminergic mechanisms are functional in peripherally chemodenervated goats and 2) determine what proportion of the inhibitory ventilatory response to DA is mediated through CB mechanisms. Our data provides evidence for both CB mediated and non-CB mediated inhibitory effects of DA on respiratory motor output in anesthetized goats.

2. MATERIALS AND METHODS

2.1. Animal Preparation

A total of 12 adult goats [39 ± 3 (mean ± SE) kg body weight] of mixed breed were used in this study. After induction of anesthesia for intubation, animals were placed in

Advances in Modeling and Control of Ventilation, edited by Hughson *et al.*
Plenum Press, New York, 1998.

dorsal recumbency under a thermostatically controlled heating blanket to maintain normal body temperature (38–40°C). Anesthesia was maintained with α-chloralose given intravenously (IV). Femoral arterial and venous catheters were implanted for monitoring systemic arterial blood pressure (ABP) and for drug administration respectively. Arterial blood samples were drawn anaerobically into heparinized syringes and immediately analyzed for pH_a, Pa_{CO2} and Pa_{O2}. A bilateral mid-cervical vagotomy and superior laryngeal nerve denervation were performed to eliminate putative inputs from aortic chemoreceptors and superior laryngeal nerve paraganglia. A C6 phrenic nerve root was dissected free from surrounding tissue low in the neck, cut distally and desheathed for neural recording. The nerve was placed on bipolar platinum hook electrodes and immersed in mineral oil to prevent dessication. Nerve activity was preamplified 10,000 times, filtered (band pass 0.01–5.0 kHz) and then further amplified. The output was visualized on an oscilloscope and polygraph chart recorder fed to an audiomonitor and recorded on tape using a modified VCR for off-line analysis. The amplified signal was rectified and integrated to obtain a moving average of peak phrenic nerve activity (PNA).

2.2. Measurements

Once surgical preparations were completed, the animals were paralyzed with pancuronium bromide (2.0 mg initial dose then 1.0 $mg \cdot hr^{-1}$ IV) to prevent disruption of the preparation and artificially ventilated. Respiratory frequency was set between 15 and 20 $breaths \cdot min^{-1}$ and tidal volume between 300 and 500 ml depending on the size of the animal. FI_{O2} was adjusted to maintain Pa_{O2} above 90 Torr. It was necessary to add CO_2 to the inspired gases so that breathing was above the apneic threshold for stable phrenic recording. Arterial blood gases were sampled frequently to ensure maintenance of blood gas and acid-base homeostasis during artificial ventilation.

2.3. Protocol

CB function was confirmed with bolus injections of sodium cyanide (NaCN; 50.0 $\mu g \cdot kg^{-1}$), which elicited a brisk augmentation of PNA. Intravenous DA bolus injections (0.1–50.0 $\mu g \cdot kg^{-1}$) and slow infusions (5.0 and 50.0 $\mu g \cdot kg^{-1} \cdot min^{-1}$; 1 $ml \cdot min^{-1}$ for 5 min) were tested in randomized order. Arterial blood samples were drawn during control periods prior to bolus injections and immediately before and in the final min of DA infusions. DA bolus injections and infusions were repeated in hyperoxia (n = 7). In view of the substantial hemodynamic effects of DA in this preparation (see Results), in 5 goats, we attempted to mimic the hypotensive effect of DA administration with the β-adrenoceptor agonist isoproterenol (ISO; 0.5–5.0 $\mu g \cdot kg^{-1}$).

Animals were then bilaterally CBD (n = 10). It was necessary to further increase FI_{CO2} after CBD to maintain stable phrenic activity. CBD was confirmed by an absence of phrenic response to IV NaCN bolus injections. DA trials were repeated in CBD goats. In 2 goats, a lingual artery was cannulated and the tip of the catheter was advanced into the common carotid artery upstream from the CB for close intra-arterial delivery of DA. This allowed for the comparison of IV versus intracarotid effects of DA on respiratory motor output following CBD. Finally, in all 10 peripherally chemodenervated animals, the DA D_2-receptor antagonist domperidone (DOM) was administered (1.0 $mg \cdot kg^{-1}$ IV) and DA trials were repeated. In 2 goats, the α-adrenoceptor antagonist phentolamine was given (1.0 $mg \cdot kg^{-1}$ IV) and DA trials were repeated.

2.4. Data and Statistical Analysis

Phrenic burst frequency and peak integrated phrenic amplitude, systolic and diastolic blood pressures and heart rate (HR) were averaged over the last 10 seconds (sec) of the control period (30 sec for 50.0 $\mu g \cdot kg^{-1}$ DA) and for 10 (or 30) sec immediately following drug administration of bolus IV injections (allowing time for circulatory delay). For DA infusions, respiratory and cardiovascular variables were averaged over 1 min immediately prior to the beginning of the infusions and for the last min of the infusions. PNA (frequency x peak amplitude) was taken as an index of ventilation. All results are expressed as mean ± SE. To facilitate comparisons between animals and between treatments respiratory data are expressed in relative terms, i.e., as percent change. Statistical significance was evaluated using Student's paired t tests with P values less than 0.05 taken as significant.

3. RESULTS

3.1. CB Intact Group

3.1.1. Dopamine Bolus Injections. Intravenous DA bolus injections (0.1–50.0 $\mu g \cdot kg^{-1}$) were examined in 10 goats with intact carotid sinus nerves (CSN) under normoxic conditions. DA administration caused dose-dependent inhibition of PNA in all animals. Typically, DA injections caused a slowing of respiratory frequency and a decrease in peak phrenic amplitude. The magnitude and duration of the DA-induced ventilatory inhibition varied from goat to goat, but ventilatory responses were dose-related in all animals often leading to phrenic apnea, particularly at the higher dose of 50.0 $\mu g \cdot kg^{-1}$. Results were highly reproducible in each experiment. Examples of the inhibitory effect of DA bolus injections on PNA are illustrated in Fig. 1 and summarized for all animals in Fig. 2A.

In addition to ventilatory inhibition, there were significant changes in ABP following DA administration. Typically, at low doses (0.1–5.0 $\mu g \cdot kg^{-1}$) and up to 10 $\mu g \cdot kg^{-1}$, DA had a significant hypotensive effect (Fig. 1A) and caused a small, but significant, increase in HR. At the higher dose of 50.0 $\mu g \cdot kg^{-1}$, DA administration either reduced ABP similar to low dose DA administration but greater in magnitude, or it caused a biphasic pressor response consisting of an initial increase in ABP which was then followed by a slow gradual decrease in ABP below control values. These changes were accompanied by a moderate tachycardia.The ventilatory and pressor responses to DA were unrelated both in time-course and magnitude. Ventilatory depression always preceded noticeable changes in ABP.

DA bolus injections were repeated in 7 goats with intact CSN during hyperoxia. Ventilatory responses to IV bolus injections of NaCN (50.0 $\mu g \cdot kg^{-1}$ IV) were approximately halved under hyperoxic conditions in these experiments. In hyperoxia, DA bolus injections caused typical dose-dependent inhibition of PNA which was not significantly different from DA responses in normoxia in the same goats. Pressor and HR responses to DA were similar during hyperoxia and normoxia.

3.1.2. Dopamine Infusions. In 9 goats, slow IV infusions of 5.0 and 50.0 $\mu g \cdot kg^{-1} \cdot min^{-1}$ DA were examined during normoxic conditions. DA infusions caused dose-dependent progressive inhibition of PNA in all animals (Fig. 2B), leading to phrenic apnea particularly at the higher dose of 50 $\mu g \cdot kg^{-1} \cdot min^{-1}$ DA (7 out of 9 goats). The inhibitory effects of DA

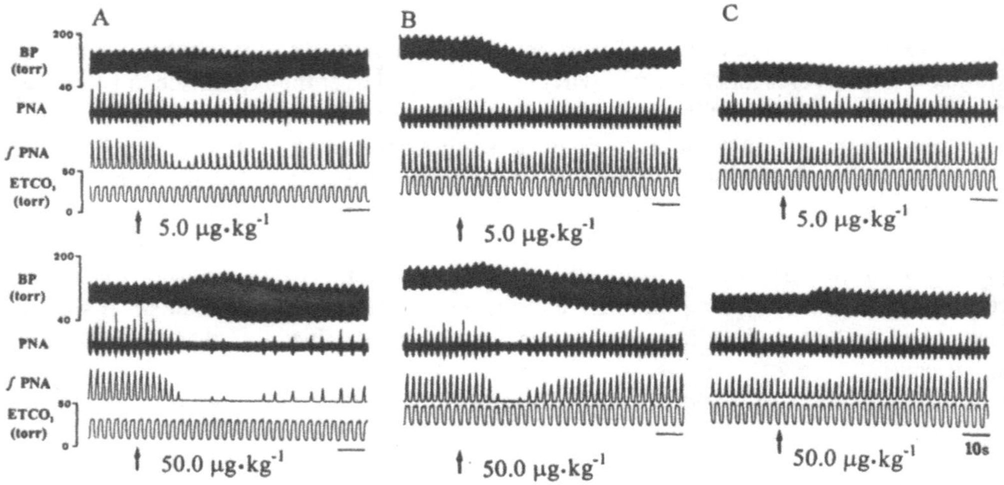

Figure 1. Ventilatory and cardiovascular effects of dopamine bolus injections in one representative goat. BP, systemic arterial blood pressure; PNA, raw phrenic nerve activity; ∫ PNA, integrated phrenic nerve activity; $ETCO_2$, end-tidal CO_2. A: Dopamine (injected IV at arrows) causes dose-dependent inhibition of phrenic nerve activity (PNA) with carotid sinus nerves intact. B: The ventilatory depressant effects of dopamine persist, but are significantly attenuated, following bilateral carotid body denervation. C: Dopamine D_2-receptor blockade with domperidone abolishes the inhibitory ventilatory effect of dopamine that persists following carotid body denervation.

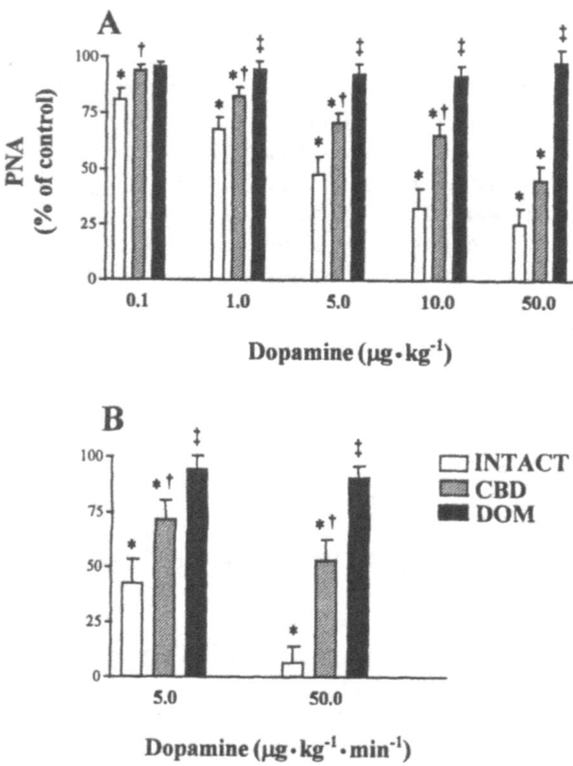

Figure 2. Effect of dopamine bolus injections (0.1–50.0 µg·kg^{-1}; A) and infusions (5.0 and 50.0 µg·kg^{-1}·min^{-1}; B) on phrenic nerve activity (PNA), expressed as a percent of pre-trial control values in 10 goats with intact carotid bodies (INTACT), following bilateral carotid body denervation (CBD) and following domperidone administration (DOM) in CBD animals. Values are means ± SE. *Significantly different ($P < 0.05$) from control PNA. †Significantly different ($P < 0.05$) from INTACT. ‡Significantly different ($P < 0.05$) from CBD.

on PNA were accompanied by substantial blood pressure changes. Typically, 5.0 $\mu g \cdot kg^{-1} \cdot min^{-1}$ DA infusion caused a slow fall in ABP which reached its nadir during the final min of the 5 min infusion period. In contrast, the 50.0 $\mu g \cdot kg^{-1} \cdot min^{-1}$ DA dose caused a biphasic pressor response consisting of an initial decrease, followed by a progressive increase in ABP. DA infusions were accompanied by a significant tachycardia. In 7 goats, DA infusions were repeated during hyperoxic conditions. Ventilatory and cardiovascular responses to DA infusions were not significantly different in hyperoxia compared to DA trials in the same animals during normoxia.

3.1.3. Isoproterenol. The effect of IV ISO bolus injections (0.5–5.0 $\mu g \cdot kg^{-1}$) were examined in 5 goats which showed typical ventilatory and cardiovascular responses to DA administration. Under control conditions, with CSN intact, low dose ISO injections (0.5–2.0 $\mu g \cdot kg^{-1}$) typically caused a decrease in ABP similar to low dose DA administration. Higher ISO doses (up to 5.0 $\mu g \cdot kg^{-1}$) caused increases in systolic pressure and decreases in diastolic pressure similar to high dose DA trials. ISO trials usually resulted in a moderate, but significant, tachycardia. In all 5 goats, however, ISO had no effect on PNA.

3.1.4. Phentolamine. In 2 CB intact goats, DA bolus injections (0.1–50.0 $\mu g \cdot kg^{-1}$) and infusions (5.0 and 50.0 $\mu g \cdot kg^{-1} \cdot min^{-1}$) were repeated following α-adrenoceptor blockade with phentolamine (1.0 $mg \cdot kg^{-1}$). α-adrenoceptor blockade was confirmed by a blunting of ventilatory and pressor responses to 1.0 $\mu g \cdot kg^{-1}$ NE administration following phentolamine treatment. Phentolamine administration did not affect inhibitory ventilatory responses to DA trials, but it caused a significant decrease in ABP. Cardiovascular responses to DA bolus injections and infusions were similar before and after phentolamine administration, except that the increase in systolic blood pressure in response to high dose DA administration before phentolamine treatment was attenuated following α-adrenoceptor blockade.

3.2. CB Denervated Group

3.2.1. Dopamine Bolus Injections. IV DA bolus injections (0.1–50.0 $\mu g \cdot kg^{-1}$) were repeated in 10 goats following bilateral CBD. Compared to DA trials in the same 10 goats before CBD, inhibitory ventilatory responses to DA were significantly attenuated in peripherally chemodenervated goats for all doses of DA with the exception of the high dose 50.0 $\mu g \cdot kg^{-1}$ DA (Fig. 2A). The durations of the inhibitory responses were also significantly attenuated after CBD. In all CBD animals, however, dose-dependent inhibition of PNA persisted in response to DA administration (Figs. 1B and 2A). Similar to CB intact trials, the magnitude of the ventilatory depression varied from goat to goat. Inhibitory responses were manifest by changes in respiratory frequency and peak phrenic amplitude. Control ABP and HR were significantly increased in all animals following bilateral CBD. However, the direction and magnitude of ABP and HR responses to DA trials were not significantly different following CBD compared to DA trials in the same animals before CBD.

3.2.2. Dopamine Infusions. Slow IV DA infusions (5.0 and 50.0 $\mu g \cdot kg^{-1} \cdot min^{-1}$) were repeated in 9 goats following CBD. Again, DA infusions caused dose-dependent inhibition of PNA. However, ventilatory responses to DA infusions were significantly attenuated after CBD (Fig. 2B). Cardiovascular responses to DA infusions were similar in time-course and magnitude to DA trials before CBD.

3.3. Domperidone Group

3.3.1. Dopamine Bolus Injections. In 10 CBD goats, IV DA bolus injections (0.1–50.0 $\mu g \cdot kg^{-1}$) were repeated, in normoxia, following pretreatment with DOM (1.0 $mg \cdot kg^{-1}$ IV). DOM administration significantly increased PNA (+56.6 ± 14.6%, n = 10) due to an increase in peak phrenic amplitude. In 2 goats, DOM had no effect on PNA. Following peripheral DA D_2-receptor blockade with DOM, the inhibitory effects of DA bolus injections on PNA, which persisted following CBD, were substantially attenuated. This effect is shown in Fig. 1C for one goat and group data are shown for all 10 goats in Fig. 2A. DOM administration significantly reduced ABP and HR compared to pre-DOM values in CBD animals. Following DOM treatment, pressor and HR responses to DA administration were significantly attenuated.

3.3.2. Dopamine Infusions. Pretreatment with DOM substantially attenuated the ventilatory depressant effects of slow IV infusions of DA (5.0 and 50.0 $\mu g \cdot kg^{-1} \cdot min^{-1}$) in 9 CBD goats (Fig. 2B). Cardiovascular responses to 5.0 $\mu g \cdot kg^{-1} \cdot min^{-1}$ DA infusions were attenuated but qualitatively similar to DA trials performed in the same animals before DOM treatment. The 50.0 $\mu g \cdot kg^{-1} \cdot min^{-1}$ DA infusion typically caused increases in systolic and diastolic pressures following DA D_2-receptor blockade. These changes were accompanied by a significant increase in HR.

4. DISCUSSION

The main findings of the present study are: 1) Exogenously administered DA elicits both CB mediated and non-CB mediated inhibitory effects on respiratory motor output in anesthetized goats. 2) The ventilatory depressant effects of DA that persist in peripherally chemodenervated goats are mediated by peripheral DA D_2-receptors, since pretreatment with the selective peripheral DA D_2-receptor antagonist DOM substantially attenuates the inhibitory effects of DA on PNA. The observation of ventilatory depressant effects of DA in the present study is consistent with previous reports from our laboratory[2,10,12] and with observations in other animal species[5,6,18,22] and human subjects[20], although an excitation of breathing has been observed following DA administration in the dog[5] and with high dose DA infusion in humans[20]. Our data further support other studies in anesthetized[1,4,6,7,14,18,22,23] and awake[2,12] animals and *in vitro* studies[21] demonstrating that DA inhibits the CB. In addition, our data clearly demonstrate a non-CB mediated inhibitory effect of DA on ventilation. In peripherally chemodenervated animals, exogenous DA administration caused dose-dependent inhibition of PNA when delivered either as IV bolus injections or slow IV infusions. Other studies have demonstrated ventilatory inhibition in CBD animals in response to DA administration[2,6,18,22] but the results of these studies are mixed and the mechanism(s) of the DA-induced hypoventilation has (have) not been elucidated.

In the present study, with CSN intact, ventilatory responses to DA bolus injections and infusions were similar during normoxia and hyperoxia. We have no information on CB activity during hyperoxia in these goats, though it appeared that CB sensitivity (at least to NaCN) was reduced during hyperoxic conditions. The persistent inhibitory effect of DA on ventilation in hyperoxia (presumably when CB activity is suppressed) is further suggestive of a non-CB site of action of DA.

In addition, our data clearly demonstrate that the ventilatory depressant effects of DA that persist in peripherally chemodenervated goats are mediated by peripheral DA D_2-

receptors. DOM administration substantially attenuated the inhibitory effects of DA on PNA in CBD animals. It is well established that DOM effectively blocks peripheral DA D_2-receptors[11,15] which have been shown to mediate CB inhibition[12,15,23]. We have previously demonstrated that DOM blocks DA-induced chemosensory inhibition[4] and DA-induced ventilatory depression[3,10,12] in the goat. Thus, the present observation of DA D_2-receptor mediated ventilatory inhibition is consistent with these findings, except that in this study DA-induced respiratory depression was mediated by inhibitory DA D_2-receptors located outside of the CB. Consistent with our finding of DA D_2-receptor mediated ventilatory depression in response to DA administration in CBD animals, Bisgard et al.[2] reported that the DA antagonist haloperidol abolishes the ventilatory inhibitory effect of DA that persists following CB excision in awake goats. Haloperidol, a butyrophenone, is effective in diminishing the effects of exogenous DA on the CB[1,7,14]. However, antagonists such as haloperidol cross the blood-brain barrier and thus have access to CNS dopaminergic receptors involved in respiratory control which may exert complex effects on respiration[16,17]. The use of DOM in the present study allowed the investigation of the effects of peripheral DA blockade without effects on CNS DA receptors since DOM does not readily cross the blood-brain barrier[11,13].

DOM administration significantly increased PNA in 8 out of 10 peripherally chemodenervated goats. In awake goats, DOM produces hyperventilation during normoxia[3,10,12] consistent with reports of a sustained increase in carotid chemoreceptor afferent activity following DOM administration in anesthetized cats[23] and goats[4] and in agreement with studies demonstrating increased carotid chemoreceptor activity following DA receptor blockade with haloperidol[1,7,14] or other DA antagonists *in vivo*[14] and *in vitro*[21]. These studies support the hypothesis that there is tonic dopaminergic inhibition of the CB during normoxia and that endogenous DA may play an important role in modulating CB function. The present finding of enhanced PNA in CBD animals following DOM administration suggests the removal of tonic inhibition from endogenous DA at some other peripheral site.

In limited trials, phentolamine was ineffective in blocking the inhibitory ventilatory effects of DA administration. This suggests that the DA-induced ventilatory depression observed in the present study was not α-adrenoceptor mediated and this is consistent with previous reports. The ventilatory depressant effects of DA in this study were accompanied by substantial changes in ABP. However, peak ventilatory and pressor responses to DA were poorly correlated and ventilatory depression invariably preceded cardiovascular changes. Furthermore, ventilatory inhibition was always observed with high dose DA administration despite a variety of pressor responses. In addition, the depressant effect of DA on ventilation was attenuated following CBD, yet pressor responses were quantitatively and qualitatively similar to control trials suggesting that the DA-induced ventilatory depression was not related to the hemodynamic effects of DA administration. Moreover, for most doses, DA administration had a significant hypotensive effect which would not be expected to produce ventilatory depression. Nevertheless, since the DA-induced phrenic response was somewhat coincident in time-course with the pressor response, IV bolus injections of ISO were given in an attempt to mimic the hemodynamic effects associated with DA administration in this preparation. β-adrenoceptor stimulation with ISO produced similar pressor responses to DA administration, but had no effect on PNA suggesting that the inhibitory effects of DA on ventilation were independent of its cardiovascular effects.

An interesting observation in the present study was that in peripherally chemodenervated animals intracarotid administration of DA produced greater inhibitory effects on breathing than comparable doses given IV. This is suggestive of an intracranial site of action of DA that is readily accessible to the arterial blood supply in this region. The presence of dopamine receptors in the CNS is well documented and studies have indicated that DA influences central respiratory regulation[16,17]. Central DA receptor stimulation with apomorphine has been shown to inhibit ventilation and DA is reported to inhibit bulbar inspiratory neurons. However, central DA blockade depresses ventilation in the goat[2] and dog[16] suggesting that there may be central excitatory effects of DA on ventilation[16,17]. A central mechanism is unlikely to explain the inhibitory effects of DA on ventilation in the present study, however, since DA and DOM do not readily cross the blood-brain barrier, suggesting a site of action that is accessible to the peripheral circulation. A candidate site may be the area postrema, a mid-line circumventricular organ that lies outside of the blood-brain barrier. For many years, the area postrema has been recognized as a chemosensitive trigger zone in the emetic reflex. However, respiratory responses to electrical and chemical stimulation of the area postrema have been reported in cats and rabbits. Gatti et al.[8] observed dose-dependent ventilatory depression in response to local application of excitatory amino acids to the area postrema. Furthermore, the area postrema has reciprocal connections with the NTS[19] and other brain regions involved in respiratory control. Interestingly, DA D_2-receptors have been identified in the area postrema and DA agonists have been shown to act at this site to produce physiological responses.

In summary, the present data clearly demonstrate that exogenous DA administration inhibits ventilation in anesthetized goats through CB and non-CB mechanisms. The ventilatory depressant effects of DA that persist in peripherally chemodenervated animals are DA D_2-receptor mediated since the inhibitory effects of DA administration were substantially attenuated following DOM treatment. The inhibitory effects of DA on respiratory motor output appear to be independent of the hemodynamic effects associated with DA administration.

ACKNOWLEDGMENTS

We wish to thank Mr. Gordon Johnson and Mr. Josue Pizarro for their excellent technical assistance. This work was supported by National Heart, Lung and Blood Institute Grants HL-53969 and HL-07654.

REFERENCES

1. Bisgard, G.E., R.A. Mitchell, and D.A. Herbert. Effects of dopamine, norepinephrine and 5-hydroxytryptamine on the carotid body of the dog. *Respir. Physiol.* 37: 61–80, 1979.
2. Bisgard, G.E., H.V. Forster, J.P. Klein, M. Manohar, and V.A. Bullard. Depression of ventilation by dopamine in goats-effects of carotid body excision. *Respir. Physiol.* 41: 379–392, 1980.
3. Bisgard, G.E., N.A. Kressin, A.M. Nielsen, L. Daristotle, C.A. Smith, and H.V. Forster. Dopamine blockade alters ventilatory acclimatization to hypoxia in goats. *Respir. Physiol.* 69: 245–255, 1987.
4. Bisgard, G.E., P.L. Janssen, and M.R. Dwinell. Excitatory and inhibitory effects of dopamine in the carotid body of goats (Abstract). *Am. J. Respir. Crit. Care Med.* 155, A297, 1997.
5. Black, A.M.S., J.H. Comroe, Jr., L. Jacobs. Species difference in carotid body response of cat and dog to dopamine and serotonin. *Am. J. Physiol.* 223: 1097–1102, 1972.
6. Cardenas, H., and P. Zapata. Dopamine-induced ventilatory depression in the rat, mediated by carotid nerve afferents. *Neurosci. Lett.* 24: 29–33, 1981.
7. Folgering, H., J. Ponte, and T. Sadig. Adrenergic mechanisms and chemoreception in the carotid body of the cat and rabbit. *J. Physiol. London* 325: 1–21, 1982.

8. Gatti, P.J., J.D. Souza, A.M.T. Da Silva, J.A. Quest, and R.A. Gillis. Chemical stimulation of the area postrema induces cardiorespiratory changes in the cat. *Brain Res.* 346: 115–123, 1985.
9. Gonzalez, C., L. Almaraz, A. Obeso, and R. Rigual. Carotid body chemoreceptors: From natural stimuli to sensory discharges. *Physiol. Rev.* 74: 829–898, 1994.
10. Janssen, P.L., K.D. O'Halloran, J. Pizarro, M.R. Dwinell, and G.E. Bisgard. Carotid body dopaminergic mechanisms are functional after acclimatization to hypoxia in goats. *Respir. Physiol.*, in press.
11. Kohli, J.D., D. Glock, and L.I. Goldberg. Selective DA 2 versus DA 1 antagonist activity of domperidone in the periphery. *Eur. J. Pharmacol.* 89: 137–141, 1983.
12. Kressin, N.A., A.M. Nielsen, R. Laravuso, and G.E. Bisgard. Domperidone-induced potentiation of ventilatory responses in awake goats. *Respir. Physiol.* 65: 169–180, 1986.
13. Laduron, P.M., and J.E. Leysen. Domperidone, a specific in vitro dopamine antagonist, devoid of in vivo central dopaminergic activity. *Biochem. Pharmacol.* 28: 2161–2165, 1979.
14. Llados, F., and P. Zapata. Effects of dopamine analogues and antagonists on carotid body chemosensors in situ. *J. Physiol. London* 274: 487–499, 1978.
15. Mir, A.K., D.S. McQueen, D.J. Pallot, and S.R. Nahorski. Direct biochemical and neuropharmacological identification of dopamine D2 receptors in the rabbit carotid body. *Brain Res.* 291: 273–283, 1984.
16. Nielsen, A.M., and G.E. Bisgard. Dopaminergic modulation of respiratory timing mechanisms in carotid-body denervated dogs. *Respir Physiol.* 53: 71–86, 1983.
17. Nielsen, A.M., and G.E. Bisgard. Differential effects on phrenic output of two dopamine agonists, apomorphine and bromocriptine. *Eur. J. Pharmacol.* 106: 69–78, 1984.
18. Nishino, T., and S. Lahiri. Effects of dopamine on chemoreflexes in breathing. *J. Appl. Physiol.* 50: 892–897, 1981.
19. van der Kooy, D., and L.Y. Koda. Organization of the projections of a circumventricular organ: the area postrema in the rat. *J. Comp. Neurol.* 219: 328–338, 1983.
20. Welsh, M.J., D.D. Heistad, and F.M. Abboud. Depression of ventilation by dopamine in man. *J. Clin. Invest.* 61: 708–713, 1978.
21. Zapata, P. Effects of dopamine on carotid chemo- and baroreceptors in vitro. *J. Physiol. London* 244: 235–251, 1975.
22. Zapata, P., and A. Zuazo. Respiratory effects of dopamine-induced inhibition of chemosensory inflow. *Respir. Physiol.* 40: 79–92, 1980.
23. Zapata, P., and F. Torrealba. Blockade of dopamine-induced chemosensory inhibition by domperidone. *Neurosci. Lett.* 51: 359–364, 1984.

EFFECT OF 8 HOURS OF ISOCAPNIC/POIKILOCAPNIC HYPOXIA ON THE VENTILATORY RESPONSE TO CO_2

Marzieh Fatemian and Peter A. Robbins

University Laboratory of Physiology
University of Oxford
Parks Road, Oxford OX1 3PT, United Kingdom

1. INTRODUCTION

The ventilatory sensitivity to CO_2 is thought to increase following a sustained exposure to hypoxia (1–5). This study was conducted to determine whether 8 h of poikilocapnic hypoxia could elicit a change in the hyperoxic ventilatory response to CO_2, and whether there were any differences between poikilocapnic and isocapnic exposures to hypoxia.

2. METHODS

Ten healthy young volunteers were studied using a purpose-built chamber where the composition of the inspired gas could be altered to maintain the end-tidal gases at their desired levels.

Three protocols were used for each subject: 1) Isocapnic hypoxia (IH): $P_{ET_{O2}}$ was held at 55 Torr and $P_{ET_{CO2}}$ at the subject's normal value, as measured before the experiment. 2) Poikilocapnic hypoxia (PH): $P_{ET_{O2}}$ was held at 55 Torr and $P_{ET_{CO2}}$ was not controlled. 3) Control (C): air breathing, neither $P_{ET_{O2}}$ nor $P_{ET_{CO2}}$ was controlled.

The ventilatory response to CO_2 was measured before (am) and after (pm) the chamber exposure by measuring the ventilation at four levels of $P_{ET_{CO2}}$. For these measurements a dynamic end-tidal forcing system was used to maintain the subject's $P_{ET_{CO2}}$ at 1.5, 4.5, 7.5 and 10.5 Torr above the normal resting value, measured at the start of the experiment. The duration of each CO_2 level was 5 minutes. $P_{ET_{O2}}$ was held at 200 Torr throughout.

Advances in Modeling and Control of Ventilation, edited by Hughson *et al.*
Plenum Press, New York, 1998.

3. RESULTS

Values for ventilation and PET_{CO2} were averaged over the last minute of each level of PET_{CO2}. The CO_2 response lines ($\dot{V}E$ vs. PET_{CO2}) before and after each protocol were obtained by linear regression and the average responses for all 10 subjects are shown in Fig. 1.

The slopes and the intercepts (PET_{CO2} at ($\dot{V}E = 0$) of the response lines were compared and tested for statistically significant changes after the hypoxic exposures. Analysis of variance on the ratio of the slopes (pm/am), with protocol as a fixed factor and subjects as a random factor revealed significant differences with respect to protocols ($p < 0.05$). Further partitioning of the variance showed that the hypoxic exposures were significantly

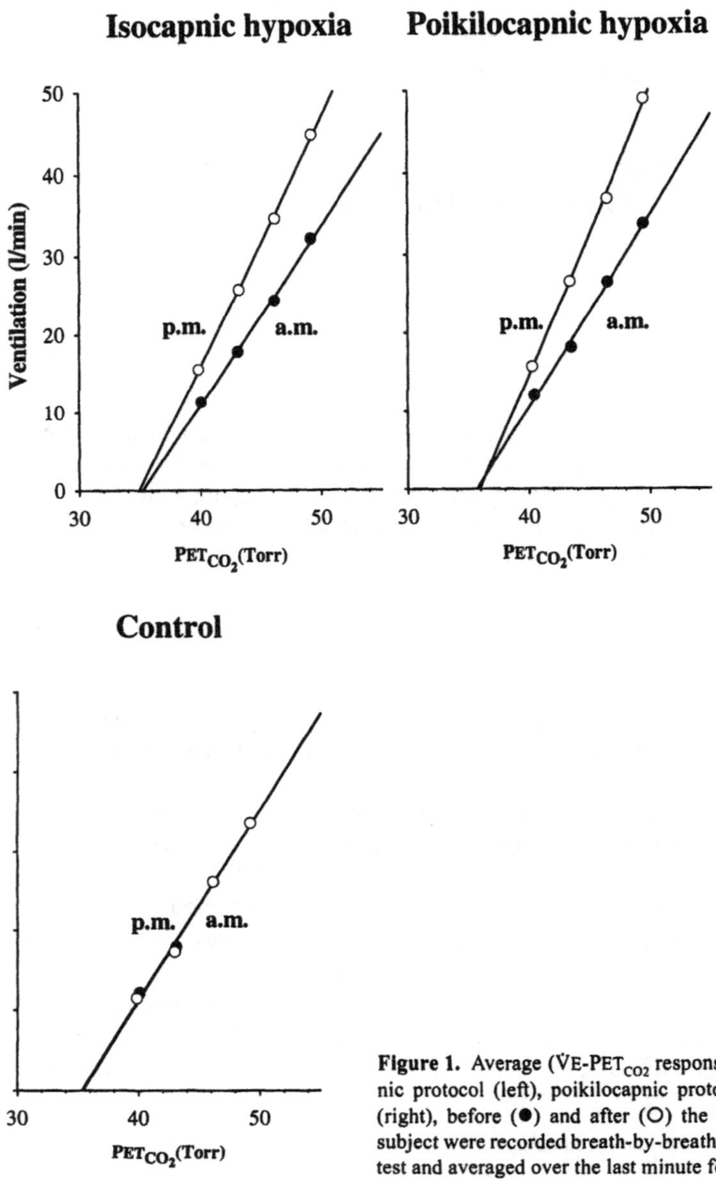

Figure 1. Average ($\dot{V}E$-PET_{CO2} response lines for all 10 subjects; isocapnic protocol (left), poikilocapnic protocol (centre) and control protocol (right), before (●) and after (○) the chamber exposure. Data for each subject were recorded breath-by-breath during the hypercapnic sensitivity test and averaged over the last minute for each level of PET_{CO2}.

different from the control case ($p < 0.01$), but no significant difference was found between the two hypoxic protocols. The shift in the intercept was very small and analysis of variance on the change in the intercept (am–pm) revealed no significant effect of protocol.

4. DISCUSSION

We conclude that 8 h of hypoxia, whether isocapnic or poikilocapnic, increases the sensitivity of the hyperoxic chemoreflex response to CO_2 in humans.

ACKNOWLEDGMENTS

This study was approved by the Central Oxford Research Ethics Committee and supported by the Wellcome Trust.

REFERENCES

1. Chiodi, H. Respiratory adaptations to chronic high altitude hypoxia. *J. Appl. Physiol.* 81–87, 1957.
2. Eger, E.I.I., R.H. Kellogg, A.H. Mines, M. Lima-Ostos, C.G. Morrill, and D.W. Kent. Influence of CO_2 on ventilatory acclimatization to altitude. *J. Appl. Physiol.* 24: 607–615, 1968.
3. Forster, H.V., J.A. Dempsey, M.L. Birnbaum, W.G. Reddan, J. Thoden, R.F. Grover, and J. Rankin. Effect of chronic exposure to hypoxia on ventilatory response to CO_2 and hypoxia. *J. Appl. Physiol.* 31: 586–592, 1971.
4. Michel, C.C., and J.S. Milledge. Respiratory regulation in man during acclimatisation to high altitude. *J. Physiol.* 168: 631–643, 1963.
5. Rahn, H., R.C. Stroud, S.M. Tenney, and J.C. Mithoefer. Adaptation to high altitude: respiratory response to CO_2 and O_2. *J. Appl. Physiol.* 6: 158–162, 1953.

4

VENTILATORY RESPONSES TO HYPOXIA AFTER 6 HOURS PASSIVE HYPERVENTILATION IN HUMANS

Xiaohui Ren and Peter A. Robbins

University Laboratory of Physiology
University of Oxford
Parks Road, Oxford OX1 3PT, United Kingdom

1. INTRODUCTION

Ventilatory acclimatization to hypoxia (VAH) is characterised by progressively increasing hyperventilation and an associated hypocapnic alkalosis. However, the mechanisms underlying acclimatization remain uncertain, and hypoxia, the initial increase in ventilation and the associated respiratory alkalosis may all play a role. The purpose of this study was to investigate the effects of hyperventilation and hypocapnia over a time course in which the initial stage of an acclimatization-like process has been observed to occur in humans.

2. METHODS

Eleven healthy subjects aged 26±10 years (mean±SD) were studied, each of whom undertook three 6-h protocols: 1) passive hyperventilation via a nasal mask with the end-tidal P_{CO_2} (PET_{CO2}) allowed to fall by 10 Torr below control (hypocapnic hyperventilation protocol, *HH*); 2) passive hyperventilation with PET_{CO2} maintained constant (eucapnic hyperventilation protocol, *EH*); 3) air breathing control (control protocol, *C*).

Before and 30 min after each protocol, the ventilatory responses to acute hypoxia (AHVR) were measured using a sequence of 6 square waves of end-tidal PO2 alternated between 100 Torr and 50 Torr, each of period 120 s, imposed by an end-tidal forcing system (1). PET_{CO2} was held constant at 1–2 Torr above control throughout.

To quantify AHVR from the data, the responses to the six square-waves of hypoxia were fitted by a single compartment model (2), and values for hypoxic sensitivity (Gp,

Advances in Modeling and Control of Ventilation, edited by Hughson *et al.*
Plenum Press, New York, 1998.

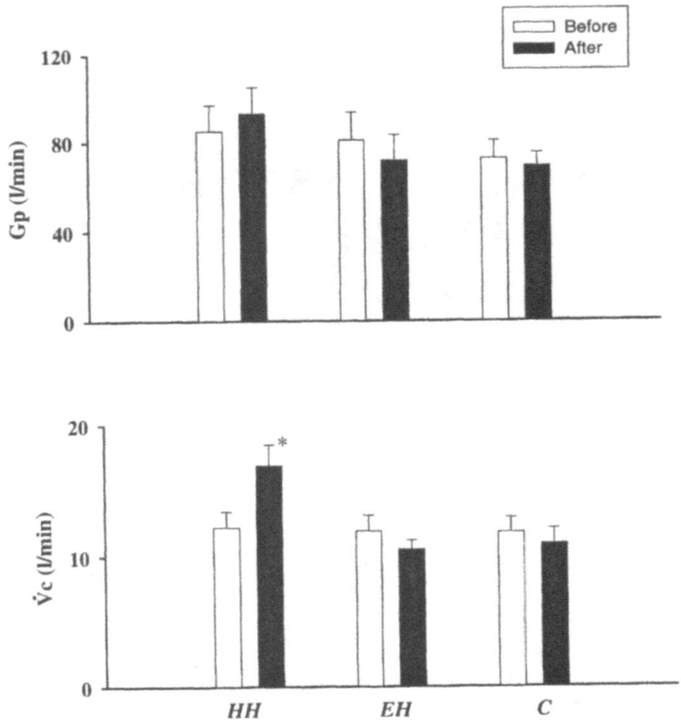

Figure 1. Averages for the hypoxic sensitivity, Gp (top panel), and for V̇c (bottom panel) for all 11 subjects for *protocol HH* (left), *protocol EH* (middle) and *protocol C* (right). White bars indicate the responses before the 6 h exposure, black bars indicate the responses after the 6 h exposure. * $p < 0.01$, ANOVA.

l/min) and the residual ventilation in the absence of an hypoxic stimulus (V̇c, l/min) were calculated for each set of data.

3. RESULTS

The ventilatory responses to hypoxia before and after each protocol are illustrated in Figure 1.

No significant effects on the value for hypoxic sensitivity Gp were detected for any of the three protocols (Analysis of Variance, ANOVA). In contrast, for V̇c, an increase in value was detected only with the *protocol HH* ($p < 0.01$, ANOVA).

4. DISCUSSION

Our results demonstrate that neither hyperventilation nor hypocapnia appears to affect the ventilatory sensitivity to hypoxia. This supports the notion that the increase in AHVR associated with VAH arises through an effect of hypoxia *per se* (3). The elevation in hypoxia-independent ventilation V̇c following 6 h of hypocapnic hyperventilation is likely to be generated by the respiratory alkalosis.

ACKNOWLEDGMENTS

The study was approved by the Central Oxford Research Ethics Committee and supported by the Wellcome Trust. X. Ren holds an ORS Award and a K. C. Wong Scholarship.

REFERENCES

1. Robbins, P. A., G. D. Swanson, and M. G. Howson. A prediction-correction scheme for forcing alveolar gases along certain time course. *J. Appl. Physiol.* 52: 1353–1357, 1982.
2. Clement, I. D., and P.A. Robbins. Dynamics of the ventilatory response to hypoxia in humans. *Respir. Physiol.* 92: 253–275, 1993.
3. Howard, L. S. G. E., and P.A. Robbins. Alterations in respiratory control during eight hours of isocapnic and poikilocapnic hypoxia in humans. *J. Appl. Physiol.* 78: 1098–1107, 1995.

VENTILATORY EFFECTS OF 8 HOURS OF ISOCAPNIC HYPOXIA WITH AND WITHOUT β-BLOCKADE

Christine Clar, Keith L. Dorrington, and Peter A. Robbins

University Laboratory of Physiology
University of Oxford
Parks Road, Oxford OX1 3PT, United Kingdom

1. INTRODUCTION

Studies of humans at high altitude show increases in cardiac output, heart rate, and plasma noradrenaline levels (2,4), suggesting an increase in sympathetic activity. Ventilation ($\dot{V}E$) as well as heart rate progressively increase during the first few hours of a hypoxic exposure, and both responses have a component that is not rapidly reversible (1,7). We investigated the hypothesis that changing sympathetic activity may be the common factor underlying those slow responses, and studied ventilatory responses during a prolonged hypoxic exposure in the presence of β-blockade as compared with control. Subjects were studied under isocapnic conditions to eliminate the confounding effect of hypocapnia developing as $\dot{V}E$ increases during the exposure.

2. METHODS

Ten healthy human volunteers were studied (6 male, 4 female). They were seated in a chamber in which end-tidal levels of O_2 and CO_2 (PET_{O2}, PET_{CO2}) could be maintained at the desired levels by computer control. Four 8h protocols were employed: (1) Isocapnic hypoxia (PET_{O2} = 50 mm Hg, PET_{CO2} = subject's normal value) with 80 mg doses of oral β-blockers (propranolol) given 8-hourly. (2) Isocapnic hypoxia, as in protocol 1, with placebo. (3) Air breathing control, with propranolol. (4) Air breathing control, with placebo.

$\dot{V}E$ was measured in the chamber at times 0, 1 h, 2 h, 4 h, 6 h, and 8 h using a mouthpiece assembly. In addition to these measurements, subjects were exposed to 5-minute spells of hyperoxia (PET_{O2} = 300 mm Hg) at t = 0, t = 4 h, and t = 8 h, using a fast gas-mixing system and mouthpiece assembly to examine any persisting effects of the hypoxic exposure. PET_{CO2} was maintained at 1–2 mm Hg above the subject's normal value for this test.

Advances in Modeling and Control of Ventilation, edited by Hughson *et al.*
Plenum Press, New York, 1998.

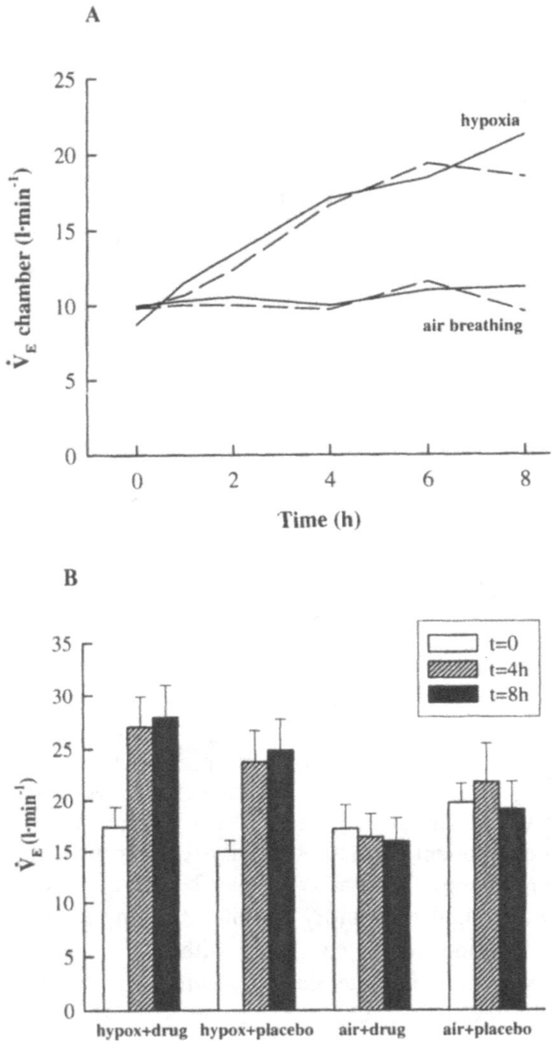

Figure 1. A: V̇E measured during 8 hours of chamber residence (solid lines, values with propranolol; broken lines, values with placebo). B: V̇E during 5 mins of hyperoxia, measured using an end-tidal forcing system. The four groups of bars represent the four experiments, and each bar in a group represents one of the measurement points.

3. RESULTS

β-Blockade did not significantly affect changes in V̇E during 8 hours of isocapnic hypoxia (ANOVA, Fig. 1A). Similarly, the progressive increase in V̇E that is seen when exposing subjects to brief spells of hyperoxia during a prolonged hypoxic exposure was not altered by β-blockade (ANOVA, Fig. 1B).

4. DISCUSSION

Our results did not provide any evidence that changes in sympathetic activity during 8 hours of isocapnic hypoxia play a role in mediating the ventilatory changes observed during that period. There was no evidence that the level and duration of hypoxia employed in this study produced the increases in sympathetic activity and noradrenaline levels that have been shown to stimulate carotid body activity in animals (5,6), or that the dose of propranolol used produced the depression of carotid body activity reported in other studies (3).

ACKNOWLEDGMENTS

The work was supported by the Wellcome Trust and approved by the Central Oxford Research Ethics Committee.

REFERENCES

1. Clar, C., M.E.F. Pedersen, M.J. Poulin, J.G. Tansley, and P.A. Robbins. Effects of 8 h of eucapnic and poikilocapnic hypoxia on middle cerebral artery velocity and heart rate in humans. *Experimental Physiology* 82: 791–802, 1997.
2. Cunningham, W.L., E.J. Becker, and F. Kreuzer. Catecholamines in plasma and urine at high altitude. *Journal of Applied Physiology* 20: 607–610, 1965.
3. Folgering, H., J. Ponte, and T. Sadig. Adrenergic mechanisms and chemoreception in the carotid body of the cat and rabbit. *Journal of Physiology* 325: 1–22, 1982.
4. Grover, R.F., J.T. Reeves, J.T. Maher, R.E. McCullough, J.C. Cruz, J.C. Denniston, and A. Cymerman. Maintained stroke volume but impaired arterial oxygenation in man at high altitude with supplemental CO_2. *Circulation Research* 38(5): 391–396, 1976.
5. Milsom, W.K., and T. Sadig. Interaction between norepinephrine and hypoxia in carotid body chemoreception in rabbits. *Journal of Applied Physiology* 55: 1893–1898, 1983.
6. O'Regan, R.G. Responses of carotid body chemosensory activity and blood flow to stimulation of sympathetic nerves in the cat. *Journal of Physiology* 315: 81–98, 1981.
7. Tansley, J.G., C. Clar, M.E.F. Pedersen, and P.A. Robbins. Human ventilatory response to acute hyperoxia during and after 8 h of both isocapnic and poikilocapnic hypoxia. *Journal of Applied Physiology* 82: 513–519, 1997.

6

MODULATION OF VENTILATORY SENSITIVITY TO HYPOXIA BY DOPAMINE AND DOMPERIDONE BEFORE AND AFTER PROLONGED EXPOSURE TO HYPOXIA IN HUMANS

Michala E. F. Pedersen, Keith L. Dorrington, and Peter A. Robbins

University Laboratory of Physiology
University of Oxford
Parks Road, Oxford OX1 3PT, United Kingdom

1. INTRODUCTION

Acclimatisation to altitude involves an increase in the ventilatory sensitivity to hypoxia (AHVR). Since low dose dopamine decreases AHVR and domperidone increases the same (1), then the increase in AHVR at altitude may be generated by a decrease in peripheral dopaminergic activity. There is evidence to support this in cats (2), but not so far in humans. We hypothesised that, if dopamine activity is decreased by prolonged hypoxia, then the effect of the blockade would also be decreased. In order to determine whether there are any changes in the sensitivity to dopamine, the study also compares the inhibitory effects on AHVR of low dose dopamine infusions with and without prior sustained hypoxia.

2. METHODS

Nine subjects were studied twice under each of three pharmacological conditions: 1) control, with no drug administered, 2) dopamine (3 μg/min/kg) and 3) domperidone (Motilin®, 40 mg). For each pharmacological condition, one of the studies was conducted without prior hypoxia and one of the studies was conducted following an 8h exposure to hypoxia (end-tidal P_{O_2}: 55 Torr).

Measurements of AHVR were undertaken with end-tidal P_{CO_2} held constant throughout at 1–2 Torr above subject's resting level. The end-tidal P_{O_2} was varied in a set

Advances in Modeling and Control of Ventilation, edited by Hughson *et al.*
Plenum Press, New York, 1998.

of 6 square waves, each of period of 120 s, alternating between 50 and 100 Torr. A single compartment model was used to estimate hypoxic sensitivity, Gp (l/min).

3. RESULTS

Dopamine decreased and domperidone increased the hypoxic sensitivity, Gp (see Fig. 1a and b). The effect of both drugs on Gp was under all conditions larger following hypoxia ($p \leq 0.05$, ANOVA). Interestingly, when expressed as a percentage of the relevant control value, the effects of dopamine and domperidone were similar before and after prolonged hypoxia (p = NS, ANOVA).

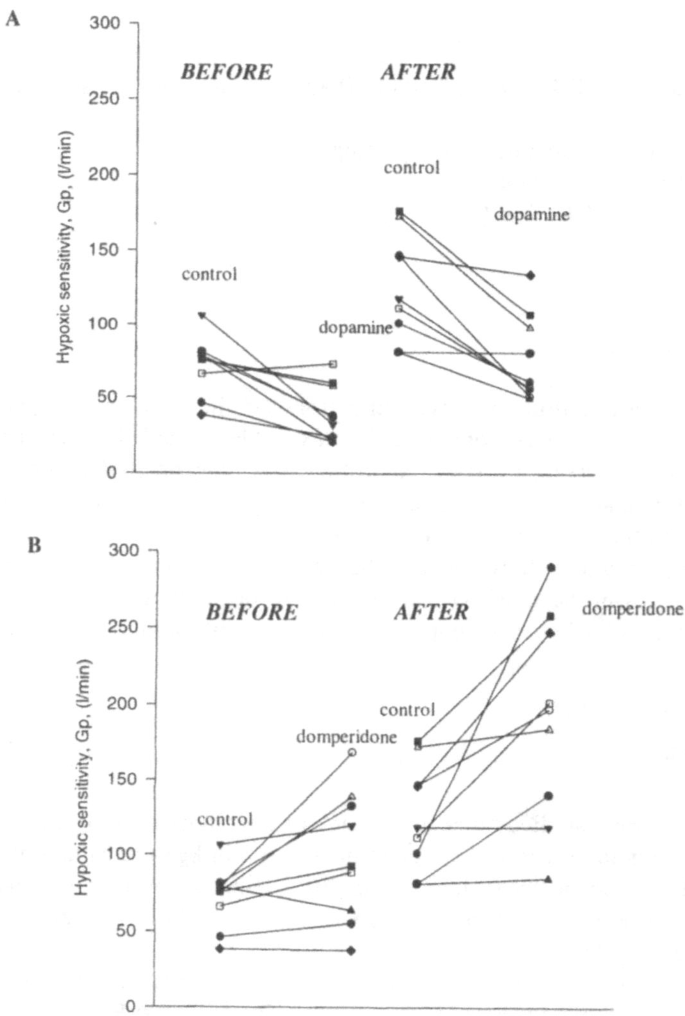

Figure 1. Effect of dopamine (A) and domperidone (B) on hypoxic sensitivity before and after 8 h of isocapnic hypoxia, 9 individual subjects.

4. DISCUSSION

The main finding of this study is that, following an 8 h period of hypoxia, the absolute effects on AHVR of low dose dopamine infusion and domperidone are increased. However, these increases are in proportion to the general increase in AHVR, and thus in relative terms the effects of low dose dopamine and domperidone are unchanged. These findings do not reflect those of Tatsumi *et al.* in the cat. This dissimilarity presumably arises either from a difference in the dose of domperidone or from a species difference.

Our results do not support the above hypothesis that the increased ventilation after prolonged hypoxia is related to a reduction in dopaminergic inhibition at the CB in humans.

ACKNOWLEDGMENTS

This study was approved by the Central Oxford Research Ethics Committee and supported by the Wellcome Trust. M.E.F. Pedersen holds a MRC student studentship and a scholarship from the Danish Research Academy.

REFERENCES

1. Bascom, D. A., I. D. Clement, K. L. Dorrington & P. A. Robbins. Effect of dopamine and domperidone on ventilation during isocapnic hypoxia in humans. *Respir. Physiol.* 85: 319–328, 1991.
2. Tatsumi, K., C. K. Pickett & J. V. Weil. Decreased carotid body hypoxic sensitivity in chronic hypoxia: role of dopamine. *Respir. Physiol.* 101: 47–57, 1995.

CHANGES IN RESPIRATORY CONTROL DURING AND AFTER 48 HOURS OF BOTH ISOCAPNIC AND POIKILOCAPNIC HYPOXIA IN HUMANS

John G. Tansley, Marzieh Fatemian, Marc J. Poulin, and Peter A. Robbins

University Laboratory of Physiology
University of Oxford
Parks Road, Oxford OX1 3PT, United Kingdom

1. INTRODUCTION

During an 8 h period of either isocapnic or poikilocapnic hypoxia, tests of the acute ventilatory response to hypoxia (AHVR) have shown that there is an increase in hypoxic sensitivity (2) accompanied by an increase in ventilation under conditions of acute hyperoxia (3). These changes were similar between the two protocols suggesting a direct effect of hypoxia *per se*.

Similar changes in respiratory control are thought to occur in the first 2–3 days of ventilatory acclimatization to altitude (VAH). In order to examine changes in respiratory control during isocapnic hypoxia over a time-scale that is similar to that of VAH we extended the period of hypoxia to 48 h. In addition we studied the 48 h recovery period following the relief of hypoxia.

2. METHODS

Ten subjects were studied. Each participated in two 48 h hypoxic exposures using a purpose-built chamber (1). In isocapnic hypoxia, IH, end-tidal P_{O2} (PET_{O2}) was maintained at 60 Torr and end-tidal P_{CO2} (PET_{CO2}) was held at the subject's control value, as measured before the experiment. In poikilocapnic hypoxia, PH, PET_{O2} was again held at 60 Torr whilst PET_{CO2} was uncontrolled. Subjects returned to the laboratory over the 48 h subsequent to the hypoxic exposure for further testing. Tests of AHVR were carried out using a dynamic end-tidal forcing system. PET_{O2} was alternated between 100 Torr and 50 Torr in a series of 6 square waves of 120 s period whilst PET_{CO2} was held 1–2 Torr above the subject's control value. A single compartment model was used to estimate the component of the response that was sensitive to acute changes in hypoxic stimulus, G_p (l min^{-1}) and the component insensitive to these changes, $\dot{V}_c$ (l min^{-1}).

Advances in Modeling and Control of Ventilation, edited by Hughson *et al.*
Plenum Press, New York, 1998.

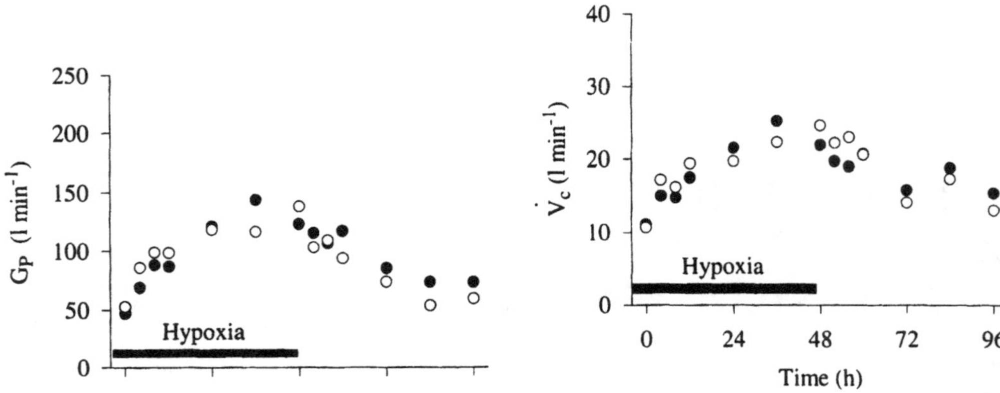

Figure 1. Mean values for model parameters G_p (l min^{-1}) and $\dot{V}_c$ (l min^{-1}). Closed symbols *protocol I*, open symbols *protocol P.*

3. RESULTS

Figure 1 shows that during the hypoxic period of both protocols there was a significant increase in both G_p and $\dot{V}_c$ ($p < 0.001$, ANOVA) and during the 48 h subsequent to the alleviation of hypoxia these parameters were observed to decrease ($p < 0.001$, ANOVA). There was no significant difference between the protocols.

4. DISCUSSION

These findings suggest that, as observed over an 8 h period, the changes in respiratory control that occur over a 48 h exposure to hypoxia are a result of the effects of hypoxia *per se*. We can conclude therefore, that the changes in respiratory control that occur over the first 2 days of ventilatory acclimatization to modest hypoxia are not a result of the concomitant hypocapnia.

ACKNOWLEDGMENTS

This study was approved by the Central Oxford Research Ethics Committee and supported by the Wellcome Trust. J.G. Tansley is a M.R.C. student.

REFERENCES

1. Howard, L.S.G.E., R.A. Barson, B.P.A. Howse, T.R. McGill, M.E. McIntyre, D.F. O'Connor, and P.A. Robbins. A chamber for controlling the end-tidal gas tensions over sustained periods in humans. *J. Appl. Physiol.* 78: 1088–1091, 1995.
2. Howard, L.S.G.E., and P.A. Robbins. Alterations in respiratory control during eight hours of isocapnic and poikilocapnic hypoxia in humans. *J. Appl. Physiol.* 78: 1098–1107, 1995.
3. Tansley, J.G., C. Clar, M.E.F. Pedersen, and P.A. Robbins. Human ventilatory response to acute hyperoxia during and after 8 h of both isocapnic and poikilocapnic hypoxia. *J. Appl. Physiol.* 82(2): 513–519, 1997.

CHEMOREFLEX EFFECTS OF LOW DOSE SEVOFLURANE IN HUMANS

Jaideep J. Pandit, Jocelyn Manning-Fox, Keith L. Dorrington, and Peter A. Robbins

University Laboratory of Physiology
University of Oxford
Parks Road, Oxford OX1 3PT, United Kingdom

1. INTRODUCTION

1.1. Effects of Low Dose Anaesthetics on the Ventilatory Response to CO_2

Knill and co-workers[6] found that doses of 0.1 MAC of volatile anaesthetics reduced the sensitivity of the response to CO_2. However, Knill assessed ventilatory responsiveness to CO_2 using Read's rebreathing method. Berkenbosch et al. showed that Read's method gave misleading results and argued in favour of a steady-state method to study ventilatory responses to CO_2[1]. Using the steady-state method, it has been found that 0.1 MAC halothane reduces the ventilation-CO_2 response slope by 30%[2], while 0.17 MAC nitrous oxide has no effect[4]. Desflurane (0.1 MAC) also does not affect the response to CO_2[5].

Interpretation of the ventilatory response to steady hypercapnia is further confounded by the non-steady effects of sustained hypercapnia. When end-tidal PCO_2 (PET_{CO2}) is held constant by values greater than 2–3 Torr above natural values, a progressive rise in ventilation (or ventilatory "drift") occurs over the period of the hypercapnic exposure[8]. Only after 2–3 hours is a true steady-state response achieved[12].

1.2. Effects of Low Dose Anaesthetics on the Ventilatory Response to Hypoxia

Knill and colleagues found that a number of anaesthetic agents at low dose profoundly depressed the acute hypoxic ventilatory response (AHVR) in humans[6]. Recent studies have largely confirmed this finding. Enflurane (0.07–0.2 MAC) depressed the response by 30–50%[7], halothane (0.1–0.2 MAC) by 50–75%[2], and isoflurane by 50%[14].

Advances in Modeling and Control of Ventilation, edited by Hughson *et al.*
Plenum Press, New York, 1998.

It is also now known that sustained isocapnic hypoxia causes a biphasic ventilatory response. Initially there is a rapid rise in ventilation (AHVR) and then a slower decline over 20–30 min (the hypoxic ventilatory decline, HVD)[17]. In anaesthetised cats the mechanism underlying HVD appears to be a depressant effect of hypoxia on the central nervous system[13]. In awake cats and awake humans, the mechanism underlying HVD is thought to be a time-dependent decline in peripheral chemoreflex activity[10]. In contrast to their depressant effects on AHVR, volatile agents do not affect the absolute magnitude of HVD[3,4,7].

1.3. Aims of Study

The aims of this study were to assess the effects of 0.1 MAC sevoflurane on (i) the spontaneous ventilation, (ii) the ventilatory response to acute hypercapnia, (iii) the ventilatory response to sustained hypercapnia, (iv) the ventilatory response to acute hypoxia, and (v) the ventilatory response to sustained hypoxia. We hypothesised that if sevoflurane resembles other volatile anaesthetics, it would greatly reduce the AHVR but have no effect on HVD.

2. METHODS

Ten healthy volunteers participated in the hypoxia protocols and eight in the hypercapnia protocols. The study was approved by the Central Oxford Research Ethics Committee. Subjects were seated in a chair and breathed through a mouthpiece whilst wearing a noseclip. End-tidal gases were controlled by dynamic end-tidal forcing, described in more detail elsewhere[9]. Sevoflurane was administered via a vaporiser which could be adjusted manually. End-tidal sevoflurane was monitored by mass spectrometry. Pulse oximeter and ECG monitoring were used.

2.1. Preliminary Protocol

Subjects undertook a preliminary protocol of 10 min of quiet air-breathing to determine ambient ventilation and end-tidal PCO_2 (PET_{CO2}). This was repeated whilst breathing sevoflurane to give an end-tidal value of 0.1 MAC.

2.2. Hypercapnia Protocols

The control hypercapnia protocol consisted of 35 min during which PET_{CO2} was held constant at 10 Torr above the natural value. PET_{O2} was held at 100 Torr throughout. The sevoflurane hypercapnia protocol was a repeat of the control protocol, but against a background of 0.1 MAC sevoflurane.

2.3. Hypoxia Protocols

The control hypoxia protocol consisted of 10 min during which the subject's PET_{O2} was held at 100 Torr, then 20 min during which it was held at 60 Torr and finally 5 min during which it was returned to 100 Torr. The PET_{CO2} was held at 10 Torr above the subject's natural value throughout. The sevoflurane hypoxia protocol was a repeat of the control protocol, but this time against a background of 0.1 MAC sevoflurane.

2.4. Data Analysis

Data were averaged into 1 min periods. The first 5 min of each protocol were excluded, leaving 30 min for analysis. The last min of the preliminary protocol was used to assess each subject's ambient PET_{CO2} and ventilation with and without sevoflurane.

2.4.1. Analysis of Hypercapnia Protocols. To assess the acute ventilatory response to hypercapnia, the ventilation at 5 min after introduction of hypercapnia was used, and compared with values from the last min of the preliminary protocol, for each of control and sevoflurane protocols. To assess the change in ventilation during sustained hypercapnia, the means of the values over the first and last 5 min of hypercapnia were used. These values yielded 10 subject means for each of these variables. These were averaged to yield the group means.

2.4.2. Analysis of Hypoxia Protocols. Four 1 min periods were used for data analysis (Fig. 1): the ventilation in the last min of euoxia before the step into hypoxia ($\dot{V}E_1$, the pre-hypoxic ventilation); the peak ventilation reached during the first 5 min of the hypoxic period ($\dot{V}E_2$, peak ventilation); the ventilation during the last min of the hypoxic period ($\dot{V}E_3$, depressed ventilation); and the minimum ventilation reached in the first 5 min after return to euoxia ($\dot{V}E_4$, post-hypoxic ventilation).

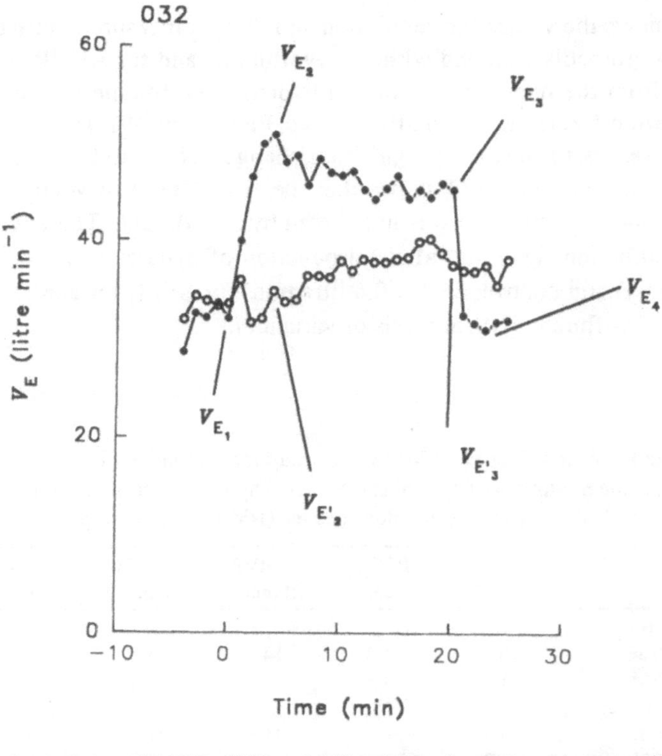

Figure 1. Method of calculation of AHVR, HVD and Off-responses for the hypoxia protocols. The results from one subject (032, without sevoflurane) are shown. Hypoxic period corresponds to times 0–20 min. ●: hypoxia protocol; O: hypercapnia protocol.

(1) The AHVR was then calculated as the difference between the peak and pre-hypoxic ventilations ($\dot{V}E_2 - \dot{V}E_1$). (2) The magnitude of the off-response was calculated as the difference between the depressed and post-hypoxic ventilations ($\dot{V}E_3 - \dot{V}E_4$). (3) Because ventilatory drift occurs in euoxic hypercapnia, it was decided to express HVD in the hypoxia protocol by controlling for the drift in the corresponding hypercapnia study in each individual subject. Figure 1 shows how this was done, using values for $\dot{V}E'_2$ and $\dot{V}E'_3$ as corresponding euoxic points for peak and depressed ventilations respectively. The formula[7]: $HVD = (\dot{V}E_2 - \dot{V}E_3) + (\dot{V}E'_3 - \dot{V}E'_2)$ was used. This technique has been employed to calculate HVD and account for ventilatory drift in a previous study[7]. Values for AHVR (ie, the on-response), HVD and off-response were obtained for each subject and from these subject means, the means for these variables for the group as a whole were obtained.

2.4.3. Statistical Analysis. The subject means for each variable were first subjected to an analysis of variance. If this indicated significant differences, the differences between the means of the variables were assessed using a paired, two-tailed t-test. Statistical significance was accepted at a value of $P < 0.05/k$ (Bonferroni's correction) where k is the number of comparisons made for that variable.

3. RESULTS

Table 1 shows the values for ventilation and PET_{CO2} measured during the last min of the preliminary protocols with and without sevoflurane, and for AHVR, HVD and off-responses taken from the hypoxia control and hypoxia sevoflurane protocols. Sevoflurane reduced ventilation by about 5%, and increased PET_{CO2} by 4%, and reduced AHVR by 20% but these were not statistically significant changes. HVD and off-responses were unaffected by sevoflurane. Figure 2 shows that the rising trend in ventilation (ventilatory drift) with sustained hypercapnia was unaffected by sevoflurane. There was no difference between the ventilations at 5 min after introduction of hypercapnia (i.e, the response to acute hypercapnia) and control: 36.1 ± 0.6 litre min^{-1} (± SEM) for control and 34.2 ± 0.5 litre min^{-1} with sevoflurane (NS, analysis of variance).

Table 1. Mean values (± SEM) for the group for ventilation ($\dot{V}E$) and PET_{CO2} from the preliminary protocol and for acute hypoxic ventilatory response (AHVR), hypoxic ventilatory decline (HVD), and off-response[a]

	$\dot{V}E$ ($l \cdot min^{-1}$)	PET_{CO2} (Torr)	AHVR ($l \cdot min^{-1}$)	HVD ($l \cdot min^{-1}$)	Off ($l \cdot min^{-1}$)
Control					
Mean	10.1	35.3	14.5	8.2	7.1
SEM	1.0	1.67	1.2	1.1	1.1
Sevoflurane					
Mean	9.6	36.6	11.6	10.6	6.3
SEM	4.82	1.06	1.6	2.8	1.4

[a]None of the differences in these variables for control and sevoflurane protocols was statistically significant: see text for further explanation.

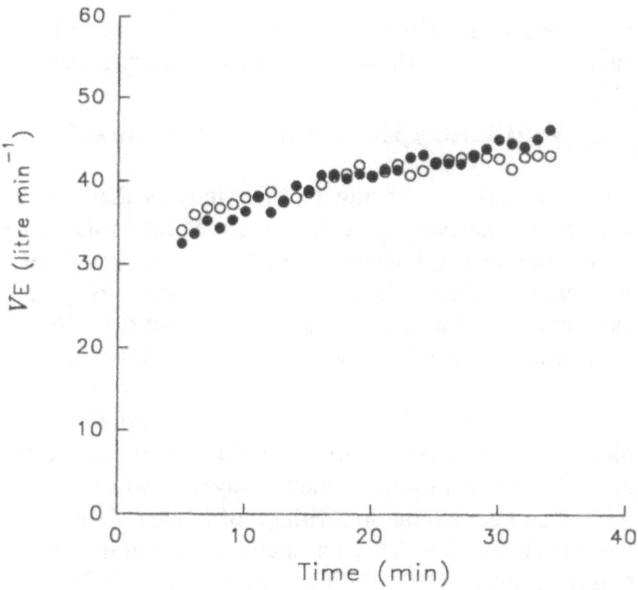

Figure 2. Graph of mean ventilation against time for the group as a whole for control (O) and sevoflurane (●) protocols with sustained hypercapnia.

4. DISCUSSION

4.1. Ventilatory Response to Hypercapnia

4.1.1. Ventilatory Response to Acute Hypercapnia. Our results are in general agreement with those of van den Elsen et al. who found that the total gain of the response to CO_2 was unaffected by sevoflurane 0.1 MAC[15,16]. Van den Elsen argued that 0.1 MAC sevoflurane selectively depresses the peripheral chemoreflex, because it reduced the gain term associated with the peripheral CO_2 chemoreflex (G_p) by 27%, but left the gain term associated with the central CO_2 chemoreflex (G_c) unaffected[15,16]. However, one problem with this approach is that if the true response to sustained hypercapnia is more complex than can be described by a two-compartment model, then it is imprecise to discuss the ventilatory response simply in terms of G_p and G_c.

4.1.2. Ventilatory Response to Sustained Hypercapnia. Reynolds and colleagues observed that ventilation rises progressively over a 30 min period of breathing 5–7% inspired CO_2[8]. Tansley and co-workers confirmed that this drift in ventilation continues for up to 2 hours when PET_{CO2} is held 6–7 Torr above natural levels[12]. The mechanisms underlying this response are unclear.

Nagyova and colleagues examined the effects of 0.2 MAC enflurane on the ventilatory response to sustained hypoxia[23]. Control protocols, in which no hypoxia was administered, were included in which PET_{CO2} was held at 5 Torr above natural values. They found that in the absence of enflurane, ventilatory drift occurred. However, in the presence of enflurane, the progressive rise in ventilation was reversed to become a progressive decline. Nagyova et al. could not exclude the possibility that the level of sedation had increased progressively throughout enflurane exposure, despite the fact that end-tidal enflurane concentrations remained constant throughout their protocols. It is also possible that there is a genuine difference between the effects of subanaesthetic enflurane and subanaesthetic sevoflurane on ventilation during sustained hypercapnia.

4.2. Ventilatory Response to Hypoxia

The striking finding of this study is that unlike other volatile agents, 0.1 MAC sevoflurane has very little depressant effect on the acute ventilatory response to hypoxia. Like other agents, however, the HVD is unaffected. Furthermore, since the magnitudes of AHVR and off-responses remained unequal during control and sevoflurane protocols, we can conclude that there is no evidence that 0.1 MAC sevoflurane affects the underlying decline in peripheral chemoreflex activity which occurs with sustained hypoxia.

4.2.1. AHVR: Comparison with Other Studies on Sevoflurane. Sarton et al. investigated the effects of 0.1 MAC sevoflurane on the response to acute (but not sustained) hypoxia[11]. The main aim of their study was to examine the effect of pain and arousal on AHVR and the interaction with sevoflurane sedation. They found that sevoflurane reduced the AHVR by 30%. This was statistically significant in their study, but both studies agree that the reduction of AHVR by sevoflurane (20–30%) is much smaller than by other volatile agents. The effects of sevoflurane on the hypoxic response at concentrations higher than 0.1 MAC have not been examined.

4.2.2. HVD and Off-Response: Physiological Significance of the Results. Two previous studies have analysed the relative magnitudes of AHVR and off-responses to examine in more detail this (lack of) effect of anaesthetics on HVD. The attempt of Nagyova and colleagues was limited by the fact that enflurane 0.2 MAC caused a progressive reduction in euoxic ventilation[7]. Dahan's group examined 0.15 MAC halothane and found that while the magnitudes of AHVR and off-responses were unequal in controls (implying a decline in peripheral chemoreflex activity), during halothane sedation they were equal[3]. They concluded that in contrast to the control state, the HVD which occurs during halothane sedation is not due to a decline in peripheral chemoreflex activity (and hence occurs by some other mechanism). Our study with sevoflurane suggests that AHVR and off-responses remain unequal, indicating that a decline in peripheral chemoreflex sensitivity still occurs in the presence of this agent.

REFERENCES

1. Berkenbosch, A., J.G. Bovill, A. Dahan, J. DeGoede, and I.C.W. Olievier. Ventilatory sensitivities from Read's rebreathing method and the steady-state method are not equal. *J. Physiol. Lond.* 411: 367–377, 1989.
2. Dahan, A., M.J.L.J. van den Elsen, A. Berkenbosch, J. DeGoede, I.C.W. Olievier, J.W. van Kleef, and J. Bovill. Effects of subanesthetic halothane on the ventilatory responses to hypercapnia and hypoxia in healthy volunteers. *Anesthesiology* 80: 727–738, 1994.
3. Dahan, A, M.J.L.J. van den Elsen, A. Berkenbosch, J. DeGoede, I.C.W. Olievier, A.G.L. Burm, and J.W. van Kleef. Influence of a subanesthetic concentration of halothane on the ventilatory response to step changes into and out of sustained isocapnic hypoxia in healthy volunteers. *Anesthesiology* 81: 850–859, 1994.
4. Dahan, A. and D.S. Ward. Effects of 20% nitrous oxide on the ventilatory response to hypercapnia and sustained isocapnic hypoxia in man. *Br. J. Anaesth* 72: 17–20, 1994.
5. Dahan, A., E. Sarton, M.J.L.J. van den Elsen, J.W. van Kleef, L. Teppema, and A. Berkenbosch. Ventilatory response to hypoxia in humans. Influences of subanesthetic desflurane. *Anesthesiology* 85: 60–68, 1996.

6. Knill, R.L. and J.L. Clement. Variable effect of anaesthetics on the ventilatory response to hypoxaemia in man. *Canad. Anaesth. Soc. J.* 29: 93–99, 1982.
7. Nagyova, B., K.L. Dorrington, M.J. Poulin, and P.A. Robbins. Influence of 0.2 minimum alveolar concentration of enflurane on the ventilatory response to sustained hypoxia in humans. *Br. J. Anaesth.* 78: 707–713, 1997.
8. Reynolds, W.J., H.T. Millhorn, and G.H. Holloman. Transient ventilatory response to graded hypercapnia in man. *J. Appl. Physiol.* 33: 47–54. 1972.
9. Robbins, P.A., G.D. Swanson, and M.G. Howson. A prediction-correction scheme for forcing alveolar gases along certain time courses. *J. Appl. Physiol.* 52: 1353–1357, 1982.
10. Robbins P.A. Hypoxic ventilatory decline: site of action. *J. Appl. Physiol.* 373–374.
11. Sarton, E., A. Dahan, L. Teppema, M.J.L.J. van den Elsen, E. Olofsen, A. Berkenbosch, and J. van Kleef. Acute pain and central nervous system arousal do not restore impaired hypoxic ventilatory response during sevoflurane sedation. *Anesthesiology* 85: 295–303, 1996.
12. Tansley, J.G., M.E.F. Pedersen, C. Clar, and P.A. Robbins. The human ventilatory response to 8h of euoxic hypercapnia followed by 8h of euoxic poikilocapnic recovery. *J. Physiol. Lond.* 497: 24*P*, 1996.
13. Van Beek, J.H.G.M., A. Berkenbosch, J. DeGoede, and I.C.W. Olievier. Effects of brain stem hypoxaemia on the regulation of breathing. *Respir. Physiol.* 57: 171–188, 1984.
14. Van den Elsen, M.J.L.J., A. Dahan, A. Berkenbosch, J. DeGoede, J.W. van Kleef, and I.C.W. Olievier. Does subanesthetic isoflurane affect the ventilatory response to acute isocapnic hypoxia in healthy volunteers? *Anesthesiology* 81: 860–867, 1994.
15. Van den Elsen M.J.L.J. *Influences of Low Dose Anesthetic Agents on Ventilatory Control in Man.* PhD Thesis, University of Leiden, 1997.
16. Van den Elsen, M., E. Sarton, L. Teppema, A. Berkenbosch, and A. Dahan. Influences of 0.1 minimum alveolar concentration of sevoflurane, desflurane and isoflurane on the dynamic ventilatory response to hypercapnia in humans. *Br. J. Anaesth.* (In press).
17. Weil J.W. and C.W. Zwillich. Assessment of ventilatory response to hypoxia: methods and interpretation. *Chest* 70 (Suppl.): 124–128, 1976.

9

DYNAMICS OF THE CEREBRAL BLOOD FLOW RESPONSE TO SUSTAINED EUOXIC HYPOCAPNIA IN HUMANS

Marc J. Poulin, Pei-Ji Liang, and Peter A. Robbins

University Laboratory of Physiology
University of Oxford
Parks Road, Oxford OX1 3PT, United Kingdom

1. INTRODUCTION

Cerebral blood flow (CBF) decreases with hypocapnia[4,6]. However, measurements of CBF a few hours after the induction of hypocapnia suggest that there is some secondary recovery of CBF over time[1,3,5]. The time course associated with this is uncertain. This study used transcranial Doppler ultrasound to assess the dynamics of the middle cerebral artery (MCA) blood flow response to 20 min of hypocapnia (PET_{CO2} = 15.0 Torr below eucapnic value).

2. METHODS

Six healthy adults were studied. Beat-by-beat values were calculated for the intensity-weighted mean velocity ($\overline{V}_{IWM}$), signal power ($\overline{P}$) which is proportional to cross-sectional area[2], and their instantaneous product, ($\overline{P \cdot V_{IWM}}$). The data were expressed as a percentage of the average value over a 5 min pre-hypocapnic period. A dynamic end-tidal forcing system was used to control PET_{CO2} and hold PET_{O2} constant at 100 Torr. The $\overline{P \cdot V}_{IWM}$ responses from six rePETitions in a single subject were fitted simultaneously to a simple model consisting of a delay, gain terms ($g_{f,on}$, $g_{f,off}$), time constants ($\tau_{f,on}$, $\tau_{f,off}$) and offsets for the on- and off-transients, and a gain term (g_s) and time constant (τ_s) for a second slower component.

3. RESULTS

An ensemble-average of the measured $\overline{P \cdot V}_{IWM}$, model fit and residuals for all repetitions for one subject is shown in Fig. 1. The pure delay between PET_{CO2} and vascular response was

Advances in Modeling and Control of Ventilation, edited by Hughson *et al.*
Plenum Press, New York, 1998.

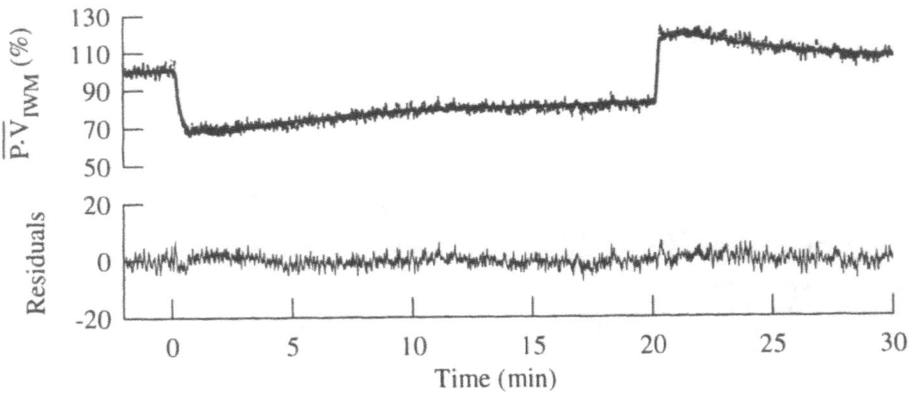

Figure 1. Response of single subject to hypocapnia. Data are points; model fit is line.

3.9 ± 0.3 s (mean ± S.E). The CBF response to hypocapnia was characterised by a significant ($p < 0.001$) slow progressive adaptation in $\overline{P \cdot V}_{IWM}$, with $g_s = 1.26 \pm 0.10\% \cdot Torr^{-1}$ and $\tau_s = 427 \pm 55$ s, that persisted throughout the hypocapnic period. The responses at the onset and relief of hypocapnia were asymmetric ($p < 0.001$, paired t-test), with $\tau_{f,on}$ (6.8 ± 0.8 s) faster than $\tau_{f,off}$ (14.3 ± 2.3 s). The gain term for the fast component of the on-transient ($g_{f,on} = 2.7 \pm 0.1\% \cdot Torr^{-1}$) was significantly smaller ($p < 0.001$) than the off-transient ($g_{f,off} = 2.9 \pm 0.1$).

4. CONCLUSION

The major finding of this study is that, after the rapid fall in $\overline{P \cdot V}_{IWM}$ at the onset of hypocapnia, there is a subsequent slow progressive increase in $\overline{P \cdot V}_{IWM}$ that continues throughout the 20 min period of hypocapnia. Additionally, there is significant asymmetry of the response to hypocapnia, characterized by a faster on-transient than off-transient.

ACKNOWLEDGMENTS

This study was approved by the Central Oxford Research Ethics Committee and was supported by the Wellcome Trust. Marc J. Poulin is supported by a Heart and Stroke Foundation of Ontario (Canada) Postdoctoral Research Fellowship.

REFERENCES

1. Albrecht, R. F., D. J. Miletich, and M. Ruttle. Cerebral effects of extended hyperventilation in unanesthetized goats. *Stroke* 18: 649–655, 1987.
2. Arts, M.G.J. and J.M.J.G. Roevros. On the instantaneous measurement of blood flow by ultrasonic means. *Med. Biol. Eng.* 10:23–34, 1972.
3. Hansen, N. B., P. T. Nowicki, R. R. Miller, T. Malone, R. G. Bickers, and J. A. Menke. Alterations in cerebral blood flow and oxygen consumption during prolonged hypocarbia. *Pediatr. Res.* 20: 147–150, 1986.
4. Kety, S. S. and C. F. Schmidt. The effects of active and passive hyperventilation on cerebral blood flow, cerebral oxygen consumption, cardiac output, and blood pressure of normal young men. *J. Clin. Invest.* 25: 107–119, 1946.
5. Raichle, M. E., J. B. Posner, and F. Plum. Cerebral blood flow during and after hyperventilation. *Arch. Neurol.* 23: 394–403, 1970.
6. Severinghaus, J. W. and N. Lassen. Step hypocapnia to separate arterial from tissue PCO_2 in the regulation of cerebral blood flow. *Circ. Res.* 20: 272–278, 1967.

EVIDENCE FOR A CENTRAL ROLE OF PROTEIN KINASE C IN MODULATION OF THE HYPOXIC VENTILATORY RESPONSE IN THE RAT

David Gozal,* Evelyne Gozal, and Gavin R. Graff

Constance S. Kaufman Pediatric Pulmonary Research Laboratory
Departments of Pediatrics, Physiology, and the Interdepartmental Neuroscience Training Program
Tulane University School of Medicine
New Orleans, Louisiana 70112

1. INTRODUCTION

Protein kinase C (PKC) has been implicated as a common mechanism in the transduction of various extracellular signals into the cell (29). PKC is ubiquitous in the central nervous system and is activated by Ca^{2+}, phospholipids and diacylglycerol (DAG) or phorbol-esters to control many physiological processes (16). The PKC family consists of three major sub-groups of isoenzymes based on their molecular structure and co-factor requirements. One sub-group comprises the classical PKC α, $\beta1$, $\beta2$, and γ isoforms, all of which share a C2 region corresponding to the Ca^{2+} binding site (23). In the other major subgroup containing the novel PKC δ, ϵ, θ, η, and μ isoforms, the C2 region is absent, and activation may occur in the absence of Ca^{2+} (14, 23, 33). Atypical isoforms such as PKC ζ, ι, and λ, also lack C2 as well as one of the repeated cysteine-rich zinc finger binding motifs within the C1 domain (24).

Immunocytochemical localization studies for PKC α, $\beta1$, $\beta2$ isoforms in rat brain have revealed heterogeneous distribution of these isoenzymes in various brain regions as well as discrete localization suggesting that individual isozymes may mediate specific functional components of neuronal activity (24, 29). A substantial body of evidence points to a critical role for PKC in both pre- and post-synaptic modulation of neuronal activity (for review see ref. 7 and 29).

* Correspondence to: David Gozal, M.D., Professor of Pediatrics and Physiology, Section of Pediatric Pulmonology, Department of Pediatrics, SL-37, Tulane University School of Medicine, 1430 Tulane Avenue, New Orleans, LA 70112. Tel: (504) 588 5601; Fax: (504) 588 5490; E-mail: dgozal@tmcpop.tmc.tulane.edu

Advances in Modeling and Control of Ventilation, edited by Hughson *et al.*
Plenum Press, New York, 1998.

Hypoxia has been shown to elicit excitatory amino acid release, intracellular Ca^{2+} increases, and enhanced degradation of phospholipids, all of which may lead to the formation of PKC activators such as DAG (2). Increased DAG membrane concentrations will induce PKC activation which can be readily assessed by examining the pattern of translocation of a particular isoform from the soluble (cytosolic) fraction to the particulate (membrane) fraction (18).

The nucleus tractus solitarii (NTS) provides the first central relay for peripheral chemoreceptor input. Within the NTS, the ventilatory response to acute hypoxia is critically dependent on excitatory amino acid-mediated neurotransmission (1, 17, 21), and more specifically on *N*-methyl-D-aspartate (NMDA) receptor activation (19, 26, 27).

The NMDA receptor has attracted significant interest due to its proposed roles in long-term potentiation, synaptogenesis, developmental structuring, and excitotoxicity. The recently cloned NMDA receptors from the rat belong to the large superfamily of ligand-gated ion channels (22). Analysis of the predicted protein sequences have uncovered characteristic structural motifs which include a large extracellular N-terminal domain, a small extracellular C-terminal domain, 4 transmembrane domains of which the second is considered to line the ion channel, and multiple potential serine/threonine phosphorylation sites within the second cytoplasmic domain (22). These serine/threonine sites could thereby provide the structural elements whereby PKC binding and activation would lead to modification and regulation of NMDA receptor activation. Indeed, Chen and Huang have demonstrated that intracellular application of PKC potentiate NMDA currents by reduction of the voltage-dependent Mg^{2+} block of ionic channels (6). Similarly, other investigators have shown a modulatory role for PKC in NMDA receptor activity suggesting that increased PKC activation will lead to potentiation of NMDA currents (30, 32).

2. PROPOSED MODEL

We hereby propose a putative model that incorporates proposed second messenger systems involved in NMDA receptor activation, and all of which possibly underlie neuronal postsynaptic activity within the NTS during acute hypoxia (Figure 1). Of note, this model does not exclude the possibility that changes in PKC activity play a role in pre-synaptic neurons and may modulate the release of a variety of neurotransmitters (for review see ref. 7). Activation of NMDA receptor is elicited by increased pre-synaptic release of glutamate via increased peripheral chemoreceptor afferent input and/or a direct effect of hypoxia on pre-synaptic neurons. Binding of glutamate to the glutamate site within the NMDA receptor may induce tyrosine and serine/threonine kinases and Ca^{2+} influx, the latter being thought to further stimulate kinase activity. Calcium calmodulin kinase II (CaCmII), and Ca^{2+}- dependent PKC isoforms could be activated by such increase in intracellular Ca^{2+}. Further Ca^{2+} release from intracellular stores via activation of phospholipase C and inositol 3-phosphate (IP3) pathways could also activate PKC, leading to its translocation and phosphorylation of the serine/threonine binding sites contained within the cytoplasmic domain of the NMDA receptor. Such phosphorylation events would increase the probability of maintaining the NMDA receptor channel in an open state and therefore further augment Ca^{2+} influx.

The activation of CaCmII would lead to increased activation of neuronal nitric oxide synthase (NOS) and NO release. We and others have previously demonstrated that nNOS plays an important excitatory role in the late phase of the biphasic response to hypoxia (8, 25). Indeed, increased NOS activity will enhance spontaneous rhythmic neural activity

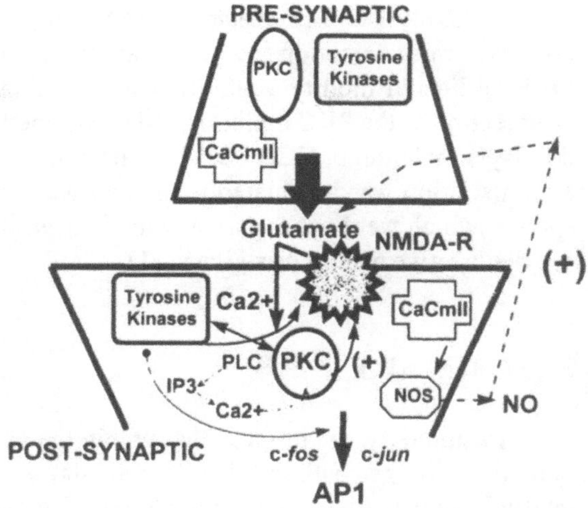

Figure 1. Putative post-synaptic signal transduction pathways which may be activated following pre-synaptic glutamate release and binding to NMDA receptor (NMDA-R). Tyrosine and PKC phosphorylation of NMDA-R will lead to calcium influx which in turn may further activate PKC and CaCmII. The latter will increase NOS activity and NO release. NO may either increase post-synaptic neuronal activity or may rapidly diffuse to the pre-synaptic neuron to enhance glutamate release. PKC activation via extracellular Ca^{2+} influx or intracellular Ca^{2+} release from IP3 stores via activation of phospholipase C (PLC) will phosphorylate serine/threonine sites in the cytoplasmic domain of NMDA-R to maintain the channel open.

within the NTS (20, 28, 31), and also reduce the magnitude of hypoxic ventilatory roll-off in a developmentally regulated fashion (9). The temporal characteristics of NOS-mediated modulation of the hypoxic ventilatory response indicate that the CaCmII-NOS pathway is more likely a slow and late component of the NMDA receptor activation in NTS neurons.

Initial evidence for an important role of PKC in cardiorespiratory control stems from work by Champagnat and Richter who demonstrated that phorbol ester-induced PKC activation is associated with increases in respiratory drive potentials (5). More recently, Haji and colleagues found that PKC inhibition reduced neuronal excitability of expiratory neurons in the ventral respiratory group of the cat (15).

In our laboratory, we initially examined the cardioventilatory responses elicited by administration of the systemically active, blood brain barrier permeable PKC inhibitor Ro 32-0432 (3, 4, 34). The two major findings of this study consisted in Ro 32-0432 inducing significant prolongations of expiratory duration, and markedly attenuating the ventilatory response to hypoxia but not to hypercapnia in unrestrained rats (12). We further examined the time course of PKC activation during hypoxia within the NTS (13). Significant isoform-selective increases in PKC translocation occurred within the NTS after 10 min hypoxic challenge with 10% O_2 balance N_2 (Figure 2). However, while in some PKC isoforms such as PKC-α, PKC-β, and PKC-δ, translocation patterns returned to normoxic levels (Figure 2), in PKC-γ increased translocation to the particulate subcellular fraction persisted despite onset of hypoxic ventilatory roll-off (13). Furthermore, NTS microinjec-

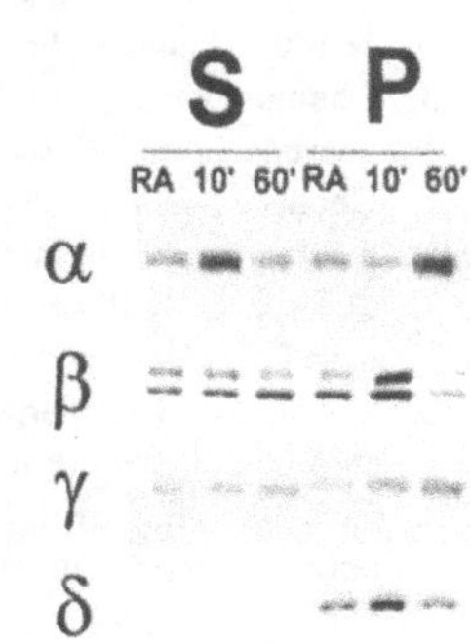

Figure 2. Western blots for PKC-α, PKC-β, PKC-γ, and PKC-δ of protein equivalents of soluble (S) and particulate (P) fractions of NTS lysates harvested from rats during normoxia (RA; left lane), at 10 min of a 10% O_2 challenge (mid-lane), and following onset of ventilatory roll-off (60′; right-lane).

tion to normoxic rats of the membrane permeable phorbol ester which activates PKC, phorbol 12-myristate 13-acetate (PMA) induced marked ventilatory enhancements (10). In contrast, microinjection of the inactive phorbol ester 4α-phorbol,12,13-didecanoate (4αPD) did not modify ventilatory measures (10). We also examined whether NTS microinjection of the PKC inhibitor {2-[1-(3-dimethylaminopropyl)-1H-indol-3-yl]-3(1H-indol-3-yl) maleimide, HCl} (BIM I) affected the hypoxic ventilatory response (11). BIM I administration was associated with significant attenuations of the hypoxic ventilatory response, which were not apparent when [2,3-*bis*(1H-Indol-3-yl)-N-methylmaleimide] (BIM V), the inactive analog was given (11).

3. SUMMARY

In summary, we propose that within the NTS, PKC activity changes provide a major second messenger pathway for the modulation of the hypoxic ventilatory response. The relative contributions and interactions of other kinases such as tyrosine kinases and CaCmII in the signal transduction cascade of the hypoxic ventilatory response within this neural structure remain to be defined.

ACKNOWLEDGMENTS

This study was supported in part by grants from the National Institute of Child Health and Development (HD-01072), the Maternal and Child Health Bureau (MCJ-229163), and the American Lung Association (CI-002-N). We are extremely grateful to Roche Products for generously providing Ro 32-0432.

REFERENCES

1. Ang, R.C., B. Hoop, and H. Kazemi. Role of glutamate as the central neurotransmitter in the hypoxic ventilatory response. *J. Appl. Physiol.* 72:1480–1487, 1992.
2. Avaldano, M.I, and N.G. Bazan. Rapid production of diacylglycerols enriched in arachidonate and stearate during early brain ischemia. *J. Neurochem.* 25:919–920, 1975.
3. Birchall, A.M., J. Bishop, D. Bradshaw, A. Cline, J. Coffey, L.H. Elliott, V.M. Gibson, A. Greenham, T.J. Hallam, W. Harris, C.H. Hill, A. Hutchings, A.G. Lamont, G. Lawton, E.J. Lewis, A. Maw, J.S. Nixon, D. Pole, J. Wadsworth, and S.E. Wilkinson. Ro 32–0432, a selective and orally active inhibitor of protein kinase C prevents T-cell activation. *J. Pharmacol. Exp. Ther.* 268:922–929, 1994.
4. Bit, R.A., P.D. Davis, L.H. Elliott, W. Harris, C.H. Hill, E. Keech, H. Kumar, G. Lawton, A. Maw, J.S. Nixon, D.R. Vesey, J. Wadsworth, and S.E. Wilkinson. Inhibitors of protein kinase C (3): Potent and highly selective bisindolylmaleimides by conformational restriction. *J. Med. Chem.* 36:21–29, 1993.
5. Champagnat, J., and D.W. Richter. Second messenger-induced modulation of the excitability of respiratory neurones. *NeuroReport* 4:861–863, 1993.
6. Chen, L., and L.Y.M. Huang. Protein kinase C reduces Mg^{2+} block of NMDA-receptor channels as a mechanism of modulation. *Nature* 356:521–523, 1992.
7. Dekker, L.V., P.N.E. De Graan, and W.H. Gispen. Transmitter release: target of regulation by protein kinase C? *Prog. Brain Res.* 89:209–233, 1991.
8. Gozal, D., J.E. Torres, Y.M. Gozal, and S.M. Littwin. Effect of nitric oxide synthase inhibition on cardiorespiratory responses in the conscious rat. *J. Appl. Physiol.* 81:2068–2077, 1996.
9. Gozal, D., E. Gozal, J.E. Torres, Y.M. Gozal, T.J. Nuckton, and P.J. Hornby. Nitric oxide modulates ventilatory responses to hypoxia in conscious developing rats. *Am J. Resp. Crit. Care Med.* 155:1755–1762, 1997.

10. Gozal, D., G.A. Holt, J.E. Torres, and E. Gozal. Protein kinase C (PKC) activation enhances cardioventilatory output in the conscious rat. *FASEB J.* 11:A350, 1997.
11. Gozal, D., G.A. Holt, J.E. Torres, and E. Gozal. Hypoxic ventilatory response is modulated by protein kinase C (PKC) activity within the nucleus tractus solitarius (NTS) of the conscious rat. *Am. J. Resp. Crit. Care Med.* 155:A299, 1997.
12. Gozal, D., G.R. Graff, J.E. Torres, S.G. Khicha, G.S. Nayak, N. Simakajornboon, and E. Gozal. Cardiorespiratory responses to systemic administration of a protein kinase C inhibitor in the conscious rat. *J. Appl. Physiol.* 1998 (In Press)
13. Gozal, D, and E. Gozal E. Hypoxic ventilatory roll-off is associated with decreases in protein kinase C activation within the nucleus tractus solitarius of the rat. *Brain Res.* 1997 (In Press)
14. Gszchwendt M., H. Leipbersperger, W. Kittstein, and F. Marks. Protein kinase C ζ and η in murine epidermis. TPA induces down regulation of PKC η but not PKC ζ. *FEBS Lett.* 307:151–155, 1992.
15. Haji, A., O. Pierrefiche, P.M. Lalley, and D.W. Richter. Protein kinase C pathways modulate respiratory pattern generation in the cat. *J. Physiol.* (London) 494:297–306, 1996.
16. Inoue M., A. Kishimoto, Y. Takai and Y. Nishizuka. Studies on a cyclic nucleotide-independent protein kinase and its proenzyme in mammalian tissues. II. Proenzyme and its activation by calcium dependent protease from the rat brain. *J. Biol. Chem.* 252:7610–7616, 1977.
17. Kazemi, H., and B. Hoop. Glutamic acid and gamma-aminobutyric acid neurotransmitters in central control of breathing. *J. Appl. Physiol.* 70:1–7, 1991.
18. Kraft, A.S., and W.B. Anderson. Phorbol esters increase the amount of Ca^{2+}, phospholipid- dependent protein kinase associated with plasma membrane. *Nature* 301:621–623, 1983.
19. Lin, J., C. Suguihara, J. Huang, D. Hehre, C. Devia, and E. Bancalari. Effect of N-methyl-D-aspartate receptor blockade on hypoxic ventilatory response in unanesthetized piglets. *J. Appl. Physiol.* 80:1759–1763, 1996.
20. Ma, S., F.M. Abboud, and R.B. Felder. Effects of L-arginine-derived nitric oxide synthesis on neuronal activity in nucleus tractus solitarius. *Am. J. Physiol.* (Regulatory, Integrative, Comp. Physiol.) 268:R487-R491, 1995.
21. Mizusawa, A., H. Ogawa, Y. Kikuchi, W. Hida, H. Kurosawa, S. Okabe, T. Takishima, and K. Shirato. In vivo release of glutamate in nucleus tractus solitarii of the rat during hypoxia. *J. Physiol. (London)* 478:55–65, 1994.
22. Moriyoshi, K., M. Masu, T. Ishii, R. Shigemoto, N. Mizuno, and S. Nakanishi. Molecular cloning and characterization of the rat NMDA receptor. *Nature* 354:31–37, 1991.
23. Nishizuka, Y. The molecular heterogeneity of protein kinase C and its implications for cellular regulation. *Nature* 334:661–665, 1988.
24. Nishizuka, Y. Intracellular signalling by hydrolysis of phospholipids and activation of protein kinase C. *Science* 258:607–614, 1992.
25. Ogawa, H., A. Mizusawa, Y. Kikuchi, W. Hida, H. Miki, and K. Shirato. Nitric oxide as a retrograde messenger in the nucleus tractus solitarii of rats during hypoxia. *J. Physiol. (London)* 486:495–504, 1995.
26. Ohtake, P.J., J.E. Torres, Y.M. Gozal, G.R. Graff, and D. Gozal. NMDA receptors mediate cardiorespiratory responses to afferent peripheral chemoreceptor input in the conscious rat. *J. Appl. Physiol.* 1998 (In press)
27. Soto-Arape, I., M.D. Burton, and H. Kazemi. Central amino acid neurotransmitters and the hypoxic ventilatory response. *Am. J. Resp. Crit. Care Med.* 151:1113–1120, 1995.
28. Tagawa, T., T. Imaizumi, S. Harada, T. Endo, M. Shiramoto, Y. Hirooka, and A. Takeshita. Nitric oxide influences neuronal activity in the nucleus tractus solitatius of rat brainstem slices. *Circ. Res.* 75:7075, 1994.
29. Tanaka, C., and Y. Nishizuka. The protein kinase C family for neuronal signaling. *Ann. Rev. Neurosci.* 17:551–567, 1994.
30. Tingley, W.G., K.W. Roche, A.K. Thompson, and R.L. Huganir. Regulation of NMDA receptor phosphorylation by alternative splicing of the C-terminal domain. *Nature* 364:70–73, 1993.
31. Torres, J.E., N.R. Kreisman, and D. Gozal. Nitric oxide modulates *in vitro* intrinsic optical signal and neural activity in the nucleus tractus solitarius of the rat. *Neurosci. Lett.* 1997; 232:175–178.
32. Urushihara, H., M. Tohda, and Y. Nomura. Selective potentiation of *N*-methyl-d-aspartate-induced current by protein kinase C in *Xenopus* oocytes injected with rat brain mRNA. *J. Biol. Chem.* 267:11697–11700, 1992.
33. Ways, D.K., P.P. Cook, C. Webster, and P.J. Parker. Effect of phorbol ester on protein kinase C ζ. *J. Biol.Chem.* 267:4799–4805, 1992.
34. Wilkinson, S.E., P.J. Parker, and J.S. Nixon. Isoenzyme specificity of bisindolylmaleimides, selective inhibitors of protein kinase C. *Biochem. J.* 294:335–337, 1993.

11

SYNAPTIC CONNECTIONS TO PHRENIC MOTONEURONS IN THE DECEREBRATE RAT

G.-F. Tian, J. H. Peever, and J. Duffin

Department of Physiology
University of Toronto
Toronto, Canada

1. INTRODUCTION

As Bianchi et al. (2) point out in their recent review of respiratory neurophysiology, the rat is becoming the "animal of choice" for experimentation. This statement is particularly true for neuroanatomical tracing experiments. However, anatomical tracing often does not identify neurons as respiratory nor does it indicate the excitatory or inhibitory nature of interconnections. This information is provided by electrophysiological experimentation, and has mostly been obtained from earlier experiments on cats. It is therefore important to discover whether such information about functional connections among respiratory neurons in cats is also true of rats, rather than to assume that the neuronal organisation is similar. We set out to discover the determinants of phrenic motoneuron membrane potential trajectories in decerebrate rats in a series of projects using electrophysiological techniques. The projects are described in "Results" in separate sections; each with a brief review of previous knowledge and our findings. We have emphasised points of difference between rats and cats.

2. METHODS

The preparation used for these experiments was the vagotomized, paralysed, ventilated and decerebrated rat, and has been fully described in previous reports (33, 34). Intracellular and extracellular recordings of neuron activity in the medulla and spinal cord were made using microelectrodes, and the projections of the recorded neurons were located using microstimulation to antidromically activate their axons within the spinal cord. Phrenic motoneurons were identified by antidromic activation from the phrenic nerve. Interconnections between neurons were detected using either cross-correlation of their action potential discharges or spike-triggered averaging of their membrane potentials. Explanations of these techniques are available in previous reports; for example (6) and (13) respectively.

Advances in Modeling and Control of Ventilation, edited by Hughson *et al.*
Plenum Press, New York, 1998.

3. RESULTS

Figure 1 shows the locations of several types of respiratory neurons recorded in these experiments, and in particular those that project to the spinal cord and the phrenic motornucleus.

3.1. Phrenic Motoneuron Membrane Potentials

In rats and cats, phrenic nerve discharge often occurs in 3 stages; see reviews (2, 26). These stages are an incrementing discharge during inspiration, a decrementing discharge during early expiration, and silence during late expiration and they correspond to those predicted

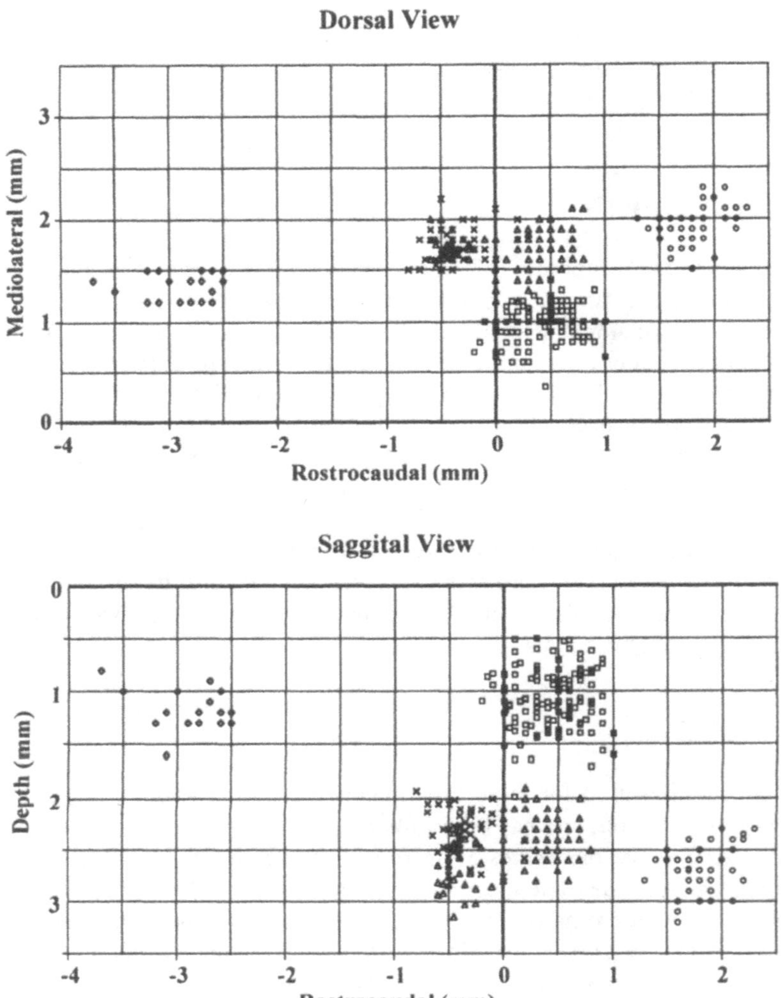

Figure 1. Locations of recorded neurons; Bötzinger-complex, bulbospinal expiratory neurons (○); dorsal-group inspiratory neurons (□); ventral-group, bulbospinal inspiratory neurons (△); ventral-group, expiratory vagal motoneurons (×) and upper-cervical inspiratory neurons (◇). Co-ordinates were estimated from micromanipulator readings and are relative to the obex at co-ordinates 0,0.

by either 3-phase (31) or 2-phase (8) models of the respiratory oscillator (32). However, previous observations of phrenic motoneuron membrane potential trajectories show only 2 stages, inspiratory and expiratory. In cats, the expiratory membrane potential trajectories usually consist of an abrupt repolarization at the onset of expiration followed by a slow hyperpolarization during expiration reaching a maximum at the end of the expiratory phase. Berger (1) also observed motoneurons (his type B) without a sudden repolarization but with the slow hyperpolarization throughout expiration, as have others (18, 23, 27).

In anaesthetised adult rats, the expiratory trajectories display a rapid repolarization at the onset of expiration followed by a slowly increasing hyperpolarization during the remainder of expiration (16) similar to previous observations in cats. The membrane potential trajectories observed in neonatal rats are quite different; during inspiration the membrane potential declines after its initial rapid rise, and during expiration the rapid repolarization at the start is absent (21).

We found evidence for a 3-stage pattern of membrane potential trajectory in intracellular recordings of 128 phrenic motoneurons in decerebrate rats (Figure 2). Their membrane potentials (mean ± SD, –59.7 ± 7.0 mV) exhibited respiratory fluctuations (12.3 ± 3.4 mV). All phrenic motoneurons depolarised during inspiration and most (105/128, 82%) hyperpolarized during expiration in two distinct stages; an abrupt repolarisation then slow depolarisation in early expiration, followed by a further hyperpolarization in late expiration, often (58/105, 55%) discharging during early expiration (29.2 ± 13.4 spikes/s). We concluded that the depolarisation and discharge of action potentials during early expiration accounts for the declining pattern of phrenic nerve discharge often observed during early expiration in anaesthetised and particularly decerebrated rats.

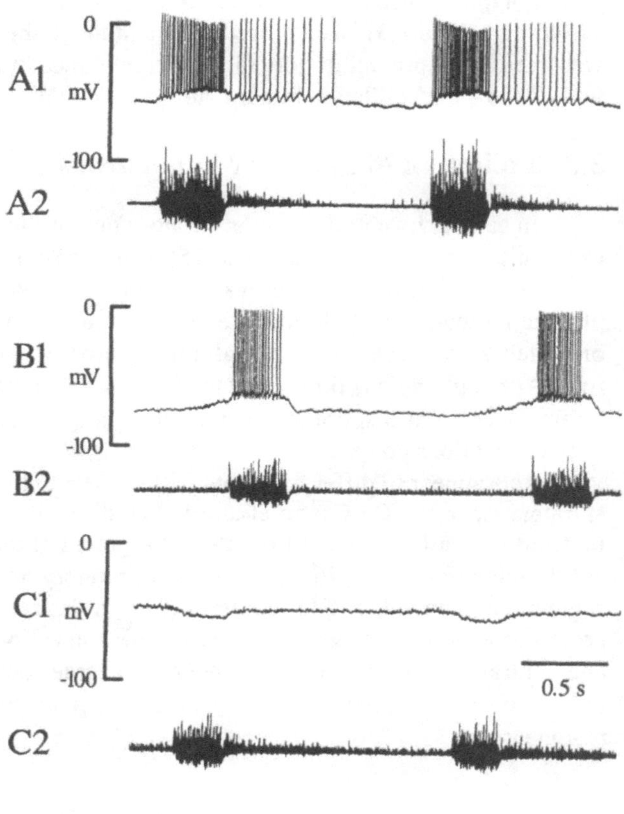

Figure 2. Membrane potential trajectories for phrenic motoneurons with two distinct expiratory phases. A: Motoneuron with discharge during early expiration. B: Motoneuron without expiratory discharge. C: Motoneuron without discharge. In each example, trace 1 is the membrane potential and trace 2 is the phrenic nerve discharge.

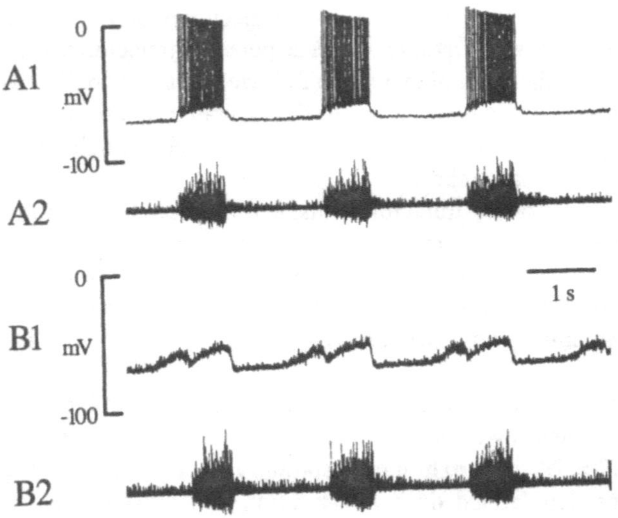

Figure 3. Reversal of inhibition in a phrenic motoneuron. A: Before iontophoresis. B: After 45 minutes of chloride iontophoresis at 10 nA. In each example, trace 1 is the membrane potential and trace 2 is the phrenic nerve discharge.

As a part of our attempts to determine the synaptic inputs responsible for the 3-stage pattern of membrane potential trajectories we used chloride iontophoresis to reverse inhibitory hyperpolarization to depolarisation in 58 phrenic motoneurons in order to observe the pattern of synaptic inhibition. In all of them, the hyperpolarizing waves during late expiration were easily reversed to depolarising waves after about 1 minute. Similar observations were made by Hayashi and Fukuda (1995) in anaesthetised rats. However, unlike these investigators we did not observe reversal waves during early expiration, despite injecting hyperpolarizing currents (5–10 nA) into some phrenic motoneurons for as long as 30–45 min (Figure 3). We concluded that although the hyperpolarization in late expiration was due to synaptic inhibition, the decrease in membrane potential during early expiration was due to a disfacilitation rather than synaptic inhibition.

3.2. Excitation of Phrenic Motoneurons

In cats, few ventral-group inspiratory neurons have been shown to monosynaptically excite either phrenic motoneurons (15) or intercostal motoneurons (24), but a majority of dorsal-group inspiratory neurons have been shown to monosynaptically excite both phrenic motoneurons (19) and intercostal motoneurons (9). These connections were demonstrated to be made via unilateral, mostly crossed, projections of these bulbospinal neurons. Although there is little evidence for monosynaptic excitation of phrenic motoneurons by upper-cervical inspiratory neurons in cats (6), Nakazono and Aoki (28) showed that such connections do exist.

Rats appear to differ from cats with respect to the spinal projections of medullary inspiratory neurons. De Castro et al. (4) found few dorsal-group inspiratory neurons could be antidromically activated from the C3 segment of the spinal cord in rats. Our attempt to antidromically activate 76 dorsal-group inspiratory neurons from the spinal cord at the C7 segment, had only 4 (5.3%) successes (3 contralateral, 1 ipsilateral). In the same study we cross-correlated dorsal-group inspiratory neuron action potentials with ipsilateral (n = 56) and contralateral (n = 20) phrenic nerve discharges but found common activation peaks in only 2 and 3 cases respectively, and no evidence for monosynaptic connections to phrenic motoneurons.

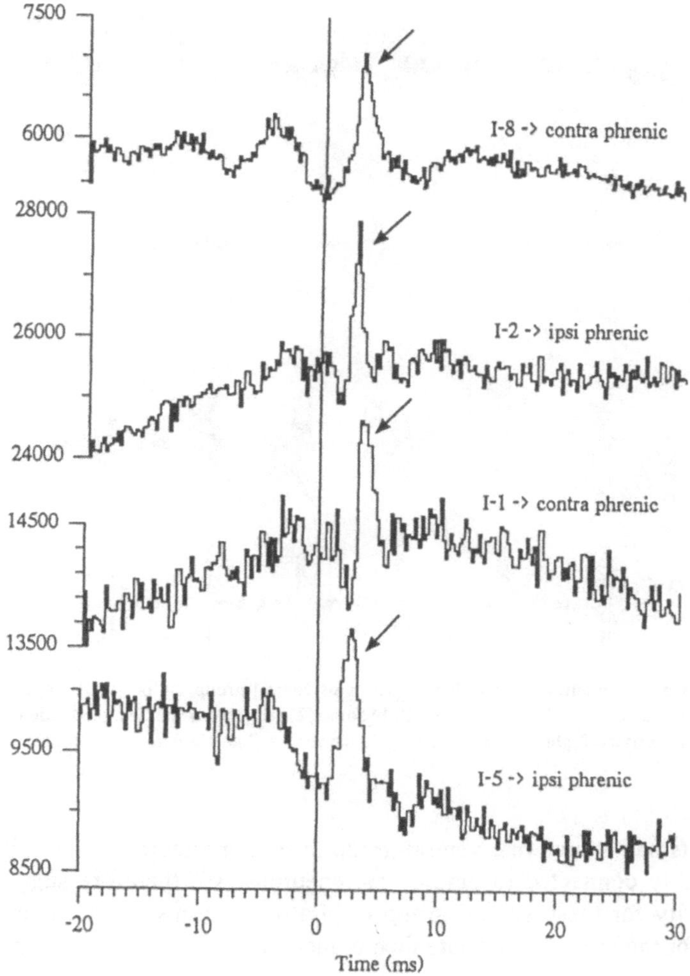

Figure 4. Examples of cross-correlation histograms computed between action potentials of ventral-group, bulbospinal inspiratory neurons and the discharge of ipsilateral and contralateral phrenic nerves. Narrow peaks at short latencies were interpreted as evidence for monosynaptic excitatory connections. Bin widths 0.2 ms; vertical axes counts/bin.

In rats, phrenic motoneurons are monosynaptically excited during inspiration by almost all ventral-group bulbospinal inspiratory neurons via predominantly bilateral connections (10, 34), and to a lesser extent by upper-cervical inspiratory neurons (33). These experiments used cross-correlation to detect connections with results such as those shown in Figure 4.

While these connections could account for the excitation of phrenic motoneurons during inspiration, the source of their excitation during early expiration was not immediately apparent. However, we recorded the pattern of extracellular activity for 137 ventral-group inspiratory neurons antidromically activated from C7 (35). Although 79% were active only during inspiration, 21% were also active during early expiration (Figure 5). While these particular neurons were not tested for connections to phrenic motoneurons,

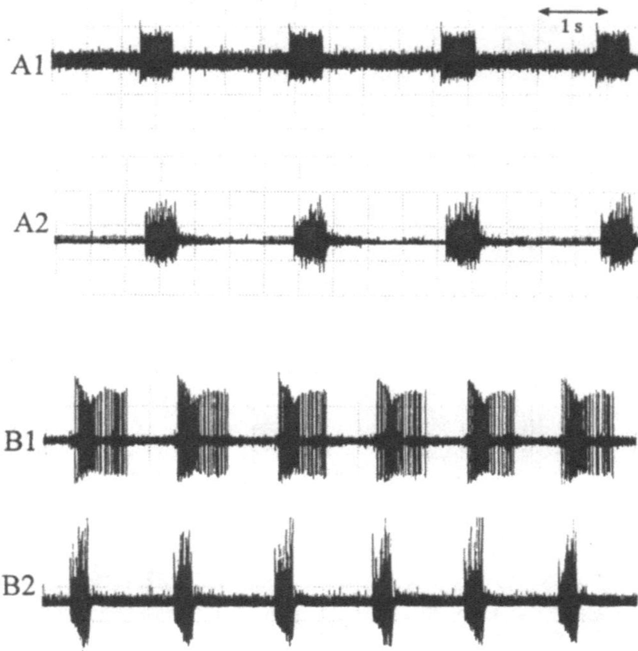

Figure 5. Examples of extracellularly recorded activities of ventral-group, bulbospinal inspiratory neurons. A: Most (79%) neurons fired only during inspiration. B: In some (21%) neurons the firing extended into early expiration. Traces: 1 neuron activity, 2 phrenic nerve activity. Time scale = 2 grid squares/s as marked.

past experiments had shown that ventral-group inspiratory neurons with bulbospinal projections invariably connected to phrenic motoneurons. We therefore suggested that the pattern of activity for these ventral-group inspiratory neurons active in early expiration could account for the excitation of phrenic motoneurons during early expiration.

3.3. Inhibition of Phrenic Motoneurons

Electrophysiological studies in cats have shown that expiratory neurons f the Bötzinger complex with an augmenting pattern of discharge have widespread bilateral collateral projections within the medulla, to both the dorsal (20) and ventral (30) respiratory groups, as well as projections to phrenic motoneurons (14) and upper-cervical inspiratory neurons (22) in the spinal cord. They monosynaptically inhibit dorsal group inspiratory neurons (25), phrenic motoneurons (23), ventral-group neurons (12, 13, 17) including the propriobulbar decrementing inspiratory neurons (7), as well as themselves (11).

However, very little is known about the Bötzinger complex in rats. Anatomical tracing techniques have confirmed its existence and possible projections to the spinal cord (5, 29), but a recent morphological study (3) found few expiratory neurons with an augmenting pattern of discharge and little evidence for spinal projections or medullary synaptic connections, although extensive ipsilateral medullary projections were observed.

We therefore sought evidence for connections to phrenic motoneurons from bulbospinal, Bötzinger complex neurons with an augmenting pattern of activity during late expiration in rats. Action potentials of 38 of these unilaterally projecting neurons were used as

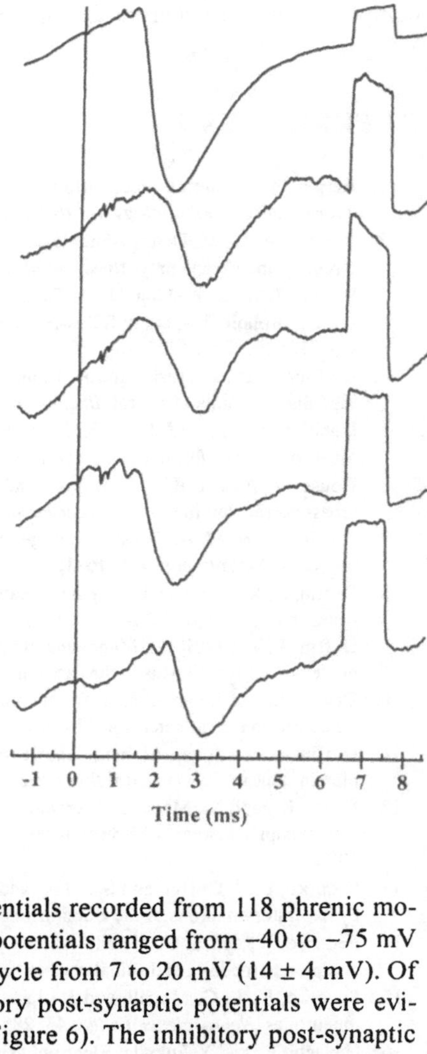

Figure 6. Examples of spike-triggered averages of phrenic motoneuron membrane potentials. The trigger pulses (0.2 ms) were derived from the action potentials of bulbospinal, Bötzinger-complex expiratory neurons 0.5 ms after their rising phase and in some traces their artefacts are visible. The averages are of 2000 sweeps, comprising four consecutive 500 sweep averages. The calibration pulse between 6.5 and 7.5 ms is 100 µV.

triggers for computing averages of intracellular potentials recorded from 118 phrenic motoneurons. Resting phrenic motoneuron membrane potentials ranged from –40 to –75 mV (–56 ± 8 mV) and fluctuations with the respiratory cycle from 7 to 20 mV (14 ± 4 mV). Of the 118 spike-triggered averages computed, inhibitory post-synaptic potentials were evident in 18 (~15%), evoked by 10 (~26%) neurons (Figure 6). The inhibitory post-synaptic potential amplitudes varied from 35 to 550 µV (105 ± 113 µV), 10–90 % fall times from 0.4 to 0.9 ms (0.63 ± 0.17 ms) and half-amplitude widths from 1.3 to 3.2 ms (2.0 ± 0.52 ms). Most (16/95, ~17%) of these averages displaying post-synaptic potentials were associated with ipsilateral trigger neurons but some (2/23, ~9%) resulted from contralateral trigger neurons. We concluded that the bulbospinal, Bötzinger complex neurons with an augmenting pattern of activity during late expiration were the source of inhibition for phrenic motoneurons during late expiration.

4. SUMMARY

Phrenic motoneuron membrane potential trajectories in decerebrate rats exhibit three stages; depolarisation during inspiration, a decreased depolarisation during early expiration and hyperpolarization during late expiration. These trajectories are a result of excitation by ventral-group medullary inspiratory neurons and upper-cervical inspiratory neurons during inspiration and the early part of expiration, and inhibition from Bötzinger-complex expiratory neurons during the late part of expiration.

REFERENCES

1. Berger, A. J. Phrenic motoneurons in the cat: Subpopulations and nature of respiratory drive potentials. *Journal of Neurophysiology* 42: 76–90, 1979.
2. Bianchi, A. L., M. Denavit-Saubié, and J. Champagnat. Central control of breathing in mammals: neuronal circuitry, membrane properties, and neurotransmitters. *Physiological Reviews* 75: 1–45, 1995.
3. Bryant, T. H., S. Yoshida, D. De Castro, and J. Lipski. Expiratory neurons of the Bötzinger complex of the rat: a morphological study following intracellular labelling with biocytin. *Journal of Comparative Neurology* 335: 267–282, 1993.
4. De Castro, D., J. Lipski, and R. Kanjhan. Electrophysiological study of dorsal respiratory neurons in the medulla oblongata of the rat. *Brain Research* 639: 49–56, 1994.
5. Dobbins, E. G., and J. L. Feldman. Brainstem network controlling descending drive to phrenic motoneurons in rat. *Journal of Comparative Neurology* 347: 64–86, 1994.
6. Douse, M. A., J. Duffin, D. Brooks, and L. Fedorko. Role of upper cervical inspiratory neurons studied by cross- correlation in the cat. *Experimental Brain Research* 90: 153–162, 1992.
7. Duffin, J., and M. A. Douse. Bötzinger expiratory neurones inhibit propriobulbar decrementing inspiratory neurones. *NeuroReport* 4: 1215–1218, 1993.
8. Duffin, J., K. Ezure, and J. Lipski. Breathing rhythm generation: focus on the rostral ventrolateral medulla. *News in Physiological Sciences* 10: 133–140, 1995.
9. Duffin, J., and J. Lipski. Monosynaptic excitation of thoracic motoneurones by inspiratory neurones of the nucleus tractus solitarius in the cat. *Journal of Physiology* 390: 415–431, 1987.
10. Duffin, J., and J. van Alphen. Bilateral connections from ventral group inspiratory neurons to phrenic motoneurons in the rat determined by cross-correlation. *Brain Research* 694: 55–60, 1995.
11. Duffin, J., and J. van Alphen. Cross-correlation of augmenting expiratory neurons of the Bötzinger complex in the cat. *Experimental Brain Research* 103: 251–255, 1995.
12. Ezure, K., and M. Manabe. Decrementing expiratory neurons of the Bötzinger complex. II. Direct inhibitory synaptic linkage with ventral respiratory group neurons. *Experimental Brain Research* 72: 159–166, 1988.
13. Fedorko, L., J. Duffin, and S. J. England. Inhibition of inspiratory neurons of the nucleus retroambigualis by expiratory neurons of the Bötzinger complex in the cat. *Experimental Neurology* 106: 74–77, 1989.
14. Fedorko, L., and E. G. Merrill. Axonal projections from the rostral expiratory neurones of the Bötzinger complex to medulla and spinal cord in the cat. *Journal of Physiology* 350: 487–496, 1984.
15. Fedorko, L., E. G. Merrill, and J. Lipski. Two descending medullary inspiratory pathways to phrenic motoneurones. *Neuroscience Letters* 43: 285–291, 1983.
16. Hayashi, F., and Y. Fukuda. Electrophysiological properties of phrenic motoneurons in adult rats. *Japanese Journal of Physiology* 45: 69–83, 1995.
17. Jiang, C., and J. Lipski. Extensive monosynaptic inhibition of ventral respiratory group neurons by augmenting neurons in the Bötzinger complex in the cat. *Experimental Brain Research* 81: 639–648, 1990.
18. Jodkowski, J. S., R. D. Guthrie, and W. E. Cameron. The activity pattern of phrenic motoneurons during the aspiration reflex: An intracellular study. *Brain Research* 505: 187–194, 1989.
19. Lipski, J., L. Kubin, and J. Jodkowski. Synaptic action of Rβ neurons on phrenic motoneurons studied with spike-triggered averaging. *Brain Research* 288: 105–118, 1983.
20. Lipski, J., and E. G. Merrill. Electrophysiological demonstration of the projection from expiratory neurones in rostral medulla to contralateral dorsal respiratory group. *Brain Research* 197: 521–524, 1980.
21. Liu, G. S., J. L. Feldman, and J. C. Smith. Excitatory amino acid-mediated transmission of inspiratory drive to phrenic motoneurons. *Journal of Neurophysiology* 64: 423–436, 1990.
22. Mateika, J. H., and J. Duffin. The connections from Bötzinger expiratory neurons to upper cervical inspiratory neurons in the cat. *Experimental Neurology* 104: 138–146, 1989.
23. Merrill, E. G., and L. Fedorko. Monosynaptic inhibition of phrenic motoneurons: a long descending projection from Bötzinger neurons. *Journal of Neuroscience* 4: 2350–2353, 1984.
24. Merrill, E. G., and J. Lipski. Inputs to intercostal motoneurons from ventrolateral medullary respiratory neurons in the cat. *Journal of Neurophysiology* 57: 1837–1853, 1987.

25. Merrill, E. G., J. Lipski, L. Kubin, and L. Fedorko. Origin of the expiratory inhibition of nucleus tractus solitarius inspiratory neurones. *Brain Research* 263: 43–50, 1983.
26. Monteau, R., and G. Hilaire. Spinal respiratory motoneurons. *Progress in Neurobiology* 37: 83–144, 1991.
27. Monteau, R., M. Khatib, and G. Hilaire. Central determination of recruitment order: intracellular study of phrenic motoneurons. *Neuroscience Letters* 56: 341–346, 1985.
28. Nakazono, Y., and M. Aoki. Excitatory connections between upper cervical inspiratory neurons and phrenic motoneurons in cats. *Journal of Applied Physiology* 77: 679–683, 1994.
29. Núñez-Abades, P. A., R. Pásaro, and A. L. Bianchi. Localization of respiratory bulbospinal and propriobulbar neurons in the region of the nucleus ambiguus of the rat. *Brain Research* 568: 165–172, 1991.
30. Otake, K., H. Sasaki, K. Ezure, and M. Manabe. Axonal projections from Bötzinger expiratory neurons to contralateral ventral and dorsal respiratory groups in the cat. *Experimental Brain Research* 72: 167–177, 1988.
31. Ramirez, J. M., and D. W. Richter. The neuronal mechanisms of respiratory rhythm generation. *Current Opinion in Neurobiology* 6: 817–825, 1996.
32. Ryback, I. A., J. F. R. Paton, and J. S. Schwaber. Modeling neural mechanisms for genesis of respiratory rhythm and pattern. III. Comparison of model performances during afferent nerve stimulation. *Journal of Neurophysiology* 77: 2027–2039, 1997.
33. Tian, G.-F., and J. Duffin. Connections from upper cervical inspiratory neurons to phrenic and intercostal motoneurons studied with cross-correlation in the decerebrate rat. *Experimental Brain Research* 110: 196–204, 1996.
34. Tian, G.-F., and J. Duffin. Spinal connections of ventral-group bulbospinal inspiratory neurons studied with cross-correlation in the decerebrate rat. *Experimental Brain Research* 111: 178–186, 1996.
35. Tian, G.-F., and J. Duffin. Synchronization of ventral-group, bulbospinal inspiratory neurons in the decerebrate rat. *Experimental Brain Research* 117: 479–487, 1997.

PHRENIC NERVE RESPONSE TO GLUTAMATE ANTAGONIST MICROINJECTION IN THE VENTRAL MEDULLA

John L. Beagle, Bernard Hoop, and Homayoun Kazemi

Pulmonary and Critical Care Unit
Medical Services
Massachusetts General Hospital
Harvard Medical School
Boston, Massachusetts 02114

1. INTRODUCTION

Effects of amino acid neurotransmitters on central respiratory drive roughly parallel their excitation or inhibition of neurons[1]. The major amino acid neurotransmitter glutamic acid is of particular interest because it stimulates ventilatory drive centrally at sites and via mechanisms within the surface of the ventral medulla (VM). Glutamate metabolism in the brain is directly related to CO_2 metabolism and fixation in the brain. Decarboxylation of glutamate via the enzyme glutamic acid decarboxylase (GAD) localized substantially in nerve endings, results in formation of the inhibitory amino acid and central respiratory depressant, γ-aminobutyric acid (GABA)[4]. Recent studies[5,6] have shown that the classic biphasic ventilatory response to acute hypoxia has a glutamatergic component, namely, the initial hyperventilatory response is mediated in part by central glutamate. Topical application of glutamatergic receptor antagonists during normoxia to the surface of the VM diminishes central ventilatory drive via reduction in phrenic nerve output[11]. The present study was therefore undertaken with multiple microinjections of the selective noncompetitive N-methyl-D-aspartate (NMDA) receptor antagonist MK-801 to localize in the VM the effects of blocking glutamate receptor on phrenic nerve output and to quantitate the effects via a model of chemoreceptor-antagonist interaction.

2. METHODS

Fifteen anesthetized (2.5% isoflurane) male Sprague-Dawley rats (300–350 g) were ventilated mechanically with 30% O_2 through a cervical tracheostomy to maintain normal

Advances in Modeling and Control of Ventilation, edited by Hughson *et al.*
Plenum Press, New York, 1998.

pH_a and P_aCO_2. Body temperature was maintained at 37.5°C by a heat lamp The cephalad end of the trachea and the esophagus were reflected rostrally and the musculature and bony surface between the tympanic bullae were removed, exposing the surface of the VM. The exposed area extended laterally to the bullae and rostrally 2 to 3 mm above the point of exit of the sixth cranial nerve. The dura was opened and fixed to the bony edges. The underlying surface of the VM remained covered by a pool of cerebrospinal fluid (CSF). Bilateral vagotomies were performed, and the phrenic nerve on one side was exposed, placed on a bipolar AgCl electrode, and grounded. Phrenic nerve activity consisting typically of bursts of about a dozen spikes (−1.0 to 1.0 mV) within 0.5 sec at burst rate of the order of 1 Hz was rectified, amplified, monitored during experiments on a chart recorder, integrated electronically and acquired at a sampling rate of 25 Hz for subsequent analysis. Individual peak heights of successive integrated neural bursts were determined from the differences between local maxima and minima, and burst-by-burst phrenic nerve output (peak burst height times frequency) was calculated.

Glutamate receptor antagonist MK-801 was prepared for microinjection as follows: MK-801 in crystalline form was dissolved in artificial CSF (140 mEq/L Na^+, 120 mEq/L Cl^-, 25 mmol/L HCO_3^-, 2.6 mEq/L K^+, 4.0 mM/L Ca^{2+}, and 2.0 mM/L Mg^{2+}) equilibrated with 95% O_2 + 5% CO_2 to pH 7.35–7.40 and PCO_2 36–42 torr. Prior to injection, pH was adjusted to 7.40 ± 0.03. Concentration of MK-801 was 8 mM, i.e., a 13.8-nL microinjectate volume contained 110 pmol MK-801. This dose was selected on the basis of earlier data from this and other laboratories which showed complete antagonism of the respiratory response at this concentration[2,4]. Artificial CSF and solutions were freshly prepared weekly. Microinjections of MK-801 were made in the rostral, intermediate and caudal chemosensitive areas of the surface of the VM at bilateral sites on a 0.25-mm rectangular grid, as diagrammed in Figure 1a.

In Figure 1a, sites are indicated at which a 10 μm diam. micropipette was placed under stereotaxic control in the chemosensitive areas of the VM with reference to coordinates taken from a standard atlas of the rat brain[10] and injected with 13.8 nL of artificial CSF containing glutamate antagonist MK-801. The micropipette was then withdrawn, moved to the corresponding contralateral site, advanced to the same depth as on the first

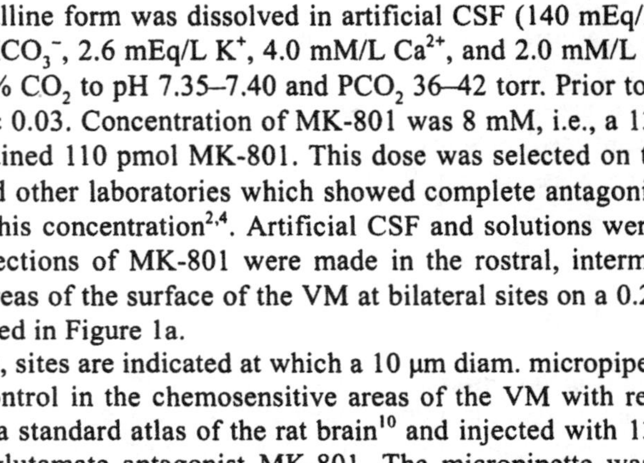

Figure 1. a. (Left) Schematic diagram of adult rat brainstem (ventral aspect) with MK-801 bilateral microinjection sites (solid circles shown on one side only). Abbreviations: C-R caudal-rostral; V VI IX X XII cranial nerves; IL interaural line; BA basilar artery; AICA anterior inferior cerebellar artery; VA vertebral artery; RA rostral area; IA intermediate area; CA caudal area. **b.** (Right) Representative measurement of phrenic nerve output (activity in arbitrary units/min) vs minutes post injection of 13.8 nL 8 mM MK-801 bilaterally. Injection site was 1.25 mm caudal to AICA, 1.25 mm lateral to midline, and 1.50 mm below the ventral medullary surface. Solid curve is calculated (cf text).

side and the injection repeated within 30 seconds of the first injection. Animals were studied during relative normoxia ($FiO_2 = 0.3$). Each animal received one to three MK-801 microinjections bilaterally. All initial microinjections were preceded by at least a ten-minute period of no intervention to assure baseline phrenic neurogram stability. Microinjection of MK-801 at specific sites in the VM caused depression of phrenic nerve output, a representative measurement of which is shown in Figure 1b.

As shown in Figure 1b, phrenic output typically decreased after bilateral microinjection, then returned to pre-injection level. The solid curve in Figure 1b is a representative calculation of phrenic nerve response based on a kinetic model of neurotransmitter chemoreception[3,4]. In brief, glutamatergic receptor sites on ventral medullary respiratory neurons, when antagonized by MK-801, reduce ventilation in proportion to the relative concentration of antagonized receptors. Antagonist molecules interact with vacant receptor sites governed by apparent binding constant K which is the ratio of dissociation to association rate constants of antagonist to receptor site. This interaction results in formation of an antagonist-blocked neural molecular complex whose ventral medullary population density (concentration) c relative to experimentally accessible total receptor concentration r determines relative phrenic nerve output. With injection of MK-801, phrenic nerve activity V is reduced from its initial value V_0 at injection time $t = 0$ in proportion to the mole fraction c/r of MK-801-antagonized receptors, i.e., $V = V_0 - V_0(c/r)$, where equilibrium c/r is represented by the classical chemical law of mass action (Michaelis-Menton kinetic), $c/r = a/(a + K)$, where a is the time-dependent MK-801 concentration at receptor sites taken as $a = jt \exp(-kt)$, where j is the binding rate of MK-801 to receptor sites, and where k is rate of MK-801 removal from injection sites. Delay times of response from end of second injection ranged from 0.0 to 3.4 minutes. All calculated curves were fitted by eye to the data and no sensitivity analysis was performed. In the present analysis, the binding rate j is taken as the product of k and the initial concentration (8 mM) of injected MK-801.

3. RESULTS

Results of 29 bilateral injections in fifteen animals are summarized in Figure 2.

Figure 2 shows mean (± SEM) binding rate j (mM/min) and apparent binding constant K (mM) vs. caudal-rostral distance (left) and vs. depth below the medullary surface (right). Binding rate j is largest (19 ± 3 mM/min) at 0.50 mm below the VM surface and

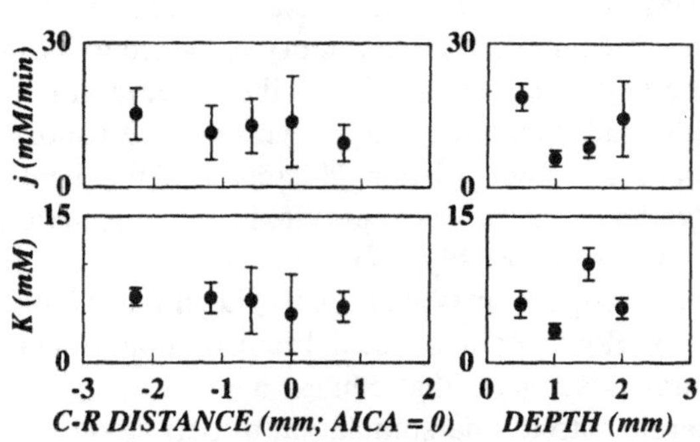

Figure 2. Binding rate j in mM/min and apparent binding constant K in mM vs caudal-rostral (C-R) distance in mm, with anterior inferior cerebellar artery taken as zero (left panels), and vs. depth in mm below ventral medullary surface.

decreases to 6 ± 2 mM/min at 1.0 mm depth ($P < 0.05$). Apparent binding constant K is 6 ± 1 mM at 0.5 mm depth and tends to a maximum of 10 ± 2 mM (n.s.) at a depth of 1.50 mm. On the left in Figure 2, depth- and laterally-averaged values of j and K are uniform to MK-801 injections at sites from 0.50 to 1.25 mm lateral to midline and over a 3.5-mm caudal-to-rostral distance centered approximately on the interface between the rostral and intermediate chemosensitive areas. That is, phrenic nerve responses to bilateral MK-801 injection show no significant differences along the caudal-rostral axis, suggesting that glutamate modulates respiration centrally via overlapping chemosensitive areas in the VM.

4. DISCUSSION

The CNS amino acid neurotransmitter glutamate has an excitatory effect on resting ventilation[1,2,9,11]. Previous work has shown that increased level of glutamate, whether by exogenous or endogenous means, stimulates ventilation[2,4]. Phrenic nerve output is also decreased after direct application of glutamate antagonist MK-801 to the surface of the VM[11]. NMDA receptor blockade with microinjections of MK-801 at specific sites in chemosensitive areas of the VM significantly decreases phrenic nerve output. Phrenic output decreased only after bilateral MK-801 microinjections (Figure 1b), which differ from effects on respiration of unilateral glutamate injection. The rate j at which MK-801 antagonizes (binds to) receptors is largest at injection sites nearest the surface of the VM and decreases with depth. Decrease in MK-801 binding rate j following bilateral microinjection was observed in the lateral portions of the rostral and intermediate areas of the VM (Figure 2, upper right). These findings support the hypothesis that glutmatergic mechanisms in the VM are important in maintaining phrenic nerve activity under normal conditions. The present results do not allow speculation about glutamatergic neuronal connections between specific chemosensitive areas of the VM and the nucleus of the solitary tract (NTS) where afferents from peripheral chemoreceptors terminate. However, anatomical studies have demonstrated neural projections between the NTS and the VM and that neural projections between the NTS and phrenic nuclei pass through VM intermediaries[8].

The importance of mechanisms of excitatory amino acid reception in central respiratory control was demonstrated by Mitra et al.[7] with topical application of N-methyl-D-aspartate (NMDA) to the intermediate chemosensitive area, which led these investigators to suggest a common respiratory interneuron pathway for the respiratory stimulatory effects of both CO_2 and NMDA. These and other results[6] lend support to our findings of the importance of the ventral medulla in the ventilatory response to hypoxia and the fact that blocking certain glutamatergic receptors abolishes the hyperventilation of hypoxia. Glutamate NMDA receptor blockade is unique to hypoxia, since the ventilatory response to acetylcholine and CO_2 remained intact[11]. Brain glutamate metabolism during hypoxia and peripheral chemodenervation suggest that in normoxic animals, peripheral chemodenervation reduces glutamate turnover, possibly reflecting a reduction in neuronal glutamatergic activity. Hypoxia alone decreases maximal glial glutamine efflux in both intact and chemodenervated animals. However, increase in brain ammonia clearance in intact but not in chemodenervated animals suggests that central hypoxia increases glutamate turnover and therefore glutamatergic activity via stimulation of peripheral chemoreceptors[5]. In the present work, values of MK-801 binding rate j are greatest close to the surface of the VM. This confirms the observation that topical application of MK-801 directly onto the intermediate chemosensitive area leads to depression of phrenic nerve output.

It should be emphasized that passive diffusion of MK-801 beyond the sites of microinjection prevent delineation of differential phrenic nerve responses which may exist between lateral portions of the rostral and intermediate chemosensitive areas. We conclude that of the neuroactive agents which affect resting ventilatory drive, bilateral microinjections of a specific antagonist of glutamate receptor into chemosensitive areas of the VM identify overlapping sites where neurotransmitter glutamate exerts its effect. These findings suggest the presence of tonic glutamergic inputs to receptors located in the rostral and intermediate zones of the VM which modulate central ventilatory drive.

ACKNOWLEDGMENTS

The authors wish to thank Drs. M.D. Burton, D.C. Johnson, I. Soto-Arape and H.J. White for assistance and valuable discussion. This work was supported in part by the U.S. Department of Health and Human Services of the Public Health Service (National Institutes of Health).

REFERENCES

1. Bianchi, A.L., M. Denavit-Saubie, and J. Champagnat. Central control of breathing in mammals: neuronal circuitry, membrane properties, and neurotransmitters. *Physiol. Rev.* 75:1–45, 1995.
2. Chiang, C.H., P. Pappagianopolous, B. Hoop, and H. Kazemi. Central cardiorespiratory effects of glutamate in dogs. *J. Appl. Physiol.* 60:2056–62, 1986
3. Hoop, B., M.R. Masjedi, V.E. Shih, and H. Kazemi. Brain glutamate metabolism during hypoxia and peripheral chemodenervation. *J. Appl. Physiol.* 69:147–54, 1990.
4. Kazemi, H., and B. Hoop. Glutamic acid and γ-aminobutyric acid neurotransmitters in central control of breathing. *J. Appl. Physiol.* 70:1–7, 1991.
5. Kazemi, H., J.L. Beagle, T. Maher, and B. Hoop. Afferent input from peripheral chemoreceptors in response to hypoxia and amino acid neurotransmitter generation in the medulla. In: *Frontiers in Arterial Chemoreception*, edited by P. Zapata, C. Eyzaguirre, and R.W. Torrance. New York: Plenum, 1996, pp. 365–9.
6. Lin, J, C. Suguihara, J. Huang, D. Hehre, C. Devia, and E. Bancalari. Effect of N-methyl-D-aspartate-receptor blockade on hypoxic ventilatory response in unanesthetized piglets. *J. Appl. Physiol.* 80:1759–1763,1996.
7. Mitra, J., N.R. Prabhakar, J.L. Overholt, and N.S. Cherniack. Respiratory effects of N-methyl-D-aspartate on the ventrolateral medullary surface. *J. Appl. Physiol.* 67:1814–1819,1989.
8. Mtui, E.P., M. Anwar, R. Gomez, D.J. Reis, and D.A. Ruggiero. Projections from the nucleus tractus solitarii to the spinal cord. *J Comp Neurol.* 337:231–52,1993.
9. Mueller, R.A., D.B.A. Lundberg, G.R. Breese, J. Hedner, T. Hedner, and J Jonason. The neuropharmacology of respiratory control. *Pharmacol. Rev.* 34:255–85, 1982.
10. Paxinos, G., and C. Watson. *The Rat Brain in Stereotaxic Coordinates*. Sydney: Academic, 1982.
11. Soto-Arape, I., M. Burton, and H. Kazemi. Central amino acid neurotransmitters and the hypoxic ventilatory response. *Am. J. Respir. Crit. Care Med.* 151:1113–20, 1995.

AXONAL PROJECTIONS FROM THE PONTINE PARABRACHIAL-KÖLLIKER-FUSE NUCLEI TO THE BÖTZINGER COMPLEX AS REVEALED BY ANTIDROMIC STIMULATION IN CATS

Son Gang, Akihiko Watanabe, and Mamoru Aoki

Department of Physiology
School of Medicine
Sapporo Medical University
South-1, West-17, Chuo-ku, Sapporo 060, Japan

1. INTRODUCTION

The pontine parabrachial and Kölliker-Fuse nuclear complex (NPB-KF) has been assumed to be the anatomical location of the pneumotaxic center (11). It contains high density of respiratory neurons with various discharge patterns (3, 6). Destruction or electrical stimulation of this area is known to produce profound changes in the respiratory rhythm (5, 24, 25). It is suggested that the NPB-KF exerts its effects by specifically modulating the activity of the medullary inspiratory 'off-switch' mechanism, which terminates inspiration and ensures the phase transition from inspiration to expiration (8, 9). However, axonal projections from the NPB-KF to the medullary structures specifically involved in the inspiratory 'off-switch' effects have not been well delineated. Previous studies by us demonstrated that the nucleus raphe magnus, a structure involved in the inspiratory 'off-switch', received strong axonal projections from the NPB-KF area (1, 21, 22). Another possible neuron group which may transmit the function of the NPB-KF is the Bötzinger complex (Böt.c.). The Böt.c. is known as a group of expiratory neurons in the vicinity of the retrofacial nucleus (14, 19). Most of those neurons have an augmenting firing pattern and widespread inhibitory connections to the medullary inspiratory premotor neurons (10, 12, 18). Recent studies by Smith *et al.* (20) and us (23) with the retrograde WGA-HRP tracing method revealed that the Böt.c. received strong projections from the lateral NPB and the KF. These studies suggest that these projections are composed of axons from the pontine respiratory and/or non-respiratory neurons which would modulate the activities of expiratory neurons in the Böt.c. The present investigation, by using the technique of antidromic activation, aimed to ascertain which types of neurons in the NPB-KF complex send efferent axonal projections to the Böt.c.

Advances in Modeling and Control of Ventilation, edited by Hughson *et al.*
Plenum Press, New York, 1998.

2. METHODS

Experiments were performed on 18 adult cats of either sex, weighing 2.7–3.8 kg. After an initial dose of ketamine (20 mg/kg, i.m.) injection, surgical anesthesia was induced with α-chloralose-urethane, (50 and 500 mg/kg, i.p., respectively). A supplementary dose (1/10 of the initial dose, i.v.) was given whenever noxious stimuli evoked changes in heart rate, blood pressure or respiratory parameters, or a withdrawal reflex. Surgical incisions and pressure points were blocked with lidocaine jelly. Surgical procedures included cannulation of trachea and femoral blood vessels (artery and vein), dissection of the C_5 root of the phrenic nerve, bilateral vagotomy and resecting the posterior rim of occipital bone. The head of the animal was fixed on a stereotaxic apparatus. The dorsal surface of the brainstem was exposed by removing the major portion of the cerebellum by suction and covered with warm paraffin oil. To prevent the brain pulsations, the animals were paralyzed with pancronium bromide (Mioblock, initial dosage of 0.5 mg/kg, supplemented by 0.1 mg/kg/hour, i.v.) and artificially ventilated. Bilateral pneumothorax was performed. Efferent phrenic nerve discharges were recorded with a bipolar silver electrode and used as an indicator of central respiratory output. The end-tidal CO_2 was monitored and kept at 5.0–6.0%. Mean arterial blood pressure was maintained at 100–130 mmHg. Rectal temperature was kept at 37°C with a heating pad.

Unit discharges in the rostral pons were extracellularly recorded with tungsten microelectrodes (shaft diameter 70 μm, tip electrically etched to 1–3 μm, impedance 6–12 MΩ) and displayed on a dual-beam memory oscilloscope. In the first 10 cases, the rostral pons was systematically explored with a series of penetrations made at 0.5 mm intervals. The exploration included a region of 2–2.5 mm lateral to the midline and 9–11.5 mm rostral to the obex. In the last 8 cases, under the guidance of our previous histological study with WGA-HRP retrograde tracing method (23), the exploration was concentrated at the lateral NPB and KF nuclei.

A tungsten microelectrode was inserted into the Böt.c. stereotaxically according to Berman's atlas (2). Expiratory unit discharges in this area were extracellularly recorded and mapped in the frontal plane to define the boundaries of the Böt.c. Then the electrode was switched to a stimulating electrode used for delivering monopolar stimulation. Test stimuli of rectangular constant current pulses (duration 0.3 ms, 2 Hz, 4–50 μA) were continuously given during searching of units in the rostral pons. When a desirable unit was recorded, a series of closely spaced penetrations (0.25 mm apart) were made within the boundaries of the Böt.c. to elicit responses in this unit. When a presumed antidromic response (an all-or-none spike with a short and invariant latency) was obtained, it was then further confirmed with the following criteria (13); 1) faithful response to high frequency stimulation (100 Hz); 2) collision of the spontaneous spike with that resulting from stimulation. Terminal arborizations of some selected antidromic units were mapped to identify their termination sites.

The stimulating and recording sites were marked by passing D.C. current (50 μA for 15 s) through the electrodes at the end of each experiment and were later identified histologically.

3. RESULTS

3.1. Respiratory Units

A total of 91 respiratory units of three major types were recorded in the rostral pons (65 inspiratory, 12 expiratory and 14 phase spanning). These respiratory units usually

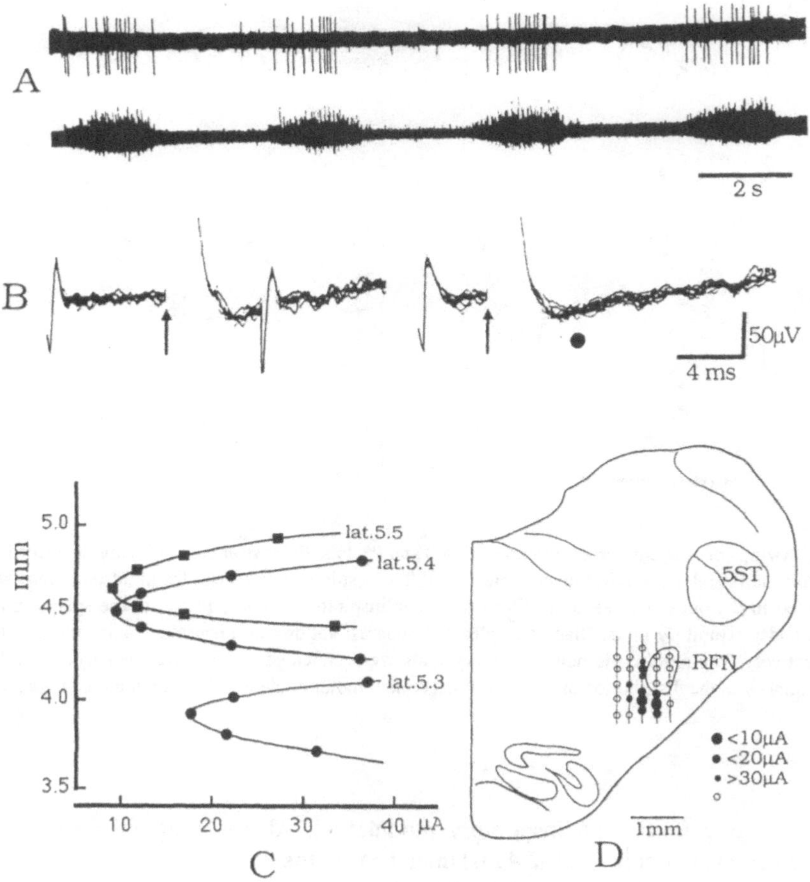

Figure 1. **A:** an inspiratory unit discharge (upper trace) recorded in the PBL. It showed an augmenting discharging pattern during the inspiratory phase of the phrenic discharge (lower trace). **B:** this unit was antidromically activated by electrical stimulation (arrow) in the Böt.c. and was confirmed by collision test. **C:** threshold vs. depth curves indicating that more than one branch were activated. Note that the latency of each curve was different from the other. **D:** drawing of a section across the Böt.c. showing the stimulating points of the curves in the C. Three low threshold stimulating points were found.

showed low discharge frequencies (< 6 Hz) and were concentrated in the lateral part of the rostral pons. All of the three types of units were tested for antidromic activation by stimulation in the region of Böt.c. (Figs. 1, 2)

1. *Inspiratory units*: 11 units were antidromically activated, comprising 16% (11/65). Among these units, 7 were recorded in the Lateral NPB (PBL). Those PBL units usually showed a phasic augmenting discharge pattern with poor signal-to-noise ratio. Their antidromic latencies were typically long, ranging from 3.8 to 5.8 ms. They might correspond to the small labeled neurons (diameter, 10–15 µm) observed in the PBL following HRP injection into the Böt.c. (23). Other two units were recorded in the KF. They showed a similar augmenting discharge pattern with a relatively low discharge frequency and better signal-to-noise ratio. Other 2 units were recorded in the lateral part of the medial NPB. They were tonically dis-

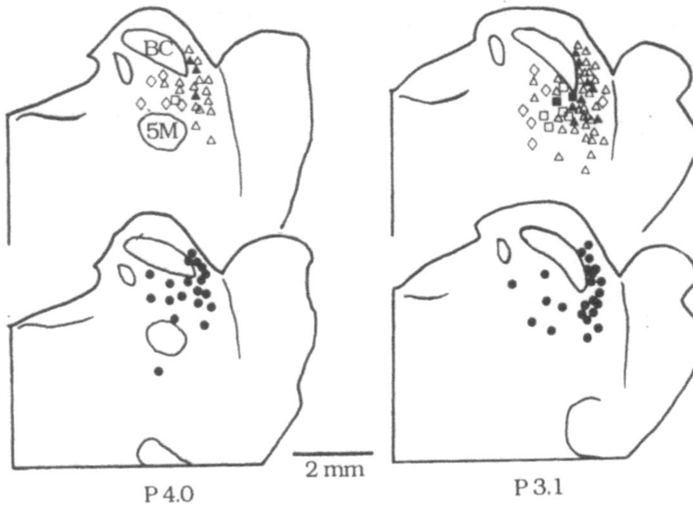

Figure 2. Drawings of sections across two levels (P4.0 and P3.1) at the rostral pons showing the recording sites of the respiratory units and the antidromic activated units. The respiratory units and the antidromic respiratory units were indicated in the upper two sections. Open triangles: inspiratory units; Squares: phase spanning respiratory units; Diamonds: expiratory units; Black triangles and squares: antidromic inspiratory units and phase spanning units, respectively. The antidromic non-respiratory units were indicated in the lower two sections (black dots). Note the similarity in the distribution patterns. 5M: trigeminal motor nucleus; BC: branchium conjunctivum.

charging units with inspiratory modulation. All the four units had shorter antidromic latencies (1.3–2.4 ms) than those of the PBL units.

2. *Phase spanning units*: All the 14 units were of expiratory-inspiratory type. They began to discharge, usually with an augmenting pattern, in the mid or late expiratory phase and extended into the inspiratory phase. Most of them were recorded in the lateral part of the medial NPB. Only 2 units were found to be antidromically activated, with latencies of 1.2 and 1.6 ms, respectively.
3. *Expiratory units*: Most expiratory units were recorded in the medial NPB (PBM). Despite of repeated tests with extensive penetrations in and around the Böt.c. region, none of them were antidromically activated.

Terminal axon arborizations of all the 13 antidromically activated units were mapped with multiple electrode penetrations (spaced at 0.25 mm apart) in the Böt.c. region (Fig. 1). At least two arborizations within the Böt.c. region were confirmed for axons of the 12 units tested. This result suggested that those antidromic units had axons terminated within the Böt.c. Since the penetrations were made mainly in the Böt.c. region, other arborizations outside the Böt.c. were not mapped.

3.2. Non-Respiratory Units

Fifty-five non-respiratory units recorded in the rostral pons were also antidromically activated by stimulation in the Böt.c. Among them, 41 were silent units or with very low discharge frequencies (0.05–0.1 Hz). They were most frequently encountered, intermingled with the inspiratory units in the PBL and KF. Actually 12 units of them were recorded simultaneously with inspiratory units. The other 14 units showed spontaneous discharges with frequencies ranging from 0.5 to 3 Hz. Two of them, recorded in the lateral PBM, showed an inspiratory modulation after asphyxia by temporarily stopping artificial respiration. The antidromic latencies of these non-respiratory units ranged from 0.8 to 6.2 ms. Terminal axonal arborizations of 11 units randomly selected were mapped. Among them, 6 units had more than two axonal arborizations in the Böt.c. region. Axons of the other 5 units seemed to be passing fibers.

4. DISCUSSION

The present study demonstrated that some respiratory and non-respiratory neurons in the pontine NPB-KF nuclear complex send descending axonal projections to the Böt.c. The existence of this descending pathway has been revealed by previous studies with retrograde WGA-HRP tracing method (20, 23). It was found that injection of WGA-HRP into the Böt.c. specifically labeled neurons in the PBL and KF nuclei. They suggested that this pathway could transmit the respiratory function of the PBL-KF nuclei. Although the PBL-KF nuclei contain the high density of respiratory neurons, it has not been known whether those respiratory neurons contributed to this pathway. The antidromic mapping technique, on the other hand, allows us to identify the axonal projections of a given type of neuron. By using this technique, we revealed that this descending projections did contain axons from the respiratory neurons, thus providing further evidence that this pathway is involved in respiratory control.

The antidromically activated respiratory and non-respiratory units were concentrated in the PBL-KF nuclei. These nuclei, especially the PBL, has not been the target of intensive exploration by previous studies with the antidromic mapping method (4). The PBL contained tremendous small sized labeled neurons after WGA-HRP injections into the Böt.c. (14). Under the guidance of these previous results, we concentrated our exploration at the lateral pontine structures including the PBL-KF and the lateral PBM. Although only a limited number of antidromic respiratory units, mainly inspiratory ones, were recorded, the actual number might be lager. In fact, we could record some inspiratory units in the PBL with presumed antidromic responses. However, the poor signal-to-noise ratio, which is characteristic of recordings from small neurons (7), made it impossible to confirm them with the collision test. Therefore, the percentage occupied by antidromically activated respiratory units might be higher than 14.3% of the present study.

It was reported that the PBL exerted facilitatory effects on inspiration (5). We also observed that electrical stimulation of the sites within the PBL, where antidromic inspiratory units were recorded, increased the amplitude of the phrenic nerve discharge and shortening of expiratory durations. On the other hand, the expiratory neurons in the Böt.c. had widespread inhibitory connections with the premotor inspiratory neurons in the medulla (12, 17). They also monosynaptically inhibited the phrenic motor neurons (17). Based on these facts, it is possible that the inspiratory facilitating effects of the PBL (and the KF) are achieved by inhibiting the activities of the Böt.c. expiratory neurons through the pathway revealed in this study. The PBL-KF nuclei and the structures immediately adjacent to the Böt.c. (sub-retrofacial nucleus and paragigantocellular nucleus) also participated in cardiovascular regulation (15, 16). The fact that the PBL-KF-Böt.c. pathway contained many axonal projections from non-respiratory neurons suggests its possible involvement in cardiovascular regulation.

REFERENCES

1. Aoki, M., Y. Fujito, Y. Kurosawa, H. Kawasaki, and I. Kosaka. Descending inputs to the upper cervical inspiratory neurons from the medullary respiratory neurons and the raphe nuclei in the cat. In: *Respiratory Muscles and Their Neuromotor Control*, edited by G. C. Sieck, S.C. Gandevia, and W. E. Cameron. New York: Alan R. Liss, 1987, p. 75–82.
2. Berman, A. L. The Brain Stem of the Cat, A Cytoarchitectonic Atlas with Stereotaxic Coordinates. University of Wisconsin Press., Madison, Wis, 1968.
3. Bertrand, F., A. Hugelin, and J. F. Vibert. A stereologic model of pneumotaxic oscillator based on spatial and temporal distributions of neuronal bursts. *J. Neurophysiol.* 37: 91–107, 1974.
4. Bianchi, A. L. and W. M. St. John. Medullary axonal projections of respiratory neurons of pontile pneumotaxic center. *Respir. Physiol.* 48: 357–373, 1982.
5. Cohen, M. I. Switching of the respiratory phases and evoked phrenic responses produced by rostral pontine electrical stimulation. *J. Physiol. Lond.* 217: 133–158, 1971.
6. Cohen, M. I., and S. C. Wang. Respiratory neuronal activity in pons of cat. *J. Neurophysiol.* 22: 33–50, 1959.
7. Connelly, C. A., H. H. Ellenberger, and J. L. Feldman. Respiratory activity in retrotrapezoid nucleus in cat. *Am. J. Physiol.* 258: L33–44, 1990.
8. Euler, C. von., and T. Trippenbach. Excitability changes of the inspiratory "off- switch" mechanism tested by electrical stimulation in nucleus parabrachialis in the cat. *Acta. Physiol. Scand.* 97: 175–188, 1976.
9. Euler, C. von., I. Marttila, J. E. Remmers, and T. Trippenbach. Effects of lesions in the parabrachial nucleus on the mechanisms for central and reflex termination of inspiration in the cat. *Acta. Physiol. Scand.* 96: 324–337, 1976.
10. Fedorko, L., and E. G. Merrill. Axonal projections from the rostral expiratory neurones of the Botzinger complex to medulla and spinal cord in the cat. *J. Physiol. Lond.* 350: 487–496, 1984.
11. Feldman, J. L. Neurophysiology of breathing in mammals. In: *Handbook of Physiology. The Nervous System. Intrinsic Regulatory Systems of the Brain.* Am. Physiol. Soc. Bethesda, MD, 1986, Sect. 1, Vol. IV, pp. 463–524.
12. Jiang, C., and J. Lipski. Extensive monosynaptic inhibition of ventral respiratory group neurons by augmenting neurons in the Bötzinger complex in the cat. *Exp. Brain. Res.* 81: 639–648, 1990.
13. Lipski, J. Antidromic activation of neurones as an analytic tool in the study of the central nervous system. *J. Neurosci. Methods.* 4: 1–32, 1981.
14. Lipski, J. and E. G. Merrill. Electrophysiological demonstration of the projection from expiratory neurones in rostral medulla to contralateral dorsal respiratory group. *Brain Res.* 197: 521–524, 1980.
15. McAllen, R. M. and R. A. Dampney. The selectivity of descending vasomotor control by subretrofacial neurons. In: *Progress in Brain Research. The Central Neural Organization of Cardiovascular Control.* edited by J. Ciriello, M. M. Caverson, and C. Polosa. Amsterdam, Elsevier, Amsterdam, 1989, p. 233–242.
16. Mraovitch, S., M. Kumada, and D. J. Reis. Role of the nucleus parabrachialis in cardiovascular regulation in cat. *Brain Res.* 232: 57–75, 1982.
17. Merrill, E. G., and L. Fedorko. Monosynaptic inhibition of phrenic motoneurons: a long descending projection from Bötzinger neurons. *J. Neurosci.* 4: 2350–2353, 1984.
18. Merrill, E. G., J. Lipski, L. Kubin, and L. Fedorko. Origin of the expiratory inhibition of nucleus tractus solitarius inspiratory neurones. *Brain Res.* 263: 43–50, 1983.
19. Otake, K., H. Sasaki, H. Mannen, and K. Ezure. Morphology of expiratory neurons of the Bötzinger complex: an HRP study in the cat. *J. Comp. Neurol.* 258: 565–579, 1987.
20. Smith, J. C., D. E. Morrison, H. H. Ellenberger, M. R. Otto, and J. L. Feldman. Brainstem projections to the major respiratory neuron populations in the medulla of the cat. *J. Comp. Neurol.* 281: 69–96, 1989.
21. Song, G., A. Mizuguchi, and M. Aoki. Axonal projections from the pontine pneumotaxic region to the nucleus raphe magnus in cats. *Respir. Physiol.* 85: 329–339, 1991.
22. Song, G., Y. Nakazono, and M. Aoki. Differential projections to the raphe nuclei from the medial parabrachial-Kölliker-Fuse (NPBM-KF) nuclear complex and the retrofacial nucleus in cats: retrograde WGA-HRP tracing. *J. Auton. Nerv. Sys.* 45: 241–244, 1993.
23. Song, G., Y. Sato, I. Kohama, and M. Aoki. Afferent projections to the Bötzinger complex from the upper cervical cord and other respiratory related structures in the brain stem in cats: retrograde WGA-HRP tracing. *J. Auton. Nerv. Sys.* 56: 1–7, 1995.
24. St. John, W.M., R. L. Glasser, and R. A. King. Apneustic breathing after vagotomy in cats with chronic pneumotaxic center lesions. *Respir. Physiol.* 12: 239–250, 1971.
25. Tang, P.C. Localization of the pneumotaxic center in the cat. *Am. J. Physiol.* 172: 645–652, 1953.

HEBBIAN COVARIANCE LEARNING

A Nexus for Respiratory Variability, Memory, and Optimization?

Daniel L. Young[1,2] and Chi-Sang Poon[1,*]

[1]Harvard-MIT Division of Health Sciences and Technology
Cambridge, Massachusetts 02139
[2]Department of Mechanical Engineering
Massachusetts Institute of Technology
Cambridge, Massachusetts 02139

1. INTRODUCTION

Respiration is a vital autonomic function that is characterized by an almost unwavering ability to maintain homeostatic blood gas and pH levels in the face of profound physiological and environmental challenges. This remarkable behavior has led many researchers to assume that respiratory control is a reflexogenic process whose sole purpose is simply to maintain nominal body gas and pH tones. While this model has born some insight into certain aspects of the respiratory system, it fails to account for more complex behaviors of the system that have been increasingly recognized.

For example, respiratory "memory" is now a widely acknowledged phenomenon (12). However, most models of the respiratory control system to date have not included such memory effects and therefore cannot elucidate its role in normal respiratory function. While some forms of respiratory memory have been elicited *in vivo* (6,15,17), more recent studies using *in-vitro* brainstem slice reparations (41) have demonstrated certain forms of long-term and short-term respiratory memory which may have a synaptic origin. These findings of synaptic plasticity in the respiratory system have many characteristics that parallel long- and short-term synaptic plasticity found in the hippocampus (7) and other brain centers (4,8). Additional studies have identified certain neurotransmitters and receptors, such as serotonin (3,16,25) and NMDA receptors (33), which appear to be involved in respiratory memory and normal respiratory function.

* Correspondence: Chi-Sang Poon, Ph.D., Harvard-MIT Division of Health Sciences and Technology, Rm 20A-126, Massachusetts Institute of Technology, Cambridge, MA 02139. Phone: (617) 258-5405; Fax: (617) 253-2514; email: cpoon@hstbme.mit.edu

Advances in Modeling and Control of Ventilation, edited by Hughson *et al.*
Plenum Press, New York, 1998.

The presence of memory and synaptic plasticity in the brainstem suggests that basic feedback/feedforward reflexogenic models are probably too simplistic to capture the essence of respiratory control. One alternative model (30,31,33) postulates that Hebbian covariance learning may participate in homeostatic respiratory control. Hebbian learning, first envisioned by Hebb (1949), describes a class of learning rules which relate changes in synaptic efficacy to pre- and post-synaptic activities. Analogous synaptic mechanisms have been found in the hippocampus (37,38), the visual cortex (14) and certain neuromuscular junctions (11). Furthermore, Hebbian covariance learning has been shown to optimally control a class of nonlinear systems (31). This approach is akin to self-tuning adaptive control described in engineering literature.

In this paper, we propose a novel model of respiratory control that employs Hebbian covariance self-tuning control. This model integrates our current knowledge of synaptic mechanisms in the brainstem with our understanding of physiological respiratory responses to various natural stimuli. The Hebbian covariance model extracts information from fluctuations in the current states of the system and hence requires continual perturbations in the system for learning to persist. Suitable excitation in the respiratory system could arise from numerous sources, e.g. intrinsic fluctuations from periodic oscillations of breathing, neuronal variability, or even chaotic respiratory dynamics (35).

By means of its elementary adaptive control law, the Hebbian controller continually drives the respiratory system to optimal operating points. Such optimal control has been proposed previously to participate in respiratory control and in homeostasis in general (31). In this manner, the Hebbian covariance model predicts that the isocapnic response to exercise is associated with an increase in the chemosensory feedback gain, thereby dispelling the need for the elusive feedforward exercise stimulus (30,31).

2. HEBBIAN LEARNING

Hebbian synaptic learning defines a class of models relating dynamic synaptic weight changes to pre- and post-synaptic activities. Such synaptic mechanisms have been hypothesized to play a fundamental role in learning and memory in the hippocampus and other brain structures (20,38). Furthermore, these elementary learning rules have been shown to be useful for various learning models, including classical conditioning (9), neural network training rules (10,27) and adaptive control (31). The general form of Hebbian learning, also called conjunctional Hebbian learning, can be expressed as:

$$\frac{dW}{dt} = k(x \cdot y), \tag{1}$$

where W is the synaptic weight; x and y represent the mean firing rate of the input and output respectively; and k defines the adaptation rate.

Hebbian covariance learning (36) is a particular incarnation of Hebbian learning which circumvents certain instability and saturation problems that plague the basic Hebbian rule. As opposed to conventional Hebbian learning, covariance learning utilizes the fluctuations of the input and output signals about their means as follows:

$$\frac{dW}{dt} = k\, \delta x \cdot \delta y, \tag{2}$$

where δx and δy are respectively the temporal variations of the pre- and post-synaptic activities about their means in a given time period. Synaptic strength is thus potentiated if changes in pre-and post-synaptic activities are positively correlated and conversely weakened if their activities are negatively correlated. Several brain regions have been shown to express synaptic dynamics that conform to this Hebbian covariance learning model (11,14,37,38).

A recent computational study has shown that adaptive self-tuning controllers utilizing Hebbian covariance learning may optimize the performance of a class of nonlinear dynamical systems (31). Such adaptive self-tuning controllers continually optimize their performance in spite of changing environmental conditions by exploiting spontaneous fluctuations in the system. In essence, such a reinforcement learning system tunes the synaptic weight by balancing the forces of the covariance of the post-synaptic and pre-synaptic activities with the autovariance of the post-synaptic activity.

3. RESPIRATORY MEMORY

Brainstem respiratory memory has received increasing attention recently, following a long period of obscurity after Gessel and co-workers first reported the presence of short-term respiratory memory over half a century ago (17). This finding was particularly intriguing because not only did it implicate memory in homeostatic control, but it was also the first evidence of "intelligent" control in the lower brain. Since this initial discovery, numerous researchers have reported other forms of respiratory memory of both long and short durations (for review, 12,31,32). For example, hypoxic respiratory depression in humans is known to persist after withdrawal of the hypoxic stimulus (6,15).

While initial reports hypothesized that respiratory memory may be due to network reverberation in medullary structures (12), recent studies indicate that these memories may infact be caused by synaptic mechanisms. One line of evidence stems from repeated carotid sinus nerve (CSN) stimulation which induces a serotonergic (5-HT)-dependent, long-term facilitation (LTF) of phrenic nerve activity (3,16). Serotonergic-dependent respiratory memory is also supported by long-term potentiation (LTP) of respiratory activity induced by stimulation of the raphe nucleus (25).

The second line of evidence for synaptic respiratory memory comes from studies of NMDA receptors which are generally considered to play a major role in the induction of synaptic plasticity (24). For instance, synaptic short-term potentiation (STP) in the nucleus tractus solitarius (NTS), the primary nucleus in the medulla which integrates afferent cardiorespiratory projections, was shown to be highly dependent on NMDA receptors (13). Furthermore, a study of genetically engineered NMDAR1 knock-out mice showed that these mutant mice, healthy at birth, died of respiratory failure within the first few days of life (33). These findings support the notion that synaptic plasticity in the brainstem, which may be of a similar nature to hippocampal plasticity (23), is critical for normal respiratory function.

Recent *in-vitro* studies using rat brainstem slice preparations have demonstrated both phasic and long-term depression (LTD) in the NTS (41). The maintenance of the LTD required elevated intracellular Ca^{2+} concentrations as evidenced by applications of the NMDA antagonist D-APV and the calcium chelator EGTA. The LTD was induced by a low frequency (1–5 Hz), 3–5 min period stimulation of the primary afferent fibers in the tractus solitarius. Such associations to intracellular Ca^{2+} and to low frequency stimulation suggest that this brainstem LTD may be of a similar origin as LTD previously described in the hippocampus and the neocortex (21). Furthermore, these findings conform to many of the predictions of the Hebbian covariance learning model, namely that weak or negative pairing of pre- and post-synaptic activities should induce synaptic depression.

4. RESPIRATORY VARIABILITY

In recent years, there has been a growing interest in studying the variabilities or fluctuations in physiologic systems as possible markers of normal or abnormal behavior. The respiratory system is one such system that exhibits various modes of variability which are associated with normal homeostatic control (31,39).

Perhaps the most significant intrinsic variability in the respiratory system stems directly from its oscillatory rhythm. Many studies have demonstrated a relationship between the respiratory pattern and fluctuations in O_2, CO_2, and pH levels in the arterial circulation (22,40) as well as in the medullary extracellular fluid (26). For example, the amplitude of arterial pH fluctuations about its mean varies inversely with respiratory frequency and directly with tidal volume (19). Such fluctuations in the systemic circulation are transmitted by the carotid body chemoreceptors with the same frequency as the respiratory rhythm (5,28). Other periodic fluctuations arise from mechanical receptors which are stimulated by the oscillatory expansion of the lungs and carried to the respiratory control center via the vagus nerve (1).

Another source of variability in the respiratory system is derived from the closed-loop dynamics of the system. For example, vagal volume feedback may drive the respiratory oscillator into chaotic regimes with breath-to-breath variabilities in peak inspiratory and end-expiratory volumes. In anesthetized rats, the respiratory rhythm becomes highly periodic with apparently less variability after bilateral vagotomy (35).

Although many reports have demonstrated the prevalence of variability in the respiratory system, few have suggested a role for such fluctuations in respiratory control. While basic feedback/feedforward control systems are often designed to minimize variability, most adaptive systems require such perturbations for continuous learning (2). As mentioned previously, Hebbian covariance learning is one example of an adaptive self-tuning controller that utilizes fluctuations in the system to acquire valuable information about the system's current performance. This dependency on variability provides another strong link between Hebbian covariance learning and normal respiratory function.

5. RESPIRATORY CONTROL MODEL

The respiratory system maintains homeostasis of blood gas concentrations and body pH levels by controlling a mechanical musculoskeletal system. Current physiological states are sensed by an array of peripheral and central receptors and relayed to the brainstem via afferent nerve fibers. Brainstem centers process these signals resulting in the controller output, the total ventilation rate, $\dot{V}_E$, thereby forming a closed-loop system.

5.1. Lungs, Receptors, and Neural Connections

The simplified respiratory model consists of the following mechanical and neural components: a lung gas exchanger, a chemoreceptor, a respiratory interneuron, and a motoneuron (Figure 1). The lungs are faced with external and internal CO_2 loads, namely inhaled CO_2 gases (P_{iCO_2}) and metabolic CO_2 production ($\dot{V}_{CO_2}$), respectively. The arterial

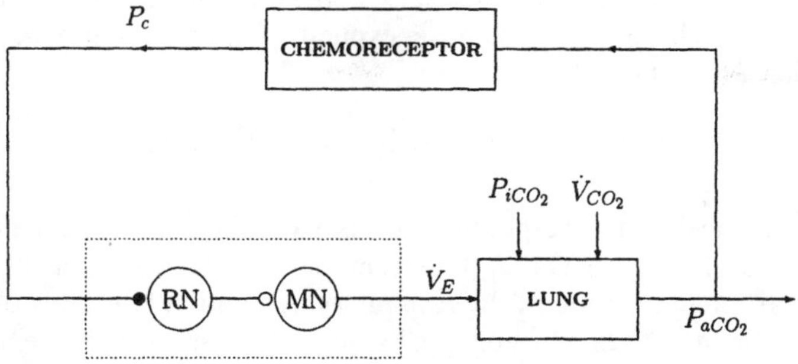

Figure 1. Self-tuning optimal regulator model of respiratory control using Hebbian covariance learning. The controller gain is adaptively regulated by synaptic plasticity in a respiratory interneuron (RN) which drives a motor neuron (MN). Synaptic potentiation and depression are governed by a Hebbian covariance rule which computes the optimal controller gain based upon correlated fluctuations in the motor outflow and chemoafferent feedback. The plastic synapse is represented by ●.

blood gas tone (P_{iCO_2}) is sensed by a representative chemoreceptor whose neural output (P_c) is relayed to a respiratory neuron (RN). The synapse mediating the afferent input to the respiratory neuron has a plastic synaptic weight (W) which defines the efficacy or strength of the synaptic transfer. The respiratory neuron projects to a motoneuron that drives the lungs. (This closed-loop model does not incorporate the medullary structures responsible for modulating the precise waveform patterns of the respiratory rhythm.)

The nonlinear lung dynamics are modeled as follows:

$$V_L \frac{dP_a}{dt} = P_{iCO_2} - P_a + \frac{K\dot{V}_{CO_2}}{\dot{V}_E}, \tag{3}$$

where P_{iCO_2} is the partial pressure of CO_2 in the inspired air; P_a is the instantaneous arterial P_{aCO_2} (with $P_a = P_{aCO_2}$ in the steady state); $\dot{V}_{CO_2}$ is the metabolic production rate of CO_2; $\dot{V}_E$ is the ventilation rate; and K and V_L are constants.

The chemoreceptor dynamics are also modeled by a first order differential equation:

$$\tau \frac{dP_c}{dt} = -P_c + \alpha(P_a - \beta) \tag{4}$$

where τ is a time constant and α and β are sensitivity and threshold parameters for chemosensitivity, respectively.

Finally, the input/output relation at the plastic synapse is given by:

$$\dot{V}_E = W(P_c, \dot{V}_E) \cdot P_c \tag{5}$$

where the synaptic weight W is a function of both the its input and output signals, P_c and $\dot{V}_E$, which will be derived in Section 5.3.

5.2. The Objective Function

To appropriately model the respiratory controller, a precise goal of the respiratory system must first be defined. One hypothesis is that the respiratory system minimizes the total cost of aberrant blood gas composition as well as the energy due to the motor act of breathing (29,30). This balancing act is made explicit with the definition of the compound objective function J (30,31):

$$J = J_c + J_m = [\alpha (P_{aCO_2} - \beta)]^2 + \ln \dot{V}_E^2 \tag{6}$$

where J_c is the chemical cost of respiration expressed as the squared deviation of the CO_2 tension from the desired level and J_m represents the mechanical cost of breathing expressed as a logarithmic function of the respiratory ventilation rate. This objective function thus defines the long-term, physiological goal of the respiratory system and has been shown to be consistent with certain respiratory responses (30,31).

5.3. Hebbian-Covariance Self-Tuning Control

Respiratory activity is known to continually fluctuate about its mean values (35,39) as well as exhibit short-term memory in response to certain afferent stimuli. Furthermore, synaptic plasticity conforming to the Hebbian covariance model has been shown to exist in the NTS, the first relay site for cardiorespiratory afferent inputs. These three respiratory characteristics support in part the hypothesized Hebbian covariance control strategy.

To obviate the discrepancies introduced by sluggishness and delays inherent in the feedback pathway, we postulate that the controller is driven by a different reward-penalty system than the physiological one, namely it utilizes a near-term objective. Hence, from Eqs. 3 and 4, we define the intermediate variable, s, given by:

$$s = \left(\frac{\tau V_L}{\dot{V}_E}\right) \ddot{P}_c + \left(\frac{V_L}{\dot{V}_E} + \tau\right) \dot{P}_c + P_c \tag{7}$$

and derive the resulting near-term objective function, Q, in discrete time:

$$Q(n) = s(n)^2 + \ln \dot{V}_E(n-2)^2. \tag{8}$$

By means of this transformation, the learning system will make short-term changes in the near-term thereby minimizing the physiological objective, J, in the long-term. In essence, the physiological long-term goal is achieved in the steady state by continual minimization of its neural correlate, Q.

We can now derive the following Hebbian covariance adaptation rule:

$$\delta W(n) = -k \left[\delta s(n) \delta \dot{V}_E(n-2) + \frac{1}{s(n)\dot{V}_E(n-2)} \delta \dot{V}_E(n-2)^2\right] \tag{9}$$

where $k > 0$ represents the learning rate. The first term on the left-hand side of Eq. 9, $\delta s(n)\delta \dot{V}_E(n-2)$, is the covariance between the controller output and a transformed measure of the chemical cost. The second term on the right hand side, $\delta \dot{V}_E(n-2)^2$, is the autovariance of the controller output. This adaptation rule continually adapts the chemosensory feedback gain by probing the environment with random fluctuations in the control output $\dot{V}_E$. By weighing the changes in the chemical costs against the changes in mechanical costs of breathing, the objective function is adaptively minimized. Lyapunov theory can be used to prove the asymptotic convergence of this adaptation rule.

By setting $\delta W = 0$ and d/dt = 0 in Eqs. 3 and 4, we obtain the following steady-state respiratory response:

$$\dot{V}_E = \alpha^2 (P_{aCO_2} - \beta) K \dot{V}_{CO_2} \tag{10}$$

This equation is the optimal response corresponding to the physiological objective function (30,31). Thus the optimization model is compatible with both the isocapnic exercise response and the hypercapnic response at rest to increased ambient CO_2.

6. COMPUTER SIMULATIONS

Simulations were carried out over a series of physiological conditions to simulate the responses during exercise and CO_2 inhalation.

6.1. Exercise Hyperpnea

The physiological response to exercise is characterized by an isocapnic increase in the ventilation rate. To date, this response has mystified many researchers because no physiological signal has been confirmed to account for the increase in respiratory output. To simulate exercise, a step change in the metabolic load ($\dot{V}_{CO_2}$) was made from rest conditions. For illustration, consider a step change in $\dot{V}_{CO_2}$ from 0.2 to 1.0 l/min. By examination of the steady-state optimal solution (Eq. 10), such a change should produce a five fold increase in the ventilation rate ($\dot{V}_E$), while P_{aCO_2} should remain at the normoxic level (~35 mmHg) in the steady state. Given that P_{aCO_2}, and thus P_c, remain constant, the increase in $\dot{V}_E$ must be due to an increase in W (see Eq. 5). By means of the adaptive Hebbian covariance rule, the controller increases ventilation by potentiating the synaptic weight in the feedback loop until the optimal solution is attained. Hence no explicit exercise stimulus is required.

6.2. CO_2 Inhalation

CO_2 inhalation was simulated by a step increase in the inspired air (P_{iCO_2}). The optimal response is to increase ventilation in proportion to the chemoafferent drive (see Eq. 10). As a result, the synaptic weight should remain constant in the steady state and the steady-state P_{aCO_2} should increase in proportion to P_{iCO_2}. Our simulations confirm this hypercapnic behavior.

The model performance closely matches the body's natural responses to exercise and exogenous CO_2 loading over a broad range of conditions. Figure 2 illustrates the final steady-state relationship between P_{aCO_2} and $\dot{V}_E$ during exercise and CO_2 inhalation. The exercise response demonstrates the system's tendency to increase ventilation and maintain a nearly constant P_{aCO_2} operating point without any additional chemical feedback. On the other hand, during increased inspired CO_2, respiratory output increases with increased chemoafferent drive and homeostasis is abolished. Figure 3 shows the feedback gain (W) in the steady state for the same series of simulations. This confirms the notion that the exercise hyperpnea response is associated with an increase in the feedback gain, while during CO_2 inhalation the gain remains unchanged. Both of these optimal responses are predicted by the steady-state solution of the near-term Hebbian covariance learning rule (Eq. 10).

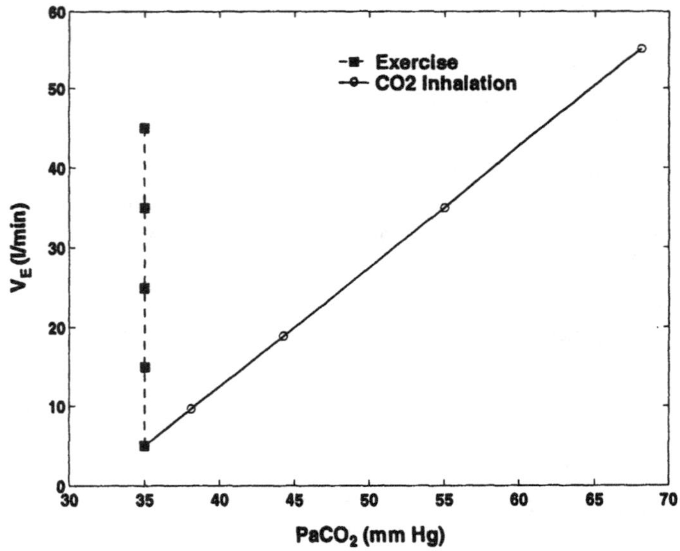

Figure 2. The optimal steady-state response of the self-tuning respiratory model under varying degrees of exercise and CO_2 inhalation. The results closely resemble actual physiologic behavior. The values for $\dot{V}_{CO_2}$ and P_{ICO_2} were varied between 0.2 and 1.8 l/min and between 0 and 65 mmHg, respectively.

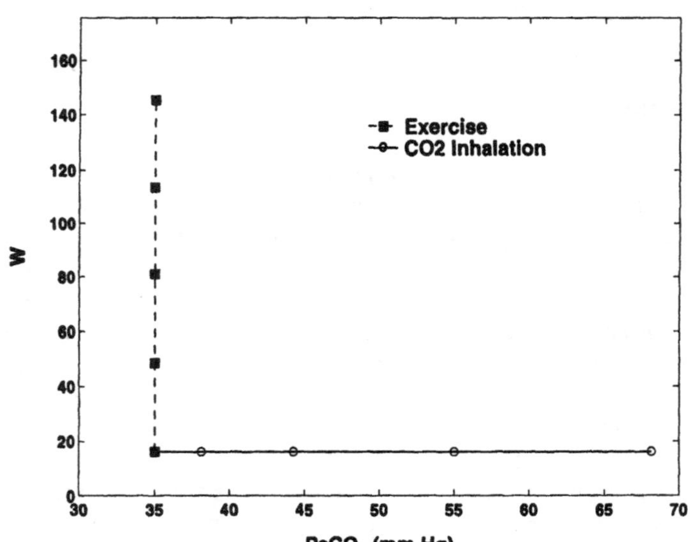

Figure 3. The steady-state relationship between arterial CO_2 (P_{aCO_2}) and the feedback gain (W) for the self-tuning respiratory model under varying degrees of exercise and CO_2 inhalation. This figure illustrates that an increase in W underlies the isocapnic exercise response while no change in W is observed during CO_2 inhalation.

7. DISCUSSION

Respiratory control and homeostatic control in general are highly specialized mechanisms necessary for life. These autonomous brainstem controllers adapt to physiological and environmental changes with amazing robustness and stability despite having both highly nonlinear environments and inherent feedback delays. Furthermore, respiratory responses to numerous perturbations appear optimal in that they minimize certain physiological costs (29,30). While reflexogenic models of respiratory control have been widely proffered, this class of models fails in many respects to explain the full spectrum of respiratory behavior including the exercise hyperpnea response.

Further motivation for the proposed model stems from recent findings of synaptic plasticity in respiratory brainstem centers. Such plasticity, which may underlie many of the reported respiratory "memory" effects (for review, 12), has many similarities to that found in the higher brain which is considered important for learning and memory. This finding suggests that the respiratory control system is not merely a feedback/feedforward control system, but that it may utilize more "intelligent" control strategies.

In this paper, we introduce a novel model of respiratory control which is compatible with many of these enigmatic characteristics of the respiratory system. Firstly, the model suggests that learning and memory are fundamental components of the respiratory control system. This notion is in accordance with the numerous memory effects which have been identified *in-vivo* and *in-vitro* as discussed in Section 3. The adaptive nature of the model differs significantly from previous reflex models of respiratory control which have assumed hard-wired connections with minimal computational abilities.

Secondly, given the abundant reports of Hebbian learning in the mammalian brain, we suggest that such a synaptic mechanism may exist within the respiratory brain centers to subserve the control objective (30,31). Several reports of synaptic plasticity in the NTS, including LTP and LTD, are consistent with this form of learning. Furthermore, analogous Hebbian covariance learning paradigms have been demonstrated to optimally control certain nonlinear system (31).

While self-tuning Hebbian covariance controllers may be shown to be stable in the sense of Lyapunov, they do require persistent excitation for continuous adaptation. In this manner, the controller derives information from fluctuations in the current states rather than from their mean values. Various sources of variability in the respiratory system, such as random fluctuations in neural patterns, oscillatory breathing patterns, and chaotic dynamics as introduced in Section 4, exist that could drive the learning process. In comparison, the cardiovascular system, another vital homeostatic control system, exhibits a high degree of chaotic dynamics in the healthy human cardiac rhythm while patients with congestive heart failure sustain a decrease in such cardiac chaos and a decrease in variability (34).

Computer simulations confirm that the Hebbian covariance learning model is capable of optimally controlling the respiratory system. The results illustrate that by implementing a basic Hebbian covariance adaptation rule, the system will respond intelligently to novel physiological and environmental perturbations. As examples, we considered the exercise and CO_2 inhalation responses. While both these conditions impose increases in arterial blood CO_2, the model responds quite differently in each case by virtue of its adaptive learning rule. In the exercise simulations, the system reacts by increasing the ventilation rate in proportion to the metabolic load ($\dot{V}_{CO_2}$). This isocapnic response is achieved by adaptively increasing the chemosensory feedback gain, thereby dispelling the need for the elusive exercise stimulus. Conversely, the CO_2 inhalation response is characterized by a gradual increase in the ventilation rate in relation to the elevated P_{aCO_2} levels. The slope of this linear curve defines the sensitivity to CO_2 and underlies the hypercapnic response.

8. CONCLUSIONS

In an attempt to account for the mounting reports of synaptic plasticity and memory, variability, and optimal behavior in the respiratory system, we have introduced a novel model for the respiratory control system. The system is modeled in the framework of classical Hebbian covariance learning, a predominant mechanism thought to underlie learning and memory in mammals. Theoretical foundations are introduced and computer simulations are used to verify the model. The results are in accordance with known physiological responses to both exercise and exogenous CO_2 loading.

ACKNOWLEDGMENTS

This work was supported by Office of Naval Research grants N00014-95-1-0414 and N00014-95-1-0863, National Science Foundation grant BCS-9216419 and National Institutes of Health grants HL52925 and HL50614. DLY is supported by a National Science Foundation graduate fellowship.

REFERENCES

1. Adrian, E. D. Afferent impulses in the vagus and their effect on respiration. *J. Physiol.* (London), 79: 332–58, 1933.
2. Åström, K. J., and B. Wittenmark. *Adaptive Control.* Addison-Wesley Publishing Company, Inc., New York, 1995.
3. Bach, K. B., and G. S. Mitchell. Hypoxia-induced long-term facilitation of respiratory activity is serotonin dependent. *Respir. Pysiol.*, 104: 251–60, 1996.
4. Baudry, M., R. F. Thompson, and J. L. Davis. *Synaptic plasticity: molecular, cellular, and functional aspects.* MIT Press, Cambridge, MA, 1993.
5. Biscoe, T. J., and M. J. Purves. Observations on the rhythmic variation in the cat carotid body chemoreceptor activity which has the same period as respiration. *J. Physiol.* (London), 190: 389–93, 1967.
6. Bisgard, G. E., and J. A. Neubauer. Peripheral and central effects of hypoxia. In A. Dempsey and A. I. Pack, editors, *Regulation of Breathing*, Ed 2. II. Lung biology in health and disease, volume 79, pages 617–668. Dekker, New York, 1995.
7. Bliss, T. V. P., and T. Lømo. Long-lasting potentiation of synaptic transmission in the dentate area of the anesthetized rabbit following stimulation of the perforant path. *J. Physiol.*, 232: 331–335, 1973.
8. Brown, T. H., P. F. Chapman, E. W. Kairiss, and C. L. Keenan. Long-term synaptic potentiation. *Science*, 242: 724–728, 1988.
9. Byrne, J. H. Cellular analysis of associative learning. *Physiol. Rev.*, 67: 329–439, 1987.
10. Carlson, A. Anti-Hebbian learning in a non-linear neural network. *Biol. Cybern.*, 64: 171–6, 1990.
11. Dan, Y., and M. M. Poo. Hebbian depression in isolated muscular synapses in vitro. *Science*, 256: 1570–3, 1992.
12. Eldridge, F. L., and D. E. Millhorn. Oscillation, gating, and memory in the respiratory control system. In N. S. Cherniack and J. G. Widdicombe, editors, *Handbook of Physiology*, volume 2, section 3, pages 93–114. American Physiological Society, Bethesda, MD, 1986.
13. England, S. J., J. E. Melton, P. Pace, and J. A. Neubauer. NMDA receptors mediate respiratory short-term potentiation in the nucleus tractus solitarius. *FASEB J.*, 6: A1826, 1992.
14. Frégnac, Y., D. Shulz, S. Thorpe, and E. Bienenstock. A cellular analogue of visual cortical plasticity. *Nature*, 333: 367–70, 1988.
15. Fregosi, R. F. Short-term potentiation of breathing in humans. *J. Appl. Physiol.*, 71: 892–9, 1991.

16. Fregosi, R. F., and G. S. Mitchell. Long-term facilitation of inspiratory intercostal nerve activity following carotid sinus nerve stimulation in cats. *J. Physiol.*(London), 477: 469–79, 1994.
17. Gesell, R., and M. A. Hamilton. Reflexogenic components of breathing. *Am. J. Physiol*, 133: 694–719, 1941.
18. Hebb, D. O. *The Organization of Behavior*. Wiley, New York, 1949.
19. Honda, Y., and M. Ueda. Fluctuations of arterial pH associated with respiratory cycle in dogs. *Jpn. J. Physiol.*, 11: 223–8, 1961.
20. Kelso, S. R., A. H. Ganong, and T. H. Brown. Hebbian synapses in hippocampus. *Proc. Natl. Acad. Sci.* USA, 83: 5326–5330, 1986.
21. Kirkwood, A., S. M. Dudek, J. T. Gold, C. D. Aizenman, and M. F. Bear. Common forms of synaptic plasticity in the hippocampus and neocortex in-vitro. *Science*, 260: 1518–1521, 1993.
22. Lewis, G., J. Ponte, and M. J. Purves. Fluctuations of PaCO2 with the same period as respiration in cat. *J. Physiol.* (London), 298: 1–11, 1980.
23. Malenka, R. C. Synaptic plasticity in the hippocampus: LTP and LTD. *Cell*, 78: 535–538, 1994.
24. Malenka, R. C., and R. A. Nicoll. NMDA-receptor-dependent synaptic plasticity: multiple forms and mechanisms. *Trends in Neural Sci.*, 16: 521–7, 1993.
25. Millhorn, D. E. Stimulation of raphe (obscurus) nucleus causes long-term potentiation of phrenic nerve activity in cat. *J. Physiol.* (London), 381: 169–79, 1994.
26. Millhorn, D. E., F. L. Eldridge, and J. P. Kiley. Oscillations of medullary extracellular fluid pH caused by breathing. *Resp. Physiol.*, 55: 193–203, 1984.
27. Minai, A. A. Covariance learning of correlated patterns in competitive networks. *Neural Comput.*, 9: 667–81, 1997.
28. Ponte, J., and M. J. Purves. Frequency response of carotid body chemoreceptors in the cat to changes of PaCO2, PaO2, and pHa. *J. Appl. Physiol.*, 37: 635–47, 1974.
29. Poon, C.-S. Ventilatory control in hypercapnia and exercise: optimization hypothesis. *J. Appl. Physiol.*, 62: 2447–2459, 1987.
30. Poon, C.-S. Adaptive neural network that subserves optimal homeostatic control of breathing. *Annals of Biomed. Engr.*, 21: 501–508, 1993.
31. Poon, C.-S. Self-tuning optimal regulation of respiratory motor output by Hebbian covariance learning. *Neural Networks*, 9: 1367–1383, 1996.
32. Poon, C.-S. Synaptic plasticity and respiratory control. In M. C. K. Khoo, editor, *Bioengineering Approaches to Pulmonary Physiology and Medicine*, pages 93–113. Plenum, New York, 1996.
33. Poon, C.-S., Y. Li, S. X. Li, and S. Tonegawa. Respiratory rhythm is altered in neonatal mice with malfunctional NMDA receptors. *FASEB J.*, 8: A389, 1994.
34. Poon, C.-S., and C. K. Merrill. Decrease of cardiac chaos in congestive heart failure. *Nature*, 389: 492–5, 1997.
35. Sammon, M. P., and E. N. Bruce. Vagal afferent activity increases dynamical dimension of respiration in rats. *J. Appl. Physiol.*, 70: 1748–1762, 1991.
36. Sejnowski, T. J. Storing covariance with nonlinearly interacting neurons. *J. Math. Biol.*, 4: 303–321, 1977.
37. Stanton, P. K. LTD, LTP, and sliding threshold for long-term synaptic plasticity. *Hippocampus*, 6: 35–42, 1996.
38. Stanton, P. K., and T. J. Sejnowski. Associative long-term depression in the hippocampus induced by Hebbian covariance. *Nature*, 339: 215–218, 1989.
39. Tobin, M. J., M. J. Mador, S. M. Guenther, R. F. Lodato, and M. A. Sackner. Variability of resting respiratory center drive and timing in health subjects. *J. Appl. Physiol.*, 65: 309–17, 1988.
40. Yamamoto, W. S. Transmission of information by the arterial blood stream with particular reference to carbon dioxide. *Biophy. J.*, 2: 143, 1962.
41. Zhou, Z., J. Champagnat, and C.-S. Poon. Phasic and long-term depression in brainstem nucleus tractus solitarius neurons: differing roles of AMPA receptor desensitization. *J. of Neurosci.*, 17: 5349–5356, 1997.

15

PERFORMANCES OF DIFFERENT CONTROL LAWS FOR AUTOMATIC OXYGEN SUPPLY FOR COPD PATIENTS

Valeri Kroumov,[1] Katsuki Yoshino,[2] and Sachio Tsukamoto[1]

[1]Okayama University of Science
Faculty of Engineering
1-1 Ridai-cho, Okayama 700, Japan
[2]Tokyo Women's Medical College
1st Department of Medicine
8-1 Kawada-cho, Shinjuku-ku, Tokyo 162, Japan

1. INTRODUCTION

Many of the patients with pulmonary insufficiency are administered to breath either pure oxygen (O_2) or high concentrations of oxygen from a mask or an intranasal tube. Depending on the level of the lung insufficiency the physician administers a certain amount of oxygen the patient have to inhale. The purpose of the oxygen therapy is to ensure that the patient's arterial partial pressure of oxygen (PaO_2) is maintained near the correct value. However, exercise is known to induce a decrease in the oxygen concentration in the blood and it is hard to predict how much PaO_2 decreases during exercise from the common pulmonary function tests such as spirometry or arterial blood gas analysis.

Furthermore, it is known that the oxygen lack becomes a powerful stimulation to respiration sometimes increasing the ventilation as much as five- to sevenfold (1). On the other hand, during the oxygen therapy, relief of the lack of oxygen occasionally causes the level of pulmonary ventilation to decrease so low that lethal levels of increasing of the carbon dioxide in the blood (hypercapnia) develop. In other words, an overdose of oxygen is as dangerous as the lack of it. So it is seriously important to keep PaO_2 at the correct level.

Because of this the automation of the process of adjusting and supplying the patient with oxygen is very much desirable. To meet this need, a number of research groups have developed and tested systems for the closed loop management of the gas levels for mechanically ventilated patients, e.g. patients under anesthesia during surgical operation (5,6,8), but as far as the authors know there is no attempt to control adaptively the oxygen supply for patients with pulmonary diseases who are on oxygen therapy.

Advances in Modeling and Control of Ventilation, edited by Hughson *et al.*
Plenum Press, New York, 1998.

Because the changing of the level of the blood oxygen is unpredictable the present study is an attempt to apply adaptive techniques to on-line estimation of the parameters of the pulmonary system model of the patient and to control the oxygen supply in order the arterial partial pressure of oxygen to maintain at the correct levels. The respiratory system is in general much dependent on equivalent volume of body tissue, blood flow rate, total volume of alveolus-factors different for every person and difficult to estimate exactly. Moreover, the dynamics of the system is affected by numerous unknown factors which does not allow to define precisely the necessary information for the synthesis of control system in the classical control theory terms. This makes the adaptive control techniques most suitable as a tool for searching for solution of the problem. In the paper the respiratory system is considered as simple as possible because of the limited knowledge at this stage. To control the amount of the supplied gas an observer (4) is used which adaptively adjusts the volume of the oxygen in order to keep PaO_2 close to the desired value.

2. DESCRIPTION OF THE CONTROL LOOP

Figure 1 illustrates the outline of the control system for PaO_2 seeing the patient as a dynamical system.

The input of the system is the amount of the inspiration gas characterized with certain fractional concentration of oxygen, and the output is the partial oxygen pressure in the blood.

The input, the fractional concentration of oxygen in inspired gas (FiO_2), is calculated using the next equation:

$$FiO_2 = \frac{\left(V_T - \frac{Q_{O_2}}{I_f I_r}\right) C_{O_{2air}} + \frac{Q_{O_2}}{I_f I_r} C_{O_{2s}}}{V_T} \tag{1}$$

where

V_T [l] = tidal volume,
Q_{O_2} [l/min] = volume of the supplied O_2,

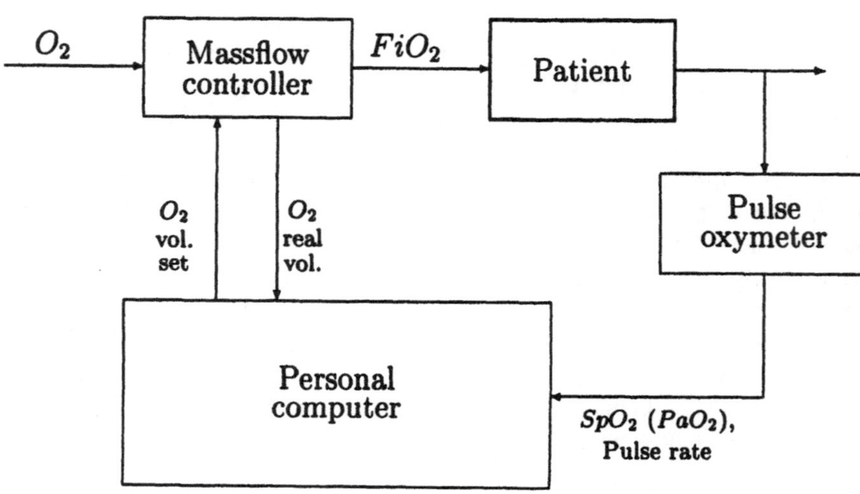

Figure 1. Block diagram of the system.

I_f [br./min] = respiration rate,
I_r = respiration/inspiration ratio,
$C_{O_{2air}}$ [%] = concentration of O_2 in air,
$C_{O_{2s}}$ [%] = concentration of the supplied O_2.

Because it is not possible to measure the partial pressure of the oxygen in the blood without a surgical intervention, the saturated oxygen in the blood (SpO_2) values obtained on-line from a pulseoxymeter are used to estimate the PaO_2 (the system output) using the Severinghaus algorithm with temperature correction and compensation for abnormal hemoglobin according to Astrup.

The personal computer receives the measurements from the pulse oxymeter and calculates the necessary amount of supply oxygen. The mass flow controller supplies oxygen according to the set value from the computer.

Special measures are taken for the safety of the control system: (a) measuring the actual amount of supplied gas, (b) monitoring of the pulse of the patient, (c) automatically checking whether the sensor is properly placed etc. In a case of abnormal readings sound and blinking colored screen alarms are activated.

3. PARAMETERS IDENTIFICATION

In order to construct the model of the respiratory system measured data from several patients in hospitalization with chronic respiratory diseases in stable clinical conditions are used. The partial oxygen pressure in the arterial blood measured when the patients were in rest in spine position while breathing normal room air was less than 75 mmHg. The studies were carried after the subjects were sufficiently informed and had given their consent. A step response of the pulmonary system of the patients was obtained by letting then to breath consequently a normal air and high concentrations of oxygen (Fig. 2).

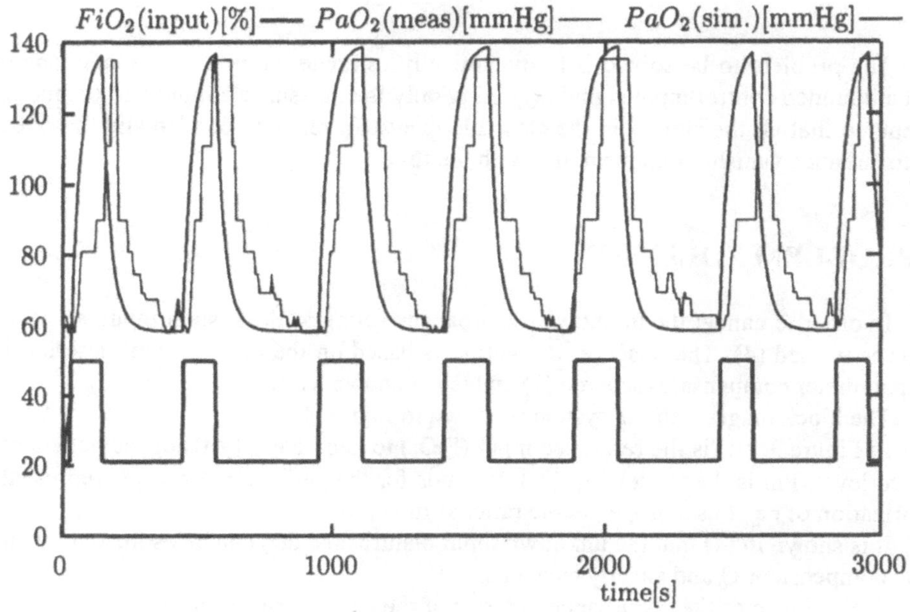

Figure 2. Simulation results.

The following first order differential equation was chosen as a model description:

$$\frac{d}{dt}y(t)+ay(t)=bu(t) \tag{2}$$

where 'a' and 'b' are unknown, y(t) is the partial pressure of O_2 in the blood, and u(t) is the FiO_2.

The unknown parameters in eq. 2 were identified using a Kreisselmeier type adaptive observer (3) with a least squares type adaptive adjusting law. Figure 2 shows the measured and simulated values of PaO_2 using the model (eq. 2) with the identified parameters substituted. It can be verified from the graphs that the simulated output matches against the measured one quite well.

4. PROBLEM STATEMENT

The parameter variations of the pulmonary system, measurement inaccuracy or other unknown factors affecting the dynamics are considered as a unknown disturbance d(t). It is assumed that the equation describing the disturbance model is known but its coefficients are unknown.

The description of the system (eq. 2) in presence of input disturbances becomes

$$\frac{d}{dt}y(t)+ay(t)=b(u(t)+d(t)) \tag{3}$$

It is supposed that the disturbance is described as

$$\frac{d^n d(t)}{dt^n}+a_n\frac{d^{n-1}d(t)}{dt^{n-1}}+\cdots+a_1\frac{d}{dt}d(t)+a_0=0 \tag{4}$$

The problem to be solved is to dynamically estimate the unknown disturbance and form a bounded control input signal u(t) using only the measurable input and output of the system, so that all the signals in the closed loop system remain bounded and the effect of the disturbance vanishes exponentially with the time.

5. PROBLEM SOLUTION

In order to cancel the unknown disturbances acting at the system input, an adaptive observer is used (4). The observer derivation is based on the most general description of two-parameter compensator scheme (7) and is not shown here.

The block diagram of the system is shown in Figure 3.

In Figure 3, r(t) is the reference input (FiO_2) to keep PaO_2 (y(t)) of the patient at the desired level, Pm is the model (eq. (2)), P stands for the patient, and N1, D1 represent the factorization of eq. 2 as a proper stable rational function.

It is shown in (4) that the unknown input disturbance d(t) can be estimated by means of the compensator Q and exactly canceled.

The estimate of the disturbance in terms of the above terms becomes:

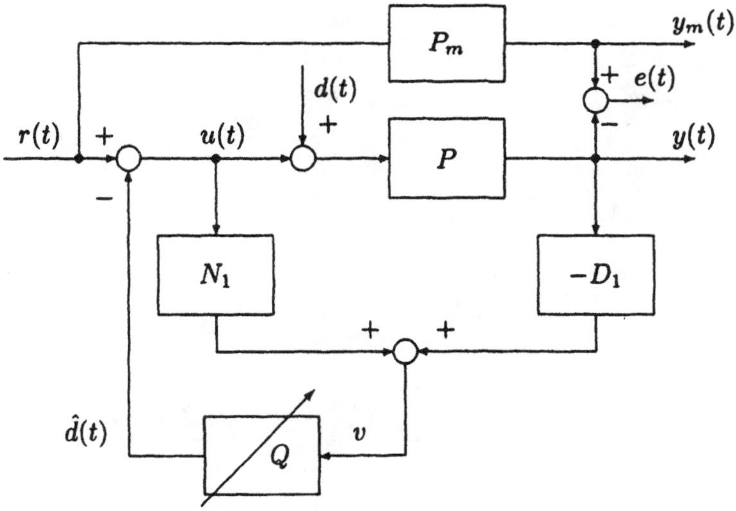

Figure 3. Block scheme of the observer-controller system.

$$\hat{d}(t) = -Q(s)(N_1(s)u(t) - D_1(s)y(t)) \tag{5}$$

6. RESULTS

A comparison between clinical test results of P control, PI control, and adaptive observer are shown here.

Figure 4 shows the data from a P control applied to a patient with pulmonary microlethiasis. It can be seen that in order to compensate the changing in the output (PaO_2)

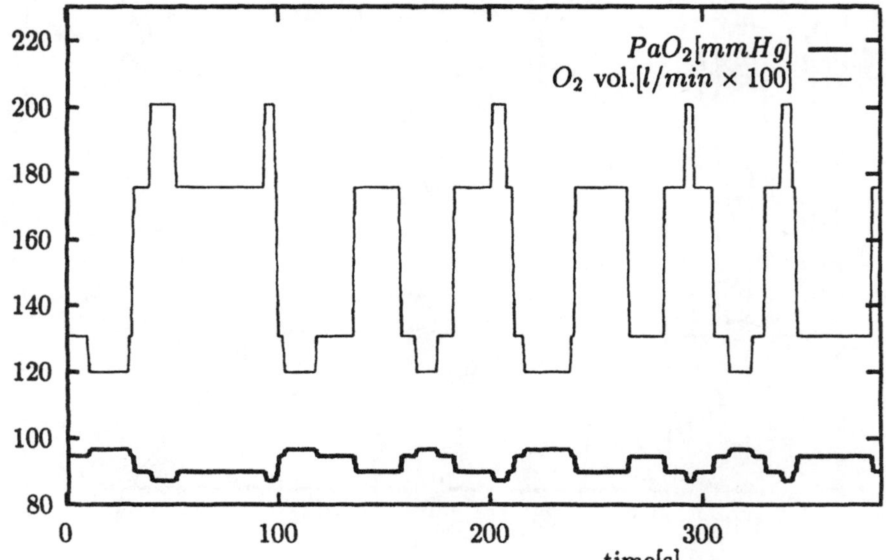

Figure 4. P control (Pulmonary microlethiasis, rest, PaO_{2set} = 95 mmHg).

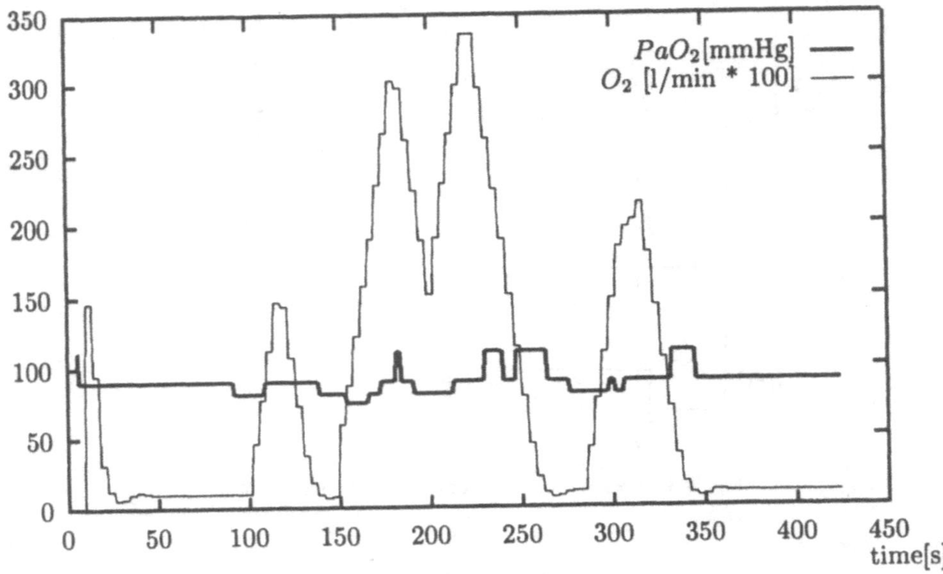

Figure 5. PI control (FLD, rest, PaO_{2set} = 81 mmHg).

the volume of the supplied oxygen becomes quite high and that there is a big overshoot in the output every time when the PaO_2 goes below the setting level. The measurements when PI control low is applied are shown on Figure 5 and it can be seen that the control input varies too much.

The result of control using the adaptive observer is shown on Figure 6. The control was applied to the same patient who cooperated for the PI control. There are big fluctua-

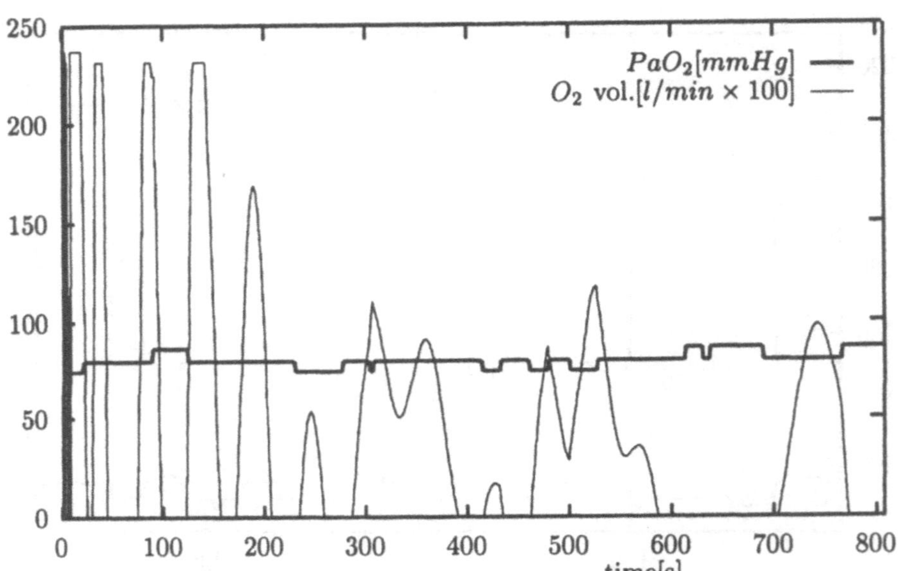

Figure 6. Adaptive control (FLD, bicycle, PaO_{2set} = 81 mmHg).

tions in the input (O_2 amount) at the initial stage, while the parameters of the observer converge (about 3 min), but after this only the necessary oxygen is supplied. Compared to the P and PI control laws the output is much more stable. Another advantage of the adaptive methods is that there is no need to perform adjustments of the parameters of the control law. In other words the proposed control using an adaptive observer can be applied to any patient without any preparations and beforehand settings. When the classical P, PI, or PID control laws are applied the control parameters have to be adjusted for every patient and quality of the control depends too much on whether the person is in rest or exercising.

It is expected that the proposed in this paper controller will decrease the discomfort from dry passages when using intranasal tube and will lead to some savings of oxygen.

7. CONCLUSIONS

A system for controlling the arterial partial pressure of oxygen in patients on long-term oxygen therapy was proposed. The system input is the flow of supplied oxygen and the output is the partial pressure of oxygen in the blood. After performing a parameter estimation of the system model, the application of an adaptive observer for estimation and further elimination of the disturbances to the patients blood oxygen level was shown. A comparison to the classical P, PI control laws was done as well. It can be concluded that the adaptive control technique is superior to P and PI control. It is expected that application of the proposed observer will contribute to the improving of the quality of life of the patients with chronic obstructive pulmonary diseases.

ACKNOWLEDGMENTS

The authors would like to thank Sanyo Electronic Industries Co., Ltd. (Okayama, Japan) for the hardware realization of the controller used in this paper. Thanks to Dr. Goto form Tokyo Women's Medical College, 1st Dept. of Medicine, who helped in data processing and measurements.

REFERENCES

1. Bullock, J., J. Boyle III, and M. B. Wang. Physiology. 3rd ed., Malvern, PA, Williams & Wilkins, 1995, pp. 257–270.
2. Kira, S., T. Petty (Eds.). Progress in Domiciliary Respiratory Care—Current Status and Perspective: Proceedings of International Symposium on Domiciliary Respiratory Care, Tokyo, Sept. 20–22. Amsterdam, Elsevier Science B. V., 1994. pp. 115–124.
3. Kreisselmeier, G. Adaptive observers with exponential rate of convergence. IEEE Trans. on Automatic Control, AC-22:2–8, 1977.
4. Kroumov, V., S. Masuda, A. Inoue, and K. Sugimoto. Proceedings of the Second Asian/Pacific International Symposium on Instrumentation, Measurement and Automatic Control, pp. 288–291, 1993.
5. Mitamura, Y., T. Mikami, H. Sugiwara, and C. Yoshimoto. An optimally controlled respirator, IEEE Trans. BME, 18:846–853, 1971.
6. Packer, J. S., T. L. Fernando, Z-M Xu, J. F. Cade, and B. Lee. 12th World Congress of IFAC, Sydney. Barton, The Institution of Engineers, Australia, 1993, 3:237–240.
7. Vidyasagar, M. Control System Synthesis: A Factorization Approach. Mass., U.S.A., The MIT Press, 1985, 146–150.
8. Wakamatsu, W. 12th World Congress of IFAC, Sydney. Barton, The Institution of Engineers, Australia, 1993, 4:467–472.

TECHNIQUES FOR ASSESSING THE SHAPE OF RESPIRATORY FLOW PROFILES FROM DATA CONTAINING MARKED BREATH-BY-BREATH RESPIRATORY VARIABILITY

Jiro Sato[1] and Peter A. Robbins[2]

[1]Department of Anesthesiology
Chiba University, Japan
[2]University Laboratory of Physiology
University of Oxford
Parks Road, Oxford OX1 3PT, United Kingdom

1. INTRODUCTION

Average respiratory flow profiles during steady breathing have been of interest as an output of the respiratory controller. However, in the process of determining average flow profiles, distortions can occur that are caused by the great variability both between breaths and within breaths. The purpose of this study is to develop a method for determining typical flow profiles which minimises such distortions.

2. METHODS AND DISCUSSION

Our method uses flow-volume loops in the process of constructing typical flow profiles. The general idea behind the method is to align the different respiratory cycles by phase of respiratory cycle before the "averaging" process is undertaken.

After any drift in the integral of the flow signal (i.e. volume) was removed, the flow and volume signals were normalised such that they had means of zero and SDs of unity. Respiratory phase could then be defined as the angle associated with the point on the normalised flow-volume diagram. For each breath, estimates for normalised flow, volume and for time at each degree of phase were obtained by linear interpolation from the values closest to either side of the phase under consideration. Data over a number of breaths could then be averaged by calculating the median values for each variable at each phase angle.

Advances in Modeling and Control of Ventilation, edited by Hughson *et al.*
Plenum Press, New York, 1998.

Because the median flows, volumes, and times are all semi-independent statistical estimates, in general it is not true that the laws of mass balance hold for these "averages", such that the volume is precisely the time integral of the flow. In order to obtain estimates that conform to the laws of mass balance, the following procedure was adopted. First, the new volume was calculated by integrating the median flow in terms of the median time. Secondly, the inspiratory and expiratory flow values were scaled proportionately so that the new inspiratory and expiratory tidal volumes correspond precisely to the original (median) tidal volume estimate. Volumes are then determined by integration of this flow time profile, and this closes the flow-volume loop.

The method was compared with that developed by Benchetrit and co-workers which reconstructed the typical flow profile from the averages of Fourier coefficients associated with individual breaths (1). The "Benchetrit" method produced a systematic distortion of the respiratory flow profile that arose from the manner in which the breaths were aligned in time. Our method reduces this distortion.

REFERENCES

1. Eisele, J.H., B. Wuyam, G. Savourey, J. Eterradossi, J.H. Bittel, and G. Benchetrit. Individuality of breathing patterns during hypoxia and exercise. J. Appl. Physiol. 72:2446–2453, 1992.

17

THE EXPIRATORY FLOW PATTERN AND THE NEUROMUSCULAR CONTROL OF BREATHING IN CATS

C. P. M. van der Grinten, C. K. van der Ent, N. E. L. Meessen, J. M. Bogaard, and S. C. M. Luijendijk

Department of Pulmonology
Maastricht University
Maastricht, The Netherlands
Department of Pediatric Pulmonology
Wilhelmina Children's Hospital
Utrecht, The Netherlands
Department of Pulmonary Diseases
University Hospital Dijkzigt
Rotterdam, The Netherlands

1. INTRODUCTION

During spontaneous breathing inspiratory muscle activity does not stop immediately at the start of expiration, but decays at a rate which can be influenced importantly by pulmonary receptors.[1] When the contribution of this decaying activity to the expiratory flow pattern is suppressed by a short end-inspiratory occlusion, the time constant of the respiratory system ($\tau_{RS} = R_{RS}/E_{RS}$) can be estimated from the expiratory flow pattern using equation 1 and P(t) = 0.[4]

$$P(t) = \dot{V}R_{RS} + VE_{RS} \tag{1}$$

P is the pressure generated by the inspiratory muscles, $\dot{V}$ is expiratory flow, V is volume, E_{RS} and R_{RS} are the elastance and resistance of the respiratory system, respectively and t is time. The integrated activity of inspiratory muscles during expiration can be described by a single exponential function.[1] Siafakas *et al.*[3] showed that inspiratory pressure is linearly related to integrated inspiratory muscle activity. Thus, P(t) during expiration can be described by $P(t = \infty) + (P(t = 0) - P(t = \infty))\, e^{-t/\tau}$ in which τ is the time constant of

Advances in Modeling and Control of Ventilation, edited by Hughson *et al.*
Plenum Press, New York, 1998.

the muscle activity. Substitution in (1) and $R_{RS} = E_{RS} \cdot \tau_{RS}$ yields equation 2, which can be solved analytically.

$$P(t=\infty)+\left(P(t=0)-P(t=\infty)\right)e^{-t/\tau}=\tau_{RS}E_{RS}\dot{V}+V(t)E_{RS} \tag{2}$$

The resulting equation for the flow is a function of three parameters: τ_{RS}, τ and the tidal volume (V_T).

$$V(t)=\frac{\dot{V}_T}{\tau_{RS}-\tau}\left(e^{-t/\tau}-e^{-t/\tau_{RS}}\right) \tag{3}$$

The derivate of equation 3 equals zero at peak expiratory flow ($\dot{V}_{peak}$). Rearrangement results in equation 4 for time to peak tidal expiratory flow (t_{PTEF}) and equation 5 for $\dot{V}_{peak}$.

$$t_{PTEF}=\frac{\tau\cdot\tau_{RS}}{\tau-\tau_{RS}}\cdot\ln\left(\tau/\tau_{RS}\right) \tag{4}$$

$$\dot{V}_{peak}=-\frac{V_T}{\tau}\cdot e^{(-t_{PTEF}/\tau)} \tag{5}$$

The aim of the study was to validate the model described above in equations 1 to 5 using data that had been gathered previously in a study by Meessen *et al.*[1]

2. METHODS

In that study 7 anaesthetized cats were used spontaneously breathing through a tracheal canula in supine position. EMG activity of the diaphragm (DIA) was measured using small, nickel-silver hooks placed in the muscle. The derived signal was amplified, rectified and integrated with a 'leaky' integrator using a time constant of 50 ms (Neurolog). Flow was assessed by a Fleisch pneumotachograph (Gould) connected on one side to the tracheal canula and on the other side to a t-piece conducting a bias flow of room air. The cervical vagi were dissected free from the carotid sheaths and placed in a pair of cooling devices. In this way the temperature of the vagi could be varied between body temperature and 4°C with an accuracy of 0.2°C.

The time constant of the decay of the integrated diaphragmatic activity during expiration was determined yielding τ_{DIA}. From the flow signal we determined t_{PTEF} and $\dot{V}_{peak}$. A large range of values for τ_{DIA} was obtained by cooling the vagus nerve, histamine injections and applying continuous positive airway pressure (CPAP) shortly after histamine injections.

Equations 4 shows that t_{PTEF} is a function of τ exclusively, when τ_{RS} is constant. Accordingly, a value for τ_{RS} can be fitted from the plots of t_{PTEF} as a function of τ for the three different experimental situations (control, histamine, histamine + CPAP). These nonlinear fits were obtained using the Levenberg-Marquardt method. Values of measured $\dot{V}_{peak}$ were compared with the values for $\dot{V}_{peak}$ predicted by equation 5.

3. RESULTS AND DISCUSSION

Part of original registrations of the flow signal (bottom) and the integrated DIA signal (top) are shown in Figure 1. In the latter signal an exponential function is fitted through part of the data showing that inspiratory activity during expiration can be characterized very well by τ_{DIA}.

From a plot of t_{PTEF} as a function of τ_{DIA}, τ_{RS} can be estimated using equation 4. This is shown in Figure 2 for the three different experimental conditions: control, histamine and histamine + CPAP. The values for τ_{RS} were 257, 240 and 190 ms, correlation coefficients 0.66, 0.81 and 0.84, and residual standard deviation 56, 56 and 45 ms for each of the three conditions, respectively. τ_{RS} values were in the range of those published for anaesthetized cats.[4] There is no increase in τ_{RS} during histamine. This was expected because bronchoconstriction would increase resistance and—with a likely unchanged E_{RS}—τ_{RS} should increase. However, pulmonary resistance as calculated from fits of pressure volume loops did not change either, although we used high doses of histamine. Thus, histamine influenced DIA probably by stimulating rapidly adapting pulmonary stretch receptors [1], without causing much bronchoconstriction. Increasing the lung volume by CPAP decreases airway resistance and lung compliance and, thus, τ_{RS}.

t_{PTEF} calculated by the model is higher than the actually measured value for low values of τ_{DIA}, especially in the lower two panels. This may be due to overestimation of τ_{DIA} at low values due to the contribution of the time constant of the integrators (50 ms).

In Figure 3 upper panel estimated $\dot{V}_{peak}$ using equation 5 is plotted as a function of the actually measured value of $\dot{V}_{peak}$, showing a considerable underestimation. There are 3

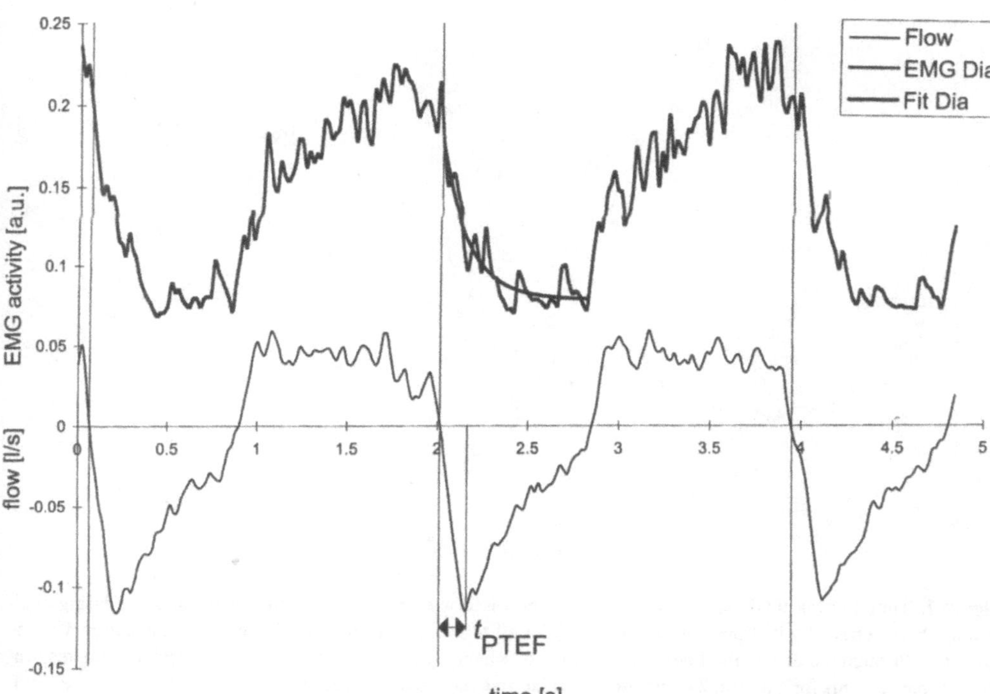

Figure 1. Original tracings of integrated diaphragmatic EMG (top) and flow (bottom). In the upper tracing a curve is fitted through a part of the EMG. In the flow tracing the calculation of t_{PTEF} is illustrated.

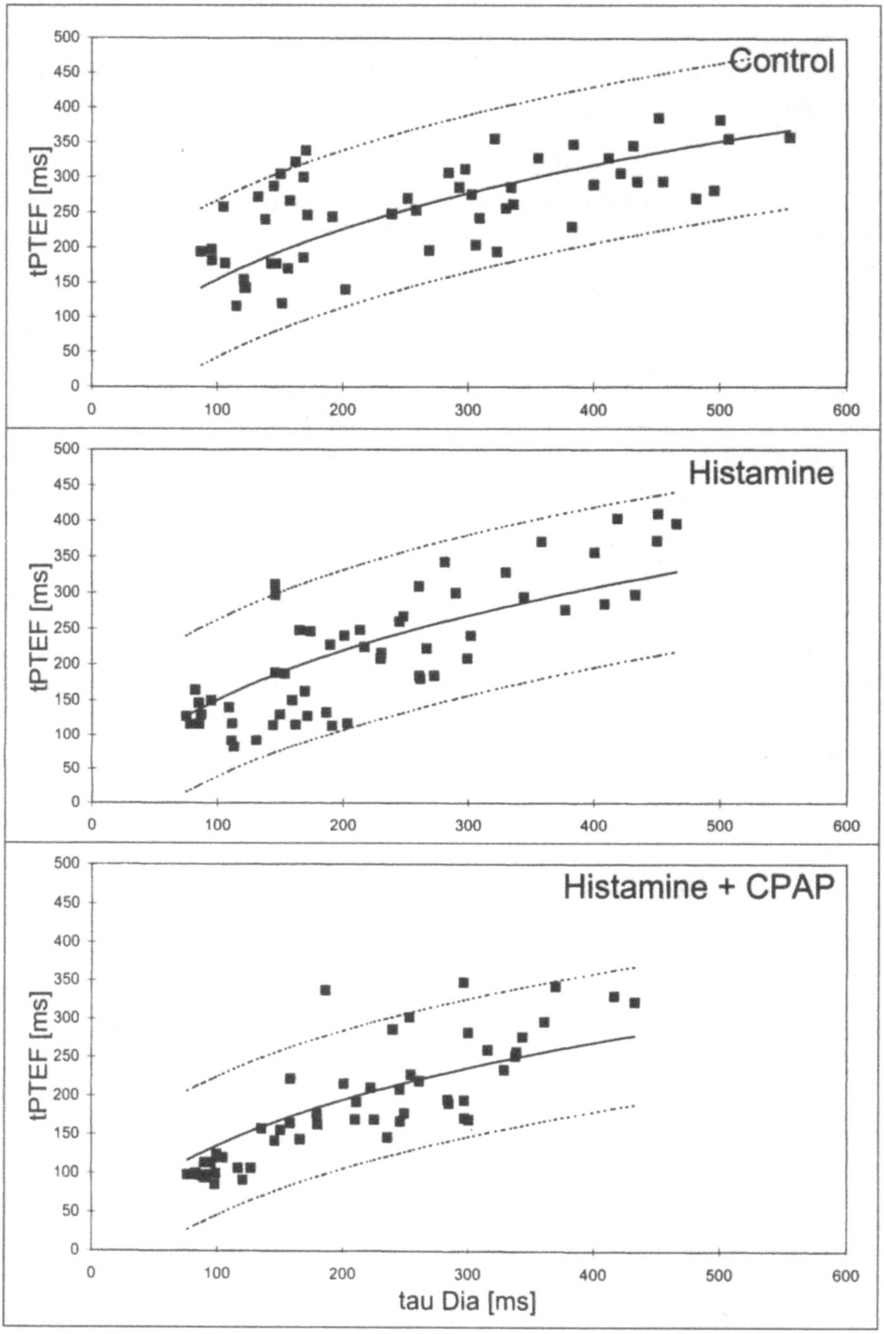

Figure 2. Time to peak tidal expiratory flow (t_{PTEF}) increased with increasing diaphragmatic activity during expiration (characterized by its time constant, tau Dia) for all three experimental conditions. Using equation 4 a line was fitted through the data (solid line), the curvature of which is characterized by the time constant of the respiratory system (257 ms for Control, 240 ms for Histamine and 190 ms for Histamine + CPAP). The dashed lines indicate two times residual standard deviation.

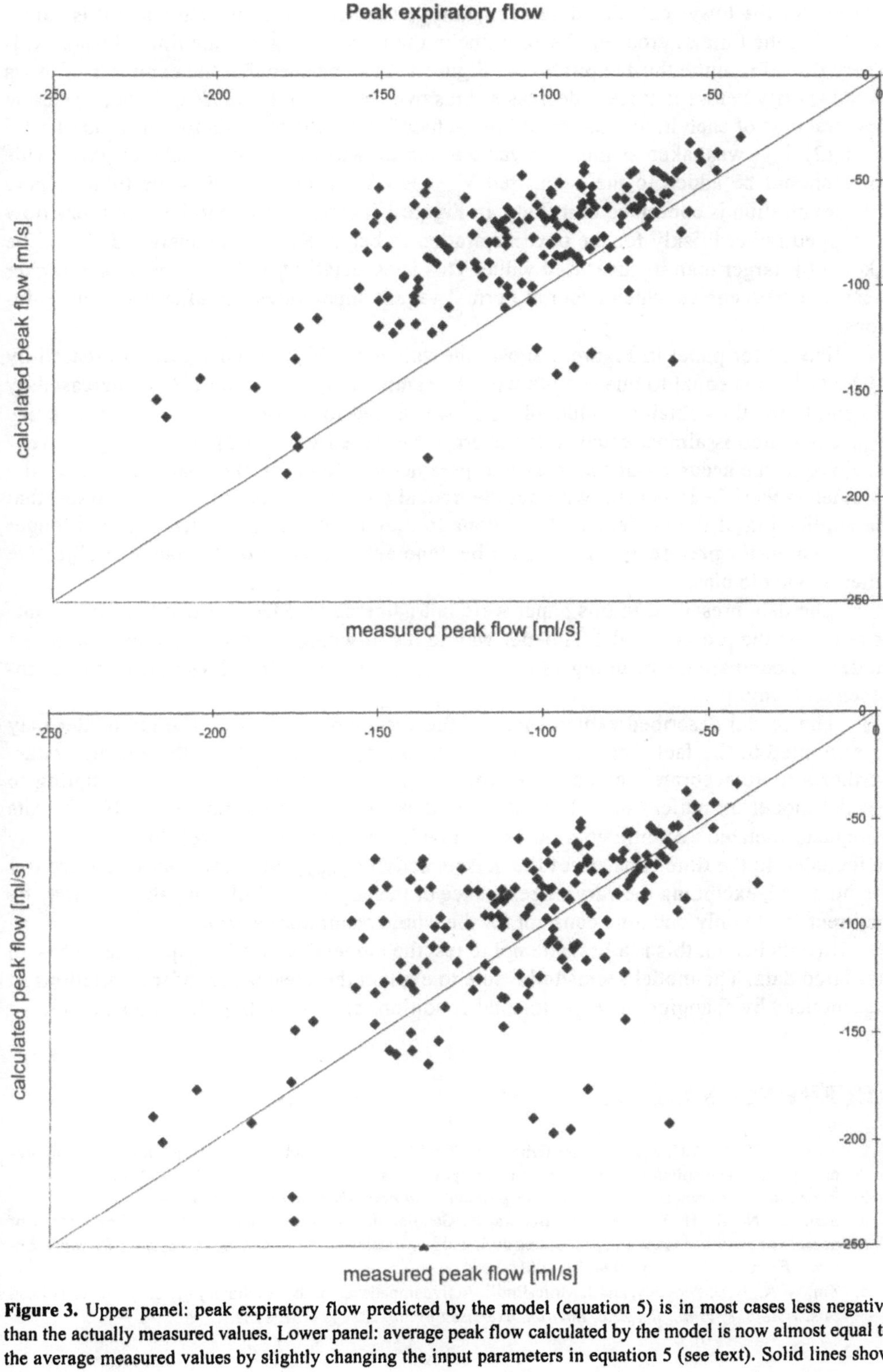

Figure 3. Upper panel: peak expiratory flow predicted by the model (equation 5) is in most cases less negative than the actually measured values. Lower panel: average peak flow calculated by the model is now almost equal to the average measured values by slightly changing the input parameters in equation 5 (see text). Solid lines show unity.

reasons for the lower calculated values: (1) t_{PTEF} is often overestimated since it is calculated from the time of crossing the zero line in the flow tracing and the time of peak expiratory flow. From the third expiration in Figure 1 it can be seen that the expiratory flow is halted shortly before it quickly decreases, thus overestimating t_{PTEF}. We did not inspect the flow tracings of each individual breath to exclude it from the analysis for this kind of artefacts. (2) $\dot{V}_{peak}$ was taken as the peak value in the flow tracing, which includes noise. This noise should be added to the calculated $\dot{V}_{peak}$. (3) V_T in equation 5 is the tidal volume when expiration is continued to infinity. In Figure 1 it can be seen that the expiratory flow is stopped rather briskly for the next inspiration to begin. So, V_T in equation 5 should be taken a bit larger than its measured value. This is especially true for the breaths at higher breathing frequencies which occur at normal vagal temperatures and after histamine infusions.

The lower panel in Figure 3 shows the values for $\dot{V}_{peak}$ when t_{PTEF} is decreased by 0.02 s (which is equal to one sample with the sample frequency used), V_T is increased by 5% and to the thus obtained value -8 $ml \cdot s^{-1}$ was added for noise correction. Now average $\dot{V}_{peak}$ calculated is almost equal to the average measured value. Apparently, $\dot{V}_{peak}$ is very sensitive to the accuracy of the measured parameters. However, the changes made to the parameters were in agreement with our theoretical considerations. It further illustrates that the application of the model should be done to individual breaths selected from a longer flow tracing. By preference, this should be done automatically on the basis of objective criteria to avoid bias.

The data presented in this paper were not gathered by Meessen *et al.*[1] with the purpose to test the present model. A better way to test it would be to fit the flow data to the model and compare the resulting values for τ_{DIA}, τ_{RS} and V_T with values obtained in an independent way.

The model described explains part of the variation in the data. The remainder may be attributed to the fact that individual cats have a τ_{RS} different from the average value. Further a more accurate method to determine t_{PTEF} would be important. It is tempting to use the model on patient data, but can it be done without precautions? Firstly, our cats were anaesthetized avoiding volitional control of respiratory manoeuvres. In part, this may be included in the time constant of the muscle activity (τ_{DIA}). Secondly, our cats were tracheotomized, excluding the varying resistance of the larynx.[2] Thirdly, in patients it may be incorrect to use only one time constant for the whole respiratory system.

In conclusion, this is a first attempt to test the model described in equations 1–5 with measured data. The model seems to be able to explain the greater part of the variations in t_{PTEF} induced by changing the experimental conditions and by cooling the vagus nerves.

REFERENCES

1. Meessen, N. E. L. ,C. P. M. van der Grinten, H. Th. M. Folgering, and S. C. M. Luijendijk. Histamine-induced end-tidal inspiratory activity and lung receptors in cats. *Eur. Respir. J.* 8: 2094–2103, 1995.
2. Rattenborg C. Laryncheal regulation of respiration. *Acta Anaesth. Scandinav.* 5: 129–140, 1961.
3. Siafakas, N. M., H. K. Chang, M. Bonora, H. Gaultier, J. Milic-Emili, and B. Duron. Time course of phrenic activity and respiratory pressures during airway occlusions in cats. *J. Appl. Physiol.: Respirat. Environ. Exercise Physiol.* 51: 99–108, 1981.
4. Zin, W. A., L. D. Pengelly, and J. Milic-Emili. Active impedance of the respiratory system in anaesthetized cats. *J. Appl. Physiol.: Respirat. Environ. Exercise Physiol.* 53: 149–157, 1982.

18

PHASE RELATIONS BETWEEN RHYTHMICAL FOREARM MOVEMENTS AND BREATHING UNDER NORMACAPNIC AND HYPERCAPNIC CONDITIONS

Dietrich Ebert, Beate Raßler, and Siegfried Waurick

Carl-Ludwig-Institute of Physiology
University of Leipzig
D-04103 Leipzig, Germany

1. INTRODUCTION

Sensory and cognitive processes, for instance, listening to texts[20] or music[9,10], or imagination of a movement[4] can influence breathing. Rhythmical limb movements were found to influence breathing during walking[1,13,15,18], running[2,3], cycling[8,12,14,22], rowing[16,21], tapping[24]. As well finger tracking[17] as eye movements[23] have been shown to be related to certain plhases of the ongoing breathing cycle. In most of these studies coordination of breathing with periodic free movements at spontaneous rates was analyzed. Some investigations, however, demonstrated the pronounced influence of movement rate on frequency and stability of coordination[6,7,18].

This study is devoted to examine coordination between breathing and tracking movements of the forearm following a sinusoidal target with systematically varied rate. A dependence of coordination on the movement rate has been found[7].

Coordination can be interpreted as the result of coupling between oscillators[5]. The theory of coupled oscillators reveals that amplitude modulations in one of the oscillators change the strength of coupling between the oscillators. The strength of coupling may reflect in the stability and the frequency of coordination phenomena.

In the present study we altered the amplitude of the respiratory oscillator increasing its chemical drive by hypercapnia (inhalation of 3.0% CO_2 in air). We expected the frequency of coordination (i.e. the frequency of constant phase-relations) and its stability (i.e. the steadiness of consecutive phase intervals between both the periodical forearm movement and breathing movements) to be modified compared to normocapnic conditions.

Advances in Modeling and Control of Ventilation, edited by Hughson *et al.*
Plenum Press, New York, 1998.

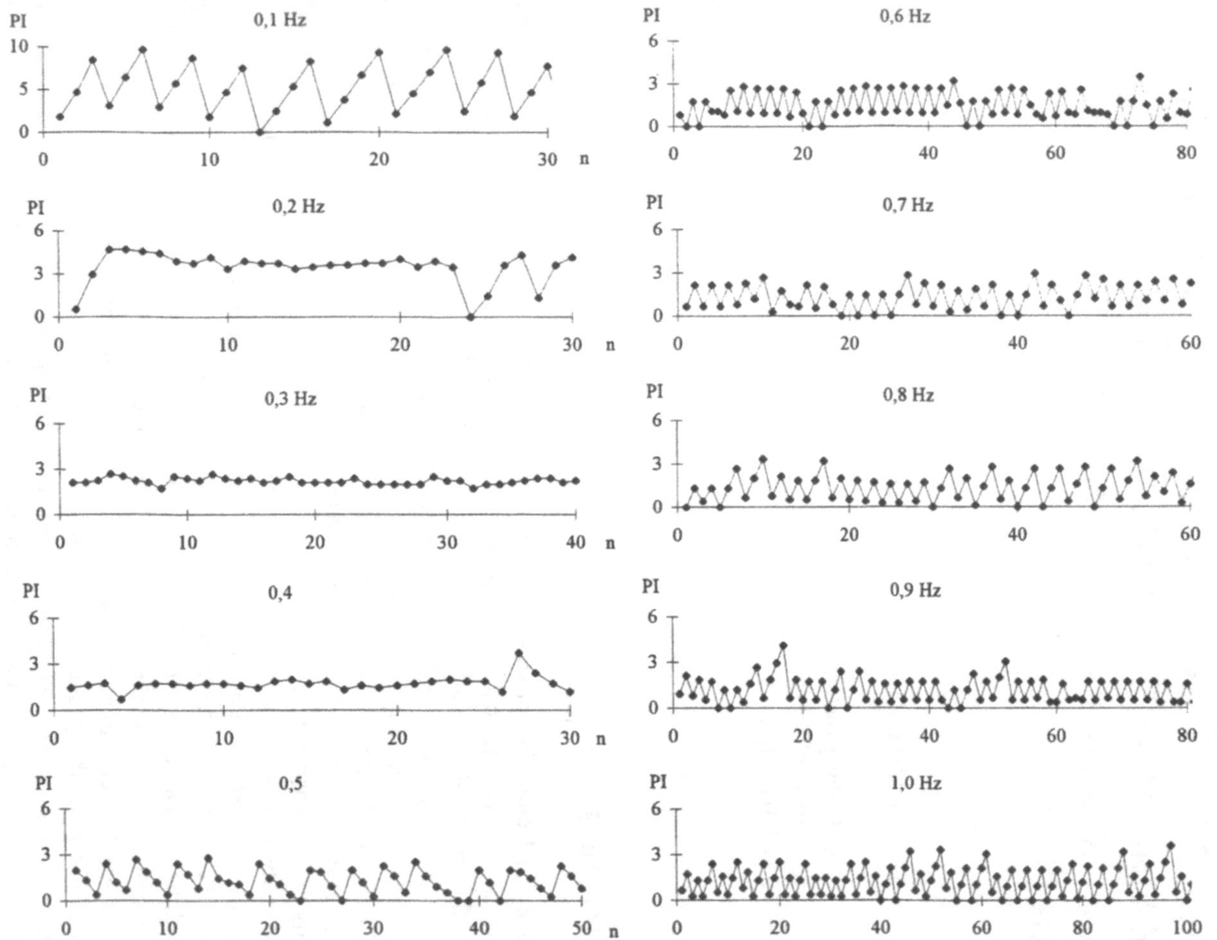

Figure 1. Left side: Sequential phase interval plots of subject 3 for tracking rates between 0.1 and 0.5 Hz (PI = phase interval, n = number of phase interval) in normocapnia. Note 1:1 coordination between 0.2 Hz and 0.4 Hz. Right side: Sequential phase interval plots of subject 3 for tracking rates between 0.6 and 1.0 Hz in normocapnia. Note 1:2 coordination between 0.6 and 0.9 Hz and 1:3 coordination at 1.0 Hz.

2. METHODS

2.1. Subjects and Experimental Set-Up

Data were analyzed from 5 subjects (3 female, 2 male, age range: 21–24 years). Subjects were asked to perform 15° elbow flexions and extensions with the preferred arm following a sinusoidal target. Movement angle was measured by a goniometer. As well arm movements as target movement were presented as light spots on a screen. Both signals were digitized at 30 Hz and stored on a disk for an off-line data analysis.

Respiration was measured simultaneously by means of a pneumotachograph (PTG) using a FLEISCH transducer fixed by a face mask.

After a period of adaptation to the measuring equipment subjects felt not disturbed any more, thus focussing their attention completely on the tracking task. Data recording was started after five to ten minutes—a period necessary to give the subjects some tracking training—and lasted about 70 s. The frequencies of the target's sine waves were varied in the following order: 0.1, 0.2, 0.3, 0.4, 0.5, 0.6, 0.7, 0.8, 0.9 and 1.0 Hz. This rate range covered the range of possible spontaneous breathing rates.

At the first day the experiments were carried out with subjects inhaling fresh air—designated as "normocapnic" condition. At the following day the experiment was repeated with the same subjects but inhaling 3% carbondioxide in the air—designated as "hypercapnic" condition. Subjects got a ten minute period of adaptation to hypercapnia, and after further five minutes of tracking practice the data recording was started again for 70 s.

2.2. Data Analysis and Statistics

For demonstration of phase coordination phase intervals (PI) were calculated (detailed description in [7]).

All zero crossings in downward direction of the breathing signal (related to inspiration onsets) were detected as breathing marker events. Likewise all zero crossings in upward direction of the arm movement signal (midphase of extension) were detected as movement marker events. Using breathing as reference signal we measured the time interval between each breathing marker event and the consecutive movement marker event. This interval was denoted as "phase interval" (PI). In order to detect phase coordination periods for each trial consecutive PIs were plotted sequentially (compare Fig. 1).

We characterized a ratio of breathing to tracking rate of 1:1 with phase intervals being almost constant (presenting in the sequential plot as a straight line of PIs parallel to the abscissa) as stable 1:1 coordination.. We define subsegments of at least three consecutive PIs as periods of stable 1:1 coordination if the corresponding differences of the consecutive phase intervals ($\Delta PI = PI_{n+1} - PI_n$) had a variation of less than 25% of the mean movement cycle duration. The number of PIs within those subsegments of coordination was calculated as percentage of the total number of all PIs in the trial. Likewise, if every second or third ΔPI in a subsegment of consecutive PIs varied less than 25% of the mean movement cycle duration, 1:2 coordination (with the ratio of breathing rate to tracking rate being 1:2) or 1:3 coordination (with the ratio of breathing rate to tracking rate being 1:3) was defined. Since the criterion for coordination being set to 25% cycle duration was weak, in comparison the same analysis was carried out a second time using a stronger selection criterion of 10% cycle duration.

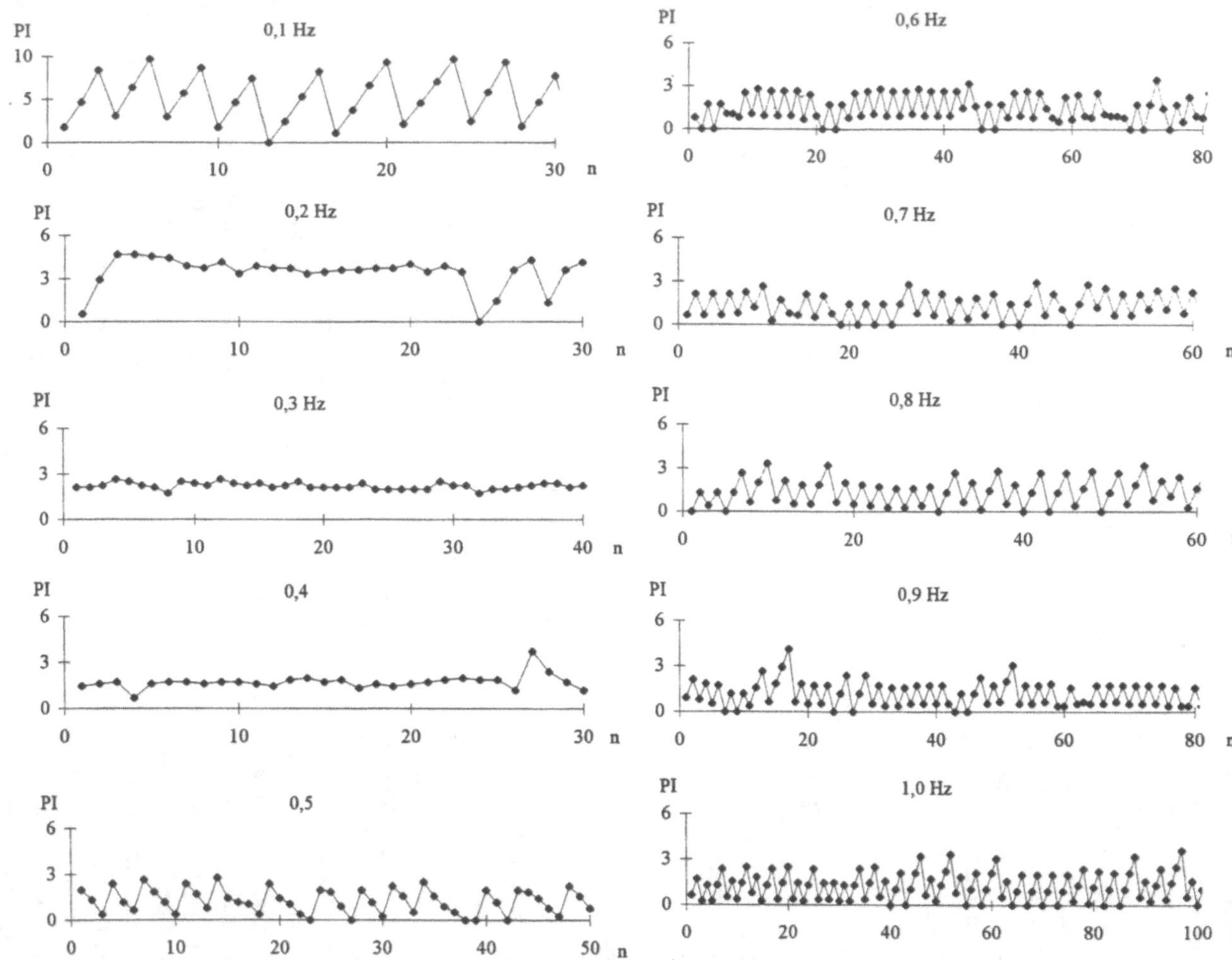

Figure 2. Left side: Sequential phase interval plots of the same subject but under hypercapnic condition. 1:1 coordination range is shifted to higher frequencies (0.3 to 0.5 Hz, some short periods also at 0.6 Hz). Right side: Also 1:2 coordination is shifted to the range between 0.7 to 1.0 Hz, 1:3 coordination is not seen in the range up to 1.0 Hz. At 0.6 unstable pattern with interspearsed 1:1 coordination is detectible, similar to 0.5 Hz under normocapnic condition (Fig. 1).

3. RESULTS

3.1. Pattern of Phase Coordination between Arm Movements and Respiration

Sequential phase interval plots revealed periods of phase coordination in all subjects. Stable 1:1 coordination was usually found in tracking movements with a tracking rate around 0.3 Hz, a rate which is slightly higher than the mean breathing rate at rest.

Fig. 1 shows typical examples of 1:n coordination between breathing and tracking movement in one subject. In plots of consecutive phase intervals the periods of 1:1 coordination were characterized by segments parallel to the abscissa (Fig. 1, left side, at 0.2, 0.3 and 0.4 Hz) with ΔPIs close to zero. In periods exhibiting 1:2 coordination PIs form two lines parallel to the abscissa (Fig. 1, right side, 0.6–0.9 Hz). 1:2 coordination between breathing and arm movements was observed at rates being twice the tracking rate of the 1:1 entrainment bands (see below). 1:3 coordination is detectible at 1.0 Hz in this subject. The variability of PI increased with increasing rate and indicates reduced coordination.

The range of target rates in a particular type of coordination (in subject 3, Fig. 2: the range from 0.2–0.5 Hz for 1:1 coordination) was termed „entrainment band". Width, but not location, of these entrainment bands varied amongst the subjects but was reproducible intraindividually. The entrainment bands of different coordination types tended to touch or to overlap each other. Group mean values of all subjects revealed in more than 50% of the recorded time a 1:1 coordination at 0.3 Hz (Fig. 4). All 1:1 entrainment bands were within the range between 0.2 and 0.6 Hz. The maximum in 1:2 coordination was found around 0.7 Hz, a rate being slightly higher than the 1st harmonic of 0.3 Hz. It occurred in about 25% of all recordings. The 1:2 entrainment band ranged from 0.5 to 1.0 Hz. The range of 1:3 coordination had its maximum at 1.0 Hz. 1:3 coordination was observed in less than 20% of the entire recording time.

3.2. Coordination under Hypercapnic Conditions

Surprisingly, the increased CO_2 content of the inspired gas led to a significant increase of coordination frequencies in all subjects at all target rates (Figs. 2, 3, 4). As can be demonstrated in subject 3 the coordination-related pattern in the sequential phase interval plot became more stable (that means, the variation of PI decreased) at the main entrainment rate.

In addition, the main entrainment rate shifted to higher rates, in most cases from 0.3 to 0.4 Hz (Fig. 2 and 3) and the rate range of the entrainment bands widened. In this case short 1:1 coordination periods were already seen at 0.2 Hz and at 0.6 Hz. All subjects exhibited these changes as demonstrated by the group mean values (Fig. 4).

Fig. 5 demonstrates breath duration (Ttot) and PIs between breathing and visually controlled forearm movements in normocapnic (left side) and hypercapnic (right side) conditions of one subject. Periods of coordination were not only characterized by steadiness of PIs reflecting in PI plots nearly parallel to the abscissa. Moreover, breathing became more regular so that the scattering of Ttot was markedly reduced compared to non-coordinated periods. In hypercapnia, periods of coordination became longer with PIs as well as Ttot being more stable than under normocapnic conditions.

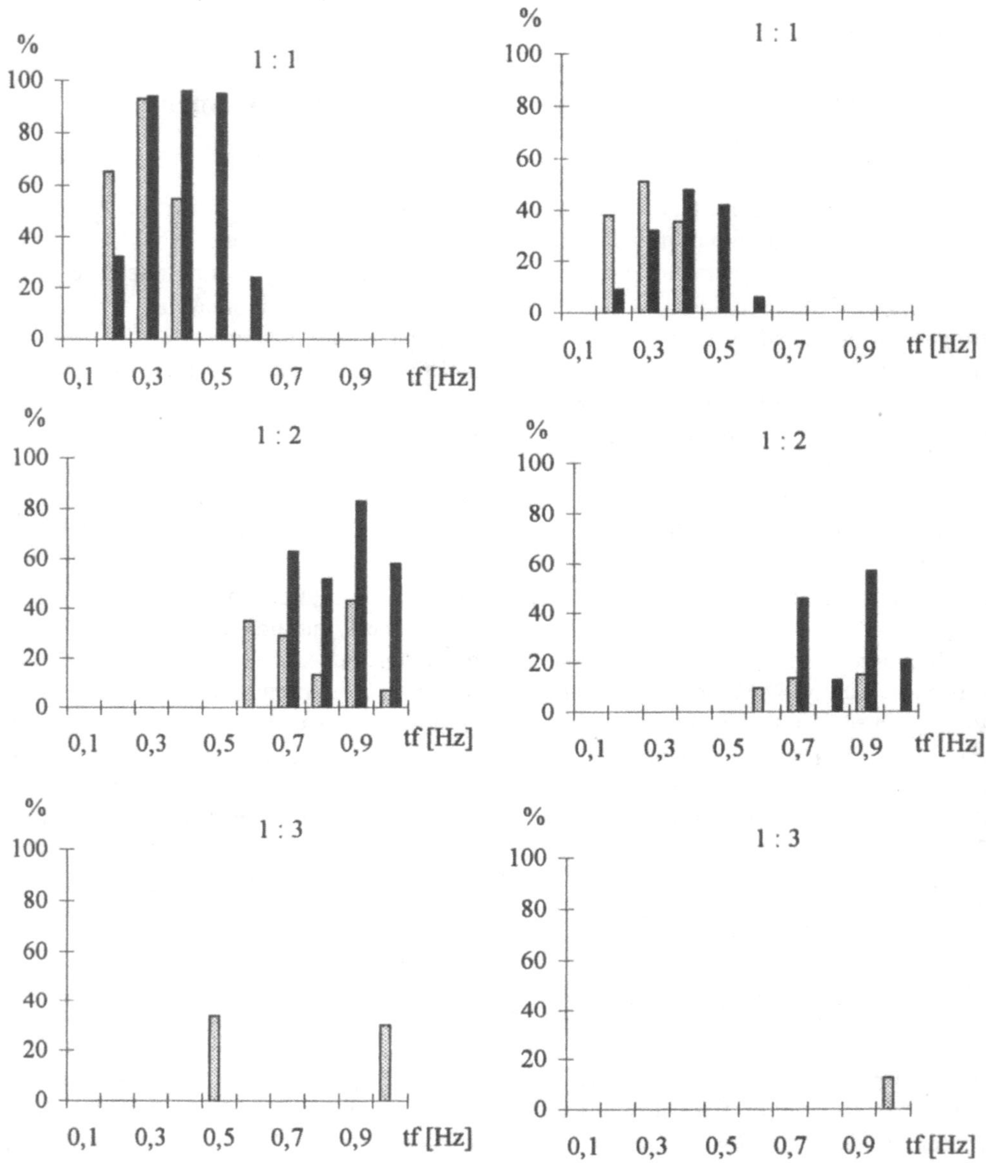

Figure 3. Percentages of PIs for 1:1 coordination, 1:2, and 1:3 coordination, in the target-frequency range between 0.1 and 1.0 Hz in subject 3. Left side: 25% criterion; right side: 10% criterion. Left columns (bright) designate normocapnic and right columns (dark) the hypercapnic conditions. Note that 1:3 coordination was not found under hypercapnic condition.

3.3. Pattern of Breathing and Breathing Rate

Under normocapnic condition in all subjects breathing at rest was fairly regular. During forearm tracking the breathing pattern remained regular, regardless whether or not breathing was coordinated with limb movements. As reported earlier[7], the variation coefficients of Ttot became significantly smaller during coordination (compare error bars in Fig. 6).

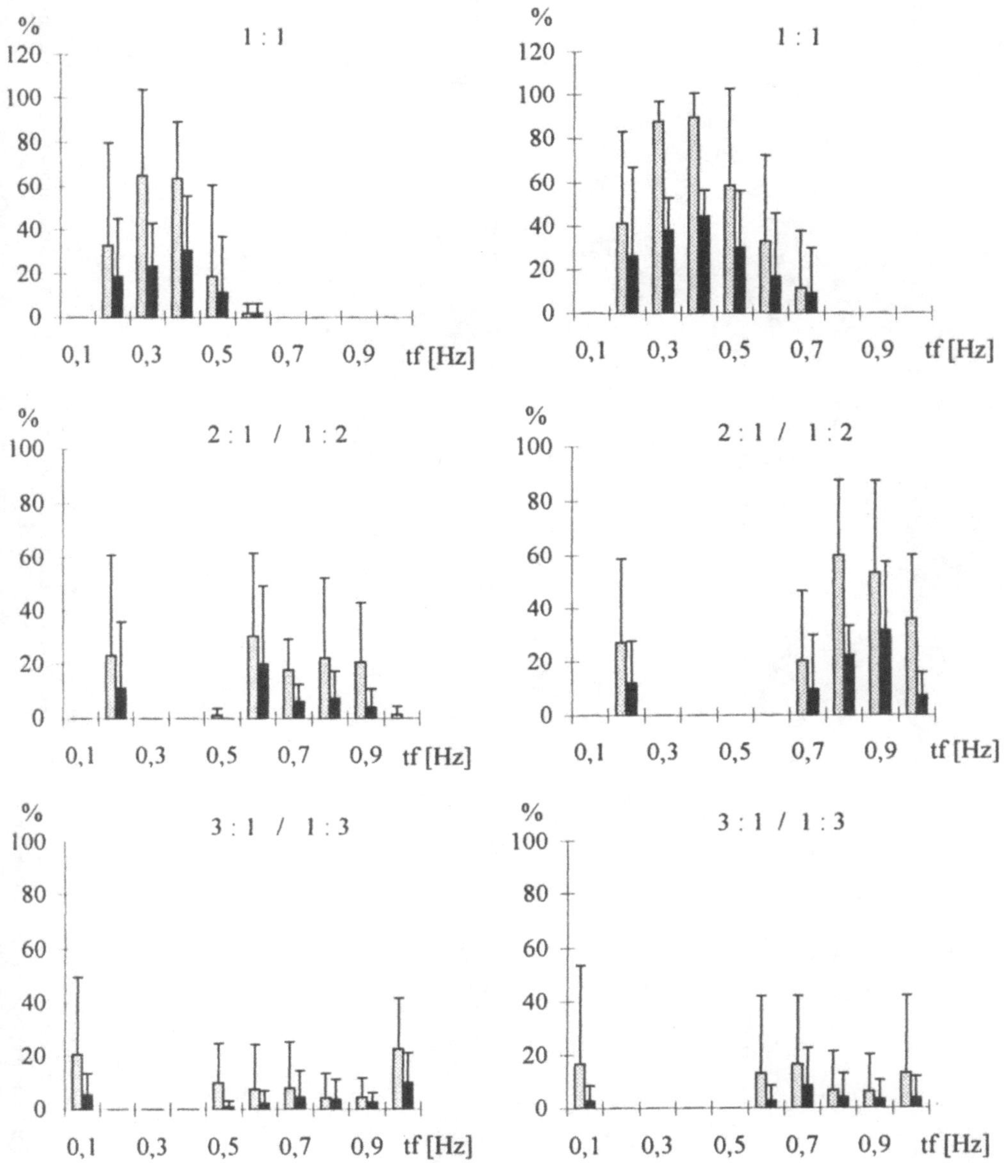

Figure 4. Mean percentages of PIs for 1:1 coordination, 2:1 (at 0.2 Hz) and 1:2, 3:1 (at 0.1 Hz) and 1:3 coordination in the target-frequency range between 0.1 and 1.0 Hz for the 25% criterion (bright columns) and the 10% criterion (dark columns). Left side: normocapnic condition, right side: hypercapnic condition. The percentages are the mean values of all normal subjects (n = 5).

Compared to breathing at rest the rate of breathing during tracking was significantly higher in all subjects ($p < 0.01$). All subjects showed a tendency towards correlation between breathing rate and target rate ($r = 0.49$, related to $R^2 = 0.24$). Under hypercapnic condition this correlation became stronger ($r = 0.79$, related to $R^2 = 0.62$). Due to the increased coordination frequency the variation of Ttot decreased, as it is demonstrated in one example in Fig. 5 as well as in the error bars of Fig. 6.

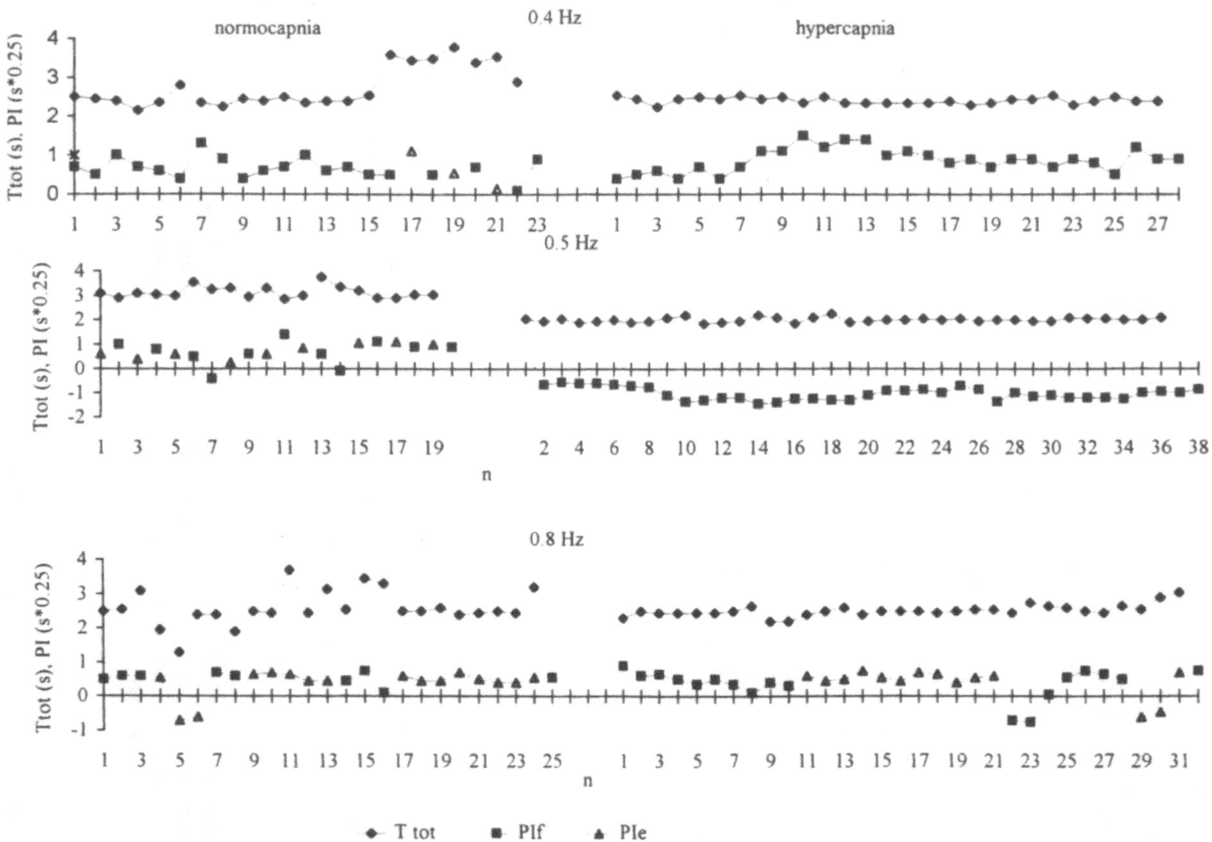

Figure 5. Segments of phase-intervals (PI) and breath durations (Ttot) of one subject, plotted breath-by-breath, at three selected target frequencies (top: 0.4 Hz, mid: 0.5 Hz, bottom:. 0.8 Hz). Abscissa: number of breath (n), Ordinate: breath duration (T_{tot}, in s), phase-interval between inspiration onset and forearm flexion onset (PIf), phase-interval between inspiration onset and forearm extension onset (PIe). For better presentation PIs are scaled in s × 0.25.

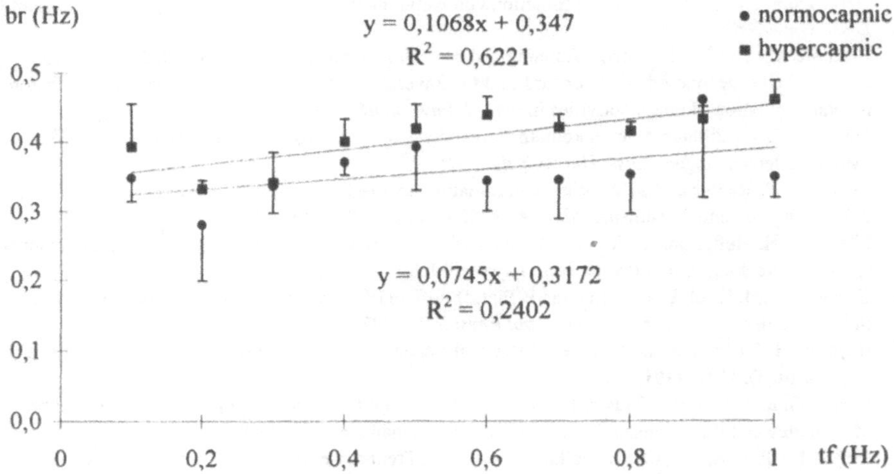

Figure 6. Group mean values (n = 5) of mean breathing rates (br, ordinate) in dependence on target frequency (tf, abscissa). Error bars indicate standard deviation. Linear regression lines are drawn. Note that breathing rates under hypercapnic condition are shifted to higher rates compared to breathing rates under normocapnic condition.

4. DISCUSSION

The present study on phase coordination between limb and breathing movements revealed that breathing is not only entrained by spontaneous movements but also by tracking movements. A clear dependence of coupling strength on target rate could be shown. Phase coordination occurred around particular rates termed "entrainment bands". From this point of view it is not sufficient to investigate coordination phenomena at only one particular rate.

The entrainment band evoking 1:1 coordination was 0.2–0.6 Hz. Dividing the respective limits of the 1:2 entrainment band by 2 yielded again the limits of the 1:1 range. This means, that coupling with a simultaneous rhythmical movement is able to change the intrinsic rate of the breathing central pattern generator (BCPG) within the limits from 0.2 to 0.6 Hz (12 to 36 breaths per minute). Not all of the subjects exhibited this bandwidth of possible entrainment—in most cases the entrainment band was smaller than 0.2–0.6 Hz. The main entrainment rate was slightly higher (0.3 Hz) than the breathing rate at rest (0.24 Hz). With 1:2 coordination the maximal percentage of coordination was observed at 0.7 Hz. Thus, the most relevant entrainment rates in our case were 0.3 and 0.7 Hz.

The hypercapnic drive on the respiratory system seems to enlarge the probability of coordinative relationships between breathing and tracking movements. We interpret the accompanied shifting of the entrainment band to higher frequencies to be due to the increased spontaneous breathing rate, caused by hypercapnia.

The increase of frequency and stability of coordination may be understood in the light of the theory of coupled oscillators: Amplifying the rhythm of one oscillator in a system of two or more coupled oscillators enhances the coupling force directed towards the other oscillator and thus, stabilizes coordination[5,11].

REFERENCES

1. Bechbache, R.R., J. Duffin. The entrainment of breathing frequency by exercise rhythm. *J. Physiol.* 272: 553–561, 1977.

2. Bernasconi, P.P.. Analyse der Koordination von Atmungsrhythmus und Bewegungsrhythmus beim Laufen. *Diss., Uni.-Zürich*, 1991.
3. Bramble, D.M. and D.R. Carrier. Running and breathing in mammals. *Science* 2/9: 251–256, 1983.
4. Decety, J., M. Jeannerod, D. Durozard and G. Baverel. Central activation of autonomic effectors during mental simulation of motor activities in man. *J. Physiol.* 461: 549–563, 1993.
5. Dörre, F. Biooszillatoren im lokomotorischen System und ihre ein- und wechselseitige Beeinflussung (Modelluntersuchungen). *Diss. Uni. Leipzig*, 1975.
6. Ebert, D., B. Raßler and S. Waurick. Coordination between spontaneous breathing and controlled sinusoidal limb movements in humans. *Pflüg. Arch.* 420, Suppl.1: R41, 160, 1992.
7. Ebert, D., H. Hefter and B. Raßler. Phase coordination between breathing and forearm movements during sinusoidal tracking. Submitted to *Resp. Physiol.* 1998.
8. Garlando, F., J. Kohl, E.A. Koller and P. Pietsch. Effect of coupling the breathing and controlled sinusoidal limb movements in humans. *Europ.J.Appl.Physiol.* 54: 497–501, 1985.
9. Haas, F., S. Distenfeld and K. Axen. Effects of perceived musical rhythm on respiratory pattern. *J. Appl. Physiol.* 61(3): 1185–1191, 1986.
10. Harrer, G. and H. Harrer. Music, Emotion and autonomic function. In: (Eds): *Music and the brain*, edit by M. Crichley and R.A. Henson, London, UK W. Heinemann med. books lim., 1977.
11. Hefter, H., P. Tass, D. Ebert, J. Volkmann and H.-J. Freund. Coupled oscillations—their relevance for normal and pathological motor control. *Pflüg. Arch.* Suppl.429/6: R 163, 604, 1995.
12. Hildebrandt, G. and F.-J. Daumann. Die Koordination von Puls und Atemrhythmus bei Arbeit. *Int.Z.angew. Physiol., Arb.Physiol.* 21: 27–48, 1965.
13. Hill, A.R., J.M. Adams, B.E. Parker, and D.F. Rochester. Short-term entrainment of ventilation to the walking cycle in humans. *J.Appl.Physiol.* 65: 570–578, 1988
14. Kohl, J., E.A. Koller and M. Jäger. Relation between pedalling- and breathing rhythm. *Europ. J. Appl. Physiol.* 47: 223–237, 1981.
15. Loring, S.H., J. Mead and T.B. Waggener. Determinants of breathing frequency during walking. *Respiration Physiol.* 82: 177–188, 1990.
16. Mahler, D.A., C.R. Shuhart, E. Brew and T.A. Stukel. Ventilatory responses and entrainment of breathing during rowing. *Medicine and Science in Sports and Exercise*, 23: 186–192, 1991.
17. Raßler, B., D. Ebert, S. Waurick and R. Junghans. Coordination between breathing and finger tracking in man. *J. Motor Behaviour* 28/1: 48–56, 1996.
18. Raßler, B. and J. Kohl. The influence of work load and speed on the coordination of breathing and walking in man. *Respiration Physiol.* 106: 317–327, 1996.
19. Raßler, B., S. Waurick and D. Ebert. Einfluß zentralnervöser Koordination auf die Steuerung von Atem- und Extremitätenmotorik des Menschen. *Biol.Cybernetics* 63: 457–462, 1990.
20. Shea, S.A., J. Walter, C. Pelley, K. Murphy and A. Guz. The effect of visual and auditory stimuli upon resting ventilation in man. *Respiration Physiol.* 68: 345–357, 1987.
21. Steinacker, J.M., M. Both and B.J. Whipp. Pulmonary mechanics and entrainment of respiration and stroke rate during rowing. *Int. J. Sports Med.*, 14, Suppl.1: S15–S19, 1993
22. Takano, N. Effects of pedal rate on respriatory responses to incremental bicycle work. *J.Physiol.* 396:389–397, 1988.
23. Waurick, S. The influence of eyetracking movements on the breathing pattern of men. *Phys. Fitness,* Praha, 443–448, 1973.
24. Wilke, J.T., R.W. Lansing and C.A. Rogers. Entrainment of respiration to repetitive finger tapping. *Physiol.Psychol.*3/4: 345–349, 1975.

19

TEMPORAL CORRELATION IN PHRENIC NEURAL ACTIVITY

Bernard Hoop, William L. Krause, and Homayoun Kazemi

Pulmonary and Critical Care Unit
Medical Services
Massachusetts General Hospital
Harvard Medical School
Boston, Massachusetts 02114

1. INTRODUCTION

Neural activity which gives rise to eupnea fluctuates in a complex manner. Apparently "noisy" variations in activity of the phrenic nerve may display a fractal scaling relationship. Fractal scaling in eupnea is the consequence of physical and chemical processes acting over short time scales at the cellular level, and which are correlated with similar processes acting simultaneously over longer time scales. Specifically, variations in phrenic neural bursts may not be independent random fluctuations or entirely due to short-range influences[4], but may exhibit temporal correlation indicative of fractal scaling. West and Deering[20] have demonstrated that fractal processes are essentially unresponsive to error and very tolerant of variability in the physiological environment. In this view, eupnea with its concomitant stability to error from a broad spectrum of inputs must have the error-tolerant properties of fractals.

Fractal scaling in phrenic neural activity (PNA) may be altered by experimental methods having quite different effects on respiratory-related neural activity, e.g., pharmacological receptor antagonism and peripheral chemodenervation. Blocking excitatory and inhibitory neurotransmission affects central and peripheral cardiorespiratory mechanisms and their underlying metabolic pathways. Carotid body denervation affects cardiorespiratory chemoreceptor neural afferent pathways. Temporal correlation may therefore be a sensitive measure of fractal scaling in eupnea and in PNA altered by peripheral chemodenervation and by blocking respiratory-related brainstem neural cell membrane ion channel activation.

The purpose of the present investigation is to determine the magnitude of temporal correlation in PNA using a novel statistical physical method capable of estimating the exponent of fractal scaling. An anesthetized rat preparation is used to study fluctuations in PNA during eupnea and during two experimental conditions imposed to alter variability in

Advances in Modeling and Control of Ventilation, edited by Hughson *et al.*
Plenum Press, New York, 1998.

PNA on different time scales: 1) removal of peripheral chemoreceptor input by carotid body denervation, and 2) blocking neural cell membrane ion channel activation by microperfusion of amino acid neurotransmitter receptor antagonists in respiratory-related areas in the brainstem. Measurements of temporal correlation in PNA are compared with a model of simulated fluctuations derived from a weighted sum of stationary and nonstationary noise on multiple time scales.

2. METHODS

Seventeen anesthetized (2.5% isoflurane) male Sprague-Dawley rats (250–350 g) were ventilated mechanically with 30% O_2 through a cervical tracheostomy to maintain $PaCO_2$ between 36 and 42 torr. Body temperature was maintained at 37.5°C by a heat lamp. The phrenic nerve on one side was exposed, placed on a bipolar AgCl electrode, and grounded. PNA consisting typically of bursts of about a dozen spikes (−1.0 to 1.0 mV) within 0.5 sec at burst rate of the order of 1 Hz was rectified, amplified, monitored during experiments on a chart recorder, integrated electronically and acquired at a sampling rate of 25 Hz for subsequent analysis. Individual peak heights of successive integrated neural bursts were determined from the differences between local maxima and minima in the sampled sequences.

In five of the seventeen animals, PNA was first measured with intact peripheral afferent innervation and then following surgical denervation of peripheral chemoreceptors (carotid bodies) in the same animal. Denervation was performed by cutting the vagosympathetic nerve trunk on each side at the level of the thyroid cartilage and stripping the adventitia bilaterally from the common, internal, and external carotid arteries and their branches in a region 0.5 mm distal and proximal to the bifurcation, severing all nerve connections. In the remaining twelve animals with intact peripheral chemoreceptors, a 10 μm diameter microperfusion pipette was placed under stereotaxic control within areas of the ventrolateral surface of the brainstem associated with respiratory-related neural control. The pipette tip was placed in the intermediate respiratory chemosensitive area of the medulla oblongata, ca. 3.0 to 4.0 mm caudal to the interaural line, 0.50 to 1.00 mm lateral to the midline, and 0.50 to 2.50 mm below the ventral surface of the medulla[2]. Antagonists of neural cell membrane ion channel receptors of the excitatory amino acid neurotransmitter glutamate, MK-801, or of the inhibitory amino acid γ-aminobutyric acid (GABA), bicuculline, were perfused at flow rates of 1.6 to 5.8 μL/min and at concentrations 5 to 40 mM for 15 to 45 minutes.

Sequences of integrated phrenic neural bursts containing typically 2×10^3 consecutive peak heights (range: 939 to 3714 peaks) recorded continuously for 30 to 90 minutes [typical mean (±SD) frequency: 0.70 ± 0.08 Hz] in all animals were analyzed for an index of temporal correlation. Such a quantitative measure of temporal correlation is a fractal scaling exponent. The scaling exponent characterizes the roughness of irregularities or the temporal correlation within the irregularities of a fractal time series. A scaling exponent for each series of phrenic burst peak heights was determined by means of detrended fluctuation analysis (DFA), described by Peng et al.[13] In this method, the cumulative (integrated) departure of peak height from mean peak height was divided into intervals of equal length n, and a least-squares line representing the linear trend in each interval was fitted to the data. The y-coordinates of the straight line segments were subtracted from the integrated values in each interval to detrend the series, and the root-mean-square fluctuation $F(n)$ was then calculated over all interval sizes.

A linear relationship between $F(n)$ and n on a double log plot indicates the presence of scaling. That is, the slope of log $F(n)$ vs. log n yields a scaling exponent α (Peng exponent) which characterizes this scaling. Different values of the Peng exponent α signify different levels of temporal correlation in fluctuation (noise) at different scales. For time series with consecutive values generated by statistically independent processes with finite variances, $\alpha = 0.5$ (uncorrelated, ordinary, or white noise), and $0.5 < \alpha < 1.0$ corresponds to processes where fluctuations in subsequent values are positively correlated (persistent noise), where $\alpha = 1.0$ corresponds to power spectral $1/f$ noise. Values of α in the range $1.0 < \alpha < 1.5$, correspond to integrated negatively correlated or anti-persistent noise, where $\alpha = 1.5$ corresponds to integrated uncorrelated (Brown) noise[13].

To validate the use of the DFA method in this study, a comparison was made with a conventional method for quantitating temporal correlation, namely, power spectral density (PSD) analysis. Positive temporal correlation is often characterized by a power law form of the power spectral density. For time series of fractional Gaussian noise[7], values of α in the range $0 < \alpha < 1.0$ are related to the negative of the exponent β of the slope of a double log plot of power vs frequency f of the power spectral density $1/f^{\beta}$ where $\beta = 2\alpha - 1$. The DFA method was compared with PSD analysis by simulating fractional Gaussian noise (fGn) as the derivative of successive random additions[6,19] with actual α-values in the range $0.05 \le \alpha \le 0.95$ at intervals of 0.05. To extend the comparison to α-values greater than 1.0, fGn simulated with actual α-values in the range $0.05 \le \alpha \le 0.50$ were integrated to yield fractional Brownian noise (also known as fractional Brownian motion) with actual α values in the range $1.05 \le \alpha \le 1.50$ which were then analyzed by each method. Mean measured vs actual α for each method are plotted in Figure 1.

In Figure 1, linear regressions of measured vs actual scaling exponent (heavy lines) are in good agreement with the line of unit slope (light line). Power spectral densities from which scaling exponents were derived for Figure 1 were calculated using rectangular weighting and interpreted as a power law of the form $1/f^{\beta}$. That is, values of β were determined from linear regressions of log power vs log frequency f over the entire frequency domain of unsmoothed data.

The results in Figure 1 for power spectral analysis agree with those of Schepers et al.[15] and with results in this laboratory[10]. Although a difference (bias) is evident in Figure 1 between measured and actual exponents for both DFA and PSD over the entire range of α, no correction has been applied for bias. This bias may be due to the chosen method of simulating fGn via successive random additions. Standard errors in α are smaller for DFA than for PSD, but DFA exhibits a greater scatter than PSD. Peng et al.[13] have characterized series of fluctuations with scaling exponent in the range $1.0 < \alpha < 1.5$ as correlation which ceases to be of power-law form. An alternative interpretation is that of integrated negatively correlated noise, by comparison with $\alpha = 1.5$ which characterizes integrated uncor-

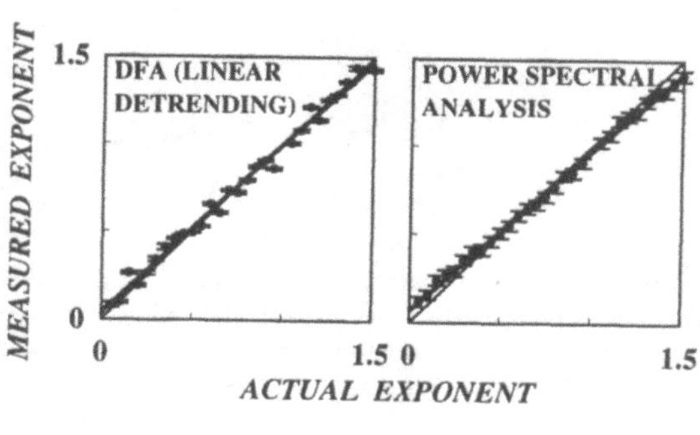

Figure 1. Mean (± SE) measured scaling exponent α vs actual α in the range $0 < \alpha < 1.5$ at increments of 0.05 in α, determined from 4096-point fractional Gaussian noise spectra. Measured α from first-order (linear) detrended fluctuation analysis (DFA) are compared with a power law form of the power spectral density, where α is related to exponent β of power spectra by $\beta = 2\alpha - 1$.

related (Brown) noise. It should be noted that the DFA of simulated noise in Figure 1 employs only linear detrending, which may not entirely account for non-stationary fluctuations in the data[18]. Higher order polynomial detrending (up to order 4) was applied to the present data. In the present analysis of phrenic nerve peak height fluctuations, there was no significant effect of higher order detrending on the Peng exponent.

A statistical physical representation of phrenic neural noise based on a multiscaled random model proposed by Hausdorff and Peng[9] was used to account for the magnitude of scaling exponent $\alpha > 1$. In the present work, the model data consisted of a weighted sum of two series of events: a stationary uncorrelated series with zero mean and its (nonstationary) integral on a time scale greater by a factor of eight. Five model series were simulated, each of 2000 points, with relative weighting of the stationary contribution ranging from 0.2 to 1.0, and with overall mean of the weighted sum chosen to approximate the typical overall mean of phrenic neural peak heights in intact animals.

3. RESULTS

Representative series of consecutive integrated phrenic nerve burst peak heights are shown in the upper three panels, and an example of a series derived from a model of phrenic neural noise is shown in the lower panel of Figure 2.

In Figure 2, series of peak heights, each on a 0–1000 scale of arbitrary units, are plotted vs peak number for an intact animal, followed by denervation of peripheral chemoreceptors in the same animal (upper two panels). The third panel shows a series of peak heights in an intact animal microperfused in the intermediate respiratory chemosensitive area of the ventral medulla with 20 mM MK-801 at a flow rate of 1.6 μL/min. A gradual decline in PNA is evident, as expected with blocking excitatory glutamate reception in the ventral medulla[2]. The fourth panel of Figure 2 shows a series of simulated peak heights using equal weighting of stationary and nonstationary uncorrelated noise on two time scales differing by a factor of eight.

Representative determinations of fractal scaling exponent are shown in Figure 3.

In Figure 3, slopes of linear regressions of log *F(n)* vs log *n* are shown by solid lines. The values of scaling exponent α (± SE) derived from these slopes are 1.197 ± 0.022 (intact); 1.098 ± 0.008 (denervated); 1.109 ± 0.008 (antagonist); 1.347 ± 0.012 (model). As

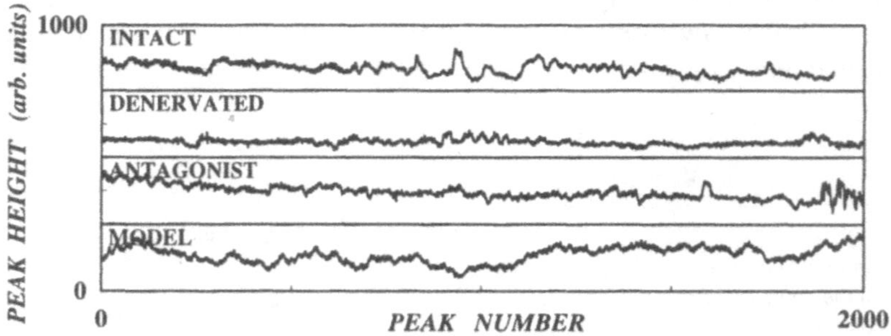

Figure 2. Upper two panels: integrated phrenic peak heights (arbitrary units) vs peak number during 60 minutes in an intact animal and in the same animal following peripheral chemoreceptor denervation, respectively. Third panel: peak heights in an intact animal during ventral medullary microperfusion of 20 mM glutamate receptor antagonist MK-801 at 1.6 μL/min. Lower panel: Model of phrenic neural noise.

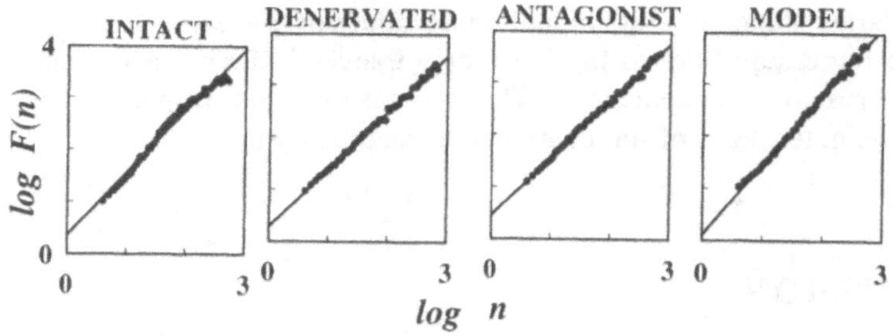

Figure 3. Log-log plots from DFA of PNA peak series in Figure 2.

illustrated in Figure 3, values of scaling exponents were determined from linear regressions over the entire domain of unsmoothed data. In the data from the intact animal illustrated in Figure 3, a distinct departure from linearity is discernible, which is reflected in the relatively larger standard error in the slope of the regression. As expected, random reordering (shuffling) the time series and recalculating α resulted in $\alpha \sim 0.5$. Reducing the sampling rate by a factor of two preserved values of $\alpha \sim 1.0$.

Mean scaling exponent α in intact animals, denervated animals, animals microperfused with antagonist, and simulated model data are summarized in Figure 4.

In Figure 4, five animals with intact peripheral chemoreceptors, phrenic peak height fluctuations are characterized by a single scaling exponent $\alpha = 1.27 \pm 0.06$ (mean ± SD). Following peripheral chemodenervation, $\alpha = 1.20 \pm 0.13$. In twelve animals with intact peripheral chemoreceptors and microperfused with either the glutamate receptor antagonist MK-801 or the GABA receptor antagonist bicuculline, phrenic peak height fluctuations are characterized by $\alpha = 1.18 \pm 0.10$. In five series of peak heights simulated with the model, $\alpha = 1.32 \pm 0.10$. Standard errors in slopes of linear regressions (α-values) for individual animals and in simulated data ranged from 0.008 to 0.023. There are no statistically significant differences between the above mean α values.

As Figure 4 illustrates, the chosen model generally represents the mean magnitude of scaling exponent in intact animals, for which $\alpha > 1.0$. Exponents in denervated animals and in animals with pharmacologically blocked glutamate and GABA receptors tend to be closer to 1.0 and more widely distributed than standard errors in individual animals would suggest. This may be due to deviation from uniform power law scaling[18]. It should be

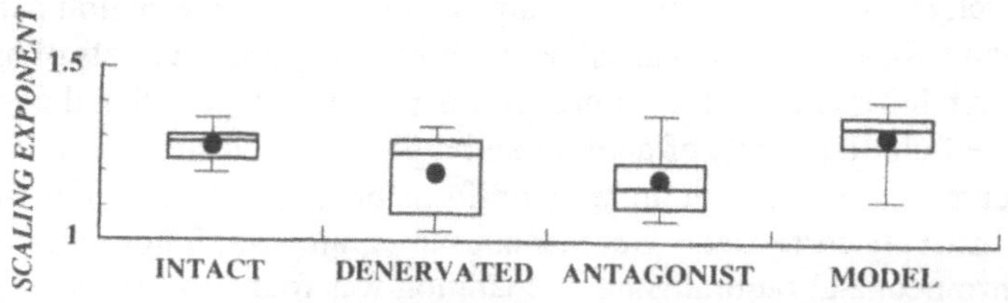

Figure 4. Box plots of measured scaling exponent α in five intact and peripherally chemodenervated animals, in 12 animals microperfused with glutamate and GABA receptor antagonist in respiratory-related chemosensitive areas of the ventral medulla, and in five simulations of a model of phrenic neural noise. The top, bottom, and line through the middle of each box in Figure 4 correspond to the 75th, 25th, and 50th percentile (median), respectively. Bars extending from the top and bottom correspond to the 90th and 10th percentile, respectively. Mean α of each group is significantly greater than 1.0 ($P < 0.05$), although there is no significant difference among groups.

noted that larger standard errors in individual values of α are due, in part, to the attempt to fit a single regression lines to log-log plots of several phrenic nerve data which clearly manifest a crossover phenomenon[13]. This was also evident in model data simulated with unequally weighted sums of stationary and nonstationary noise.

4. DISCUSSION

Statistical dependence between discrete events in PNA implies that eupnea may be self-similar with characteristics of correlated noise[1]. The magnitude of temporal correlation among fluctuations in tidal volume and respiratory rate can serve as a sensitive measure of self-similarity[4]. Phrenic neural activity underlying the depth and frequency of breathing is generated by fluctuations in conditional processes consisting of serial and parallel sequences of neuronal events at many sites and extending over a wide range of times. These time scales include central and peripheral events at the cellular level, including cell membrane ion channel activation and switching times, and peripheral events, including neural transmission times and systemic neuromodulatory and circulatory transit times[3]. Temporal correlation in PNA is therefore the consequence of fluctuations in numerous physical and chemical processes acting over short times at the cellular level and neural and circulatory processes acting simultaneously over long times, which is reflected in the exponent of fractal scaling.

Scaling in PNA may be viewed as a superposition of correlated fluctuations during eupnea arising from multiple mechanisms, including chemoreceptive and vasopressive neural and circulatory transit times operating on multiple time scales. Peng, Hausdorff, Goldberger and colleagues have shown that fluctuations in interbeat intervals of the cardiac cycle and in stride interval of human gait have long-term correlation[8,13,18]. These investigators suggest that such correlation serves as an organizing principle for complex, non-linear processes that generate fluctuations on a wide range of time scales. Bruce[4] observes that correlation in breathing which lasts for several breaths may be explained as additive uncorrelated or white noise if it is associated with integrated neural mechanisms with long time constants. However, breath-to-breath correlation may not necessarily follow an exponential decay characteristic of linear filtering of white noise. Furthermore, PNA reflects inputs to the central pattern generator acting on different time scales via multiple peripheral and central mechanisms.

Methods of nonlinear geometry have suggested chaotic behavior during eupnea[5,12], and during selective stimulation[14] of neural signals to and from the central pattern generator. On the other hand, Szeto and colleagues[16] have shown that time intervals of breathing bursts in fetal lambs exhibit fractal scaling, a property dependent on gestational age. These breath series exhibit self-similar bursts of activity and a power law distribution of the interbreath intervals. Fluctuations in tidal volume at uniform breathing rate in the fully innervated mammal are positively correlated[11]. Spontaneous integrated respiratory-related neural activity in the in vitro neonatal rat brainstem preparation was found to consist of periodic bursts with well-defined frequency and amplitude. However, fluctuations in this preparation (extended neural bursts and bursts within bursts) exhibit positive temporal correlation during stimulation with the central respiratory neurotransmitter, acetylcholine[10]. In human subjects breathing hyperoxic and hypoxic gas mixtures at rest, a large percentage of spectral power is not harmonic but exhibits a power law dependence[17]. Collectively, these studies demonstrate that fluctuations during eupnea consist of fractal noise.

The question of whether scaling in fractal noise during eupnea is uniform under any given experimental alteration of the anesthetized rat preparation can only be resolved with data series which include simultaneous measurements of other variables which may contribute to fluctuations in PNA. It is possible that in the present preparation, nonstationarity in phrenic activity derives from temporal variations in circulating acid-base quantities and blood gas contents, as well as from effects of mechanical ventilation and temperature cycling of the preparation. In addition, determination of individual peak heights of integrated neural bursts by means of a peak detection algorithm which selects local absolute maximum and minimum data point values may introduce an ordinary noise contribution. Finally, a simple multi-scaled noise model can account for temporal correlation among fluctuations in PNA during eupnea, characterized by scaling in PNA with a Peng exponent ≥ 1.0.

Within limitations of the experimental procedure and the methods used in the present analysis, we conclude that fluctuations in peak height of integrated phrenic nerve bursts exhibit persistent positive temporal correlation consistent with power spectral $1/f$ noise, and possibly including nonstationary contributions. The results suggest that noise in PNA associated with central respiratory pattern generation is scale-invariant, i.e., lacks a characteristic time scale. Furthermore, scaling in PNA remains stable after removal of peripheral chemoreceptor afferent input and after blocking respiratory-related brainstem neural cell membrane ion channel activation, two principal mechanisms underlying control of breathing.

ACKNOWLEDGMENTS

This work was supported in part by grants from the U.S. Department of Health and Human Services of the Public Health Service. One of us (WLK) was supported by USPHS Training Grant HL-07874. The authors thank Mr. J.L. Beagle for technical assistance, Drs. M.D. Burton and D.C. Johnson for helpful discussion, Dr. D.C. Johnson for the peak detection algorithm, Dr. C.-K. Peng for providing his code for detrended fluctuation analysis, and Drs. A.L. Goldberger, J.M. Hausdorff, C.-K. Peng, and B.J. West for reviewing the manuscript.

REFERENCES

1. Bassingthwaighte, J.B., L.S. Liebovitch, and B.J. West, *Fractal Physiology*. New York: Oxford, 1994.
2. Beagle, J.L., B. Hoop, and H. Kazemi. Phrenic nerve response to glutamate antagonist microinjection in the ventral medulla. In: *Advances in Control and Modeling of Ventilation*, edited by R. Hughson, D.A. Cunningham, and J. Duffin., New York: Plenum, 1998, (this volume).
3. Bianchi, A.L., M. Denavit-Saubie, and J. Champagnat. Central control of breathing in mammals: neuronal circuitry, membrane properties, and neurotransmitters. *Physiol. Rev.* 75, 1–45, 1995.
4. Bruce, E.N. Temporal variations in the pattern of breathing. *J. Appl. Physiol.* 80: 1079–1087, 1996.
5. Donaldson, G.C. The chaotic behavior of resting human respiration. Respir. Physiol. 88: 313–321, 1992.
6. Feder, J. *Fractals*. New York: Plenum, 1988, pp. 180–181.
7. Flandrin, P. On the spectrum of fractional Brownian motions. *IEEE Trans. Infor. Theor.* 35: 197–199, 1989.
8. Hausdorff, J.M., C.-K. Peng, Z. Ladin, J.Y. Wei, and A.L. Goldberger. Is walking a random walk? Evidence for long-range correlations in stride interval of human gait. *J. Appl. Physiol.* 78: 349–358, 1995.
9. Hausdorff, J.M., and C.-K. Peng. Multi-scaled randomness: a source of $1/f$ noise in biology. *Physical Review E* 54: 2154–2157, 1996.
10. Hoop, B., M.D. Burton, H. Kazemi, and L.S. Liebovitch. Correlation in stimulated respiratory neural noise. *CHAOS* 5: 609–612, 1995.

11. Hoop, B., M.D. Burton, and H. Kazemi. Fractal noise in breathing. In: *Bioengineering Approaches to Pulmonary Physiology and Medicine*, edited by M.C.K. Khoo, New York: Plenum, 1996, pp. 161–173.
12. Hughson, R.L., Y. Yamamoto, J.-O. Fortrat, R. Leask, and M.S. Fofana. Possible fractal and/or chaotic breathing patterns in resting humans. In: *Bioengineering Approaches to Pulmonary Physiology and Medicine*, edited by M.C.K. Khoo, New York: Plenum, 1996, pp. 187–196.
13. Peng, C.-K., S. Havlin, H.E. Stanley, and A.L. Goldberger. Quantification of scaling exponents and crossover phenomena in nonstationary hearbeat time series. *CHAOS* 5: 82–87, 1995.
14. Sammon, M., J.R. Romaniuk, and E.N. Bruce. Bifurcations of the respiratory pattern produced with phasic vagal stimulation in the rat. *J. Appl. Physiol.* 75: 912–926, 1993.
15. Schepers, H.E., J.H.G.M van Beek, and J.B. Bassingthwaighte. Four methods to estimate the fractal dimension from self-affine signals. *IEEE Eng. Med Biol Mag.* 11(2): 57–64, 71, 1992.
16. Szeto, H.H, P.Y. Cheng, J.A. Decena, Y. Cheng, D. Wu, and G. Dwyer. Fractal properties in fetal breathing dynamics. *Am. J. Physiol.* 263: R141–R147, 1992.
17. Tuck, S.A., Y. Yamamoto, and R.L. Hughson. The effects of hypoxia and hyperoxia on the 1/f nature of breath-by-breath ventilatory variability. In: *Modelling and Control of Ventilation*, edited by S.J.G. Semple and L. Adams. New York: Plenum, 1995, pp. 297–302.
18. Viswanathan, G.M, C.-K. Peng, H.E. Stanley, and A.L. Goldberger. Deviations from uniform power law scaling in nonstationary time series. *Physical Review E* 55: 845–849, 1997.
19. Voss, R.F. Random fractal forgeries. In: *Fundamental Algorithms in Computer Graphics*, edited by R.A. Earnshaw, Berlin: Springer, pp. 805–835, 1985.
20. West, B.J. and W. Deering. Fractal physiology for physicists: Levy Statistics. *Physics Reports* 246: 2–100, 1994.

METHODS OF ASSESSING RESPIRATORY IMPEDANCE DURING FLOW LIMITED AND NON-FLOW LIMITED INSPIRATIONS

S. A. Tuck and J. E. Remmers

Faculty of Medicine
University of Calgary
Calgary, Alberta T2N 4N1, Canada

1. INTRODUCTION

We are interested in characterizing the mechanical load on the respiratory system of obese pigs during wakefulness and sleep. These animals exhibit sleep-disordered breathing, manifest by characteristics similar to the human obstructive sleep apnea/hypopnea syndrome including inspiratory flow limitation (FL). During inspiratory FL, airflow becomes dissociated from the driving pressure; under these conditions, traditional measures such as resistance may not be suitable for describing mechanical load. Therefore, the purpose of this study was to mathematically describe the resistive pressure-airflow relationship during FL and non-FL inspirations. To accomplish this, two models of the resistive pressure-inspiratory airflow relationship during FL and two models for non-FL inspirations were compared for their ability to fit experimental data. The parameters of these models were then correlated with airway resistance.

2. METHODS

Two obese Vietnamese pot-bellied pigs were studied. These pigs were 21 and 20 months old and weighed 103 and 118 kg respectively. Under anaesthesia, the animals were chronically instrumented with wire electrodes placed between the dura and the skull to record EEG, wire electrodes secured in a neck muscle to record EMG, and a balloon placed in the intrapleural space to measure intrapleural pressure. The intrapleural balloon was constructed from thin sheets of silastic rubber to form a 3 × 3 cm square, and was attached to a length of silastic tubing, which was tunnelled subcutaneously to exit from between the shoulder blades of the animal. Non-invasive instrumentation included a piezoelectric strip (Night Watch eye sensor, Healthdyne Technologies) taped to the snout to record nose

Advances in Modeling and Control of Ventilation, edited by Hughson *et al.*
Plenum Press, New York, 1998.

twitch, and a pneumotachograph (3700 Hans Rudolph) attached to a custom-fit, non-obstructing facemask to measure airflow ($\dot{V}$).

The pigs slept in a raised box with plexiglass sides. The pneumotachograph was connected to a differential pressure transducer (MP-45-15, Validyne). The intrapleural balloon was inflated with 1 ml of air and connected to a differential pressure transducer (MP-45-32, Validyne) and referred to mask pressure. The signals were amplified and filtered by a polygraph (Model 7D, Grass Instruments) and recorded on FM tape (7DS, Racal). The airflow and pressure signals were digitized at a sampling frequency of 100 Hz using a personal computer, A/D board (CIO-AD16, Computer Boards) and commercial software (Datapac, Run Technologies).

Nose twitch was used in lieu of EOG measurements to determine sleep state of the animal in conjunction with EEG and EMG, according to published criteria for pigs (5). Twelve non-FL inspirations during wakefulness and twelve FL inspirations during NREM sleep were analysed for each pig. Resistive pressure (P) was calculated by subtracting lung elastic pressure, derived from compliance at points of zero flow, from intrapleural pressure. Flow limitation was defined as an increase in resistive pressure with no change or a decrease in airflow for greater than half of the inspiration. Inspirations were considered non-flow limited if no flow limitation was evident throughout inspiration.

The resistive pressure-airflow curves of the non-FL inspirations were fit with a linear equation (Eq. 1) and a second order equation (Eq. 2) using linear and nonlinear regression respectively.

$$P = m\dot{V} + b \tag{1}$$

$$P = K_1\dot{V} + K_2\dot{V}^2 \tag{2}$$

The resistive pressure-airflow curves of the FL inspirations were fit with a second order equation (Eq. 2) and a rectangular hyperbolic equation (Eq. 3) using nonlinear regression.

$$\dot{V} = (\alpha P)/(\beta + P) \tag{3}$$

For the rectangular hyperbolic equation, α describes the asymptote for peak flow, and β, a shape factor, corresponds to the pressure at $\alpha/2$. Goodness of fit of each equation was calculated for each inspiration by correlating the observed and predicted values of the dependent variable at each sampling point to obtain a correlation coefficient, using the Pearson product moment correlation method.

Resistance ($P/\dot{V}$) was calculated for each inspiration by two methods. Resistance was calculated at mid-inspiration (R_{mid}), and by averaging resistance at each sampling point throughout inspiration (R_{avg}). R_{mid} and R_{avg} of the twelve FL and twelve non-FL inspirations were correlated with the model parameters to obtain a correlation coefficient for each pig, using the Pearson product moment correlation method.

3. RESULTS

3.1. Non-Flow Limited Inspirations

The resistive pressure-airflow relationship during non-FL inspirations appeared linear (Figure 1). Sampled values clustered in the region of high $\dot{V}$ and high P, with fewer

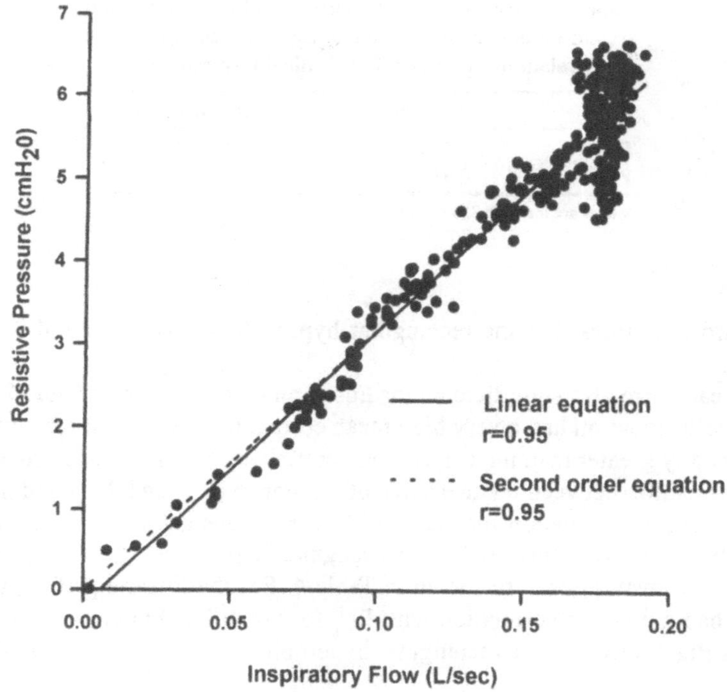

Figure 1. The application of the linear and second order equations to a typical non-flow limited inspiration.

sampling points at low $\dot{V}$ which represents the beginning and end of inspiration, where $\dot{V}$ and P are changing rapidly. In Figure 1, the regressions calculated using Eqs. 1 and 2 are also shown. In this example, both equations described the data adequately. The mean parameter estimates of Eqs. 1 and 2 for the twelve non-FL inspirations of each pig are shown in Figure 2. For the linear equation (Eq. 1), the y-intercept, b, was zero or close to zero for both pigs. The mean parameter estimates for the second order equation (Eq. 2), K_1 and K_2, had substantially larger standard deviations than the parameter estimates for the linear equation.

The mean correlation coefficients for Eqs. 1 and 2 are shown in Table 1. Both equations had high mean correlation coefficients for both pigs, with no significant difference between the correlation coefficients of the two equations.

3.2. Flow Limited Inspirations

For the FL inspirations, the resistive pressure-airflow relationship was curvilinear (Figure 3). At the end of inspiration, resistive pressure did not follow the same path compared to the start of inspiration, ie. for a given flow, resistive pressure was greater at the end of inspiration than at the start of inspiration. This end-inspiratory region was excluded from the curve-fitting, which typically involved the exclusion of 5 to 10 sampling points. In Figure 3, the regressions calculated using Eqs. 2 and 3 are also shown. The rectangular hyperbolic equation (Eq. 3) closely followed the shape of the resistive pressure-airflow relationship, but the second order equation (Eq. 2) fit the data poorly. The mean parameter estimates of Eqs. 2 and 3 for the twelve FL inspirations of each pig are shown in Figure 4. For the second order equation, K_1 was negative and K_2 was very large for both pigs, with

Table 1. Mean correlation coefficients for the linear and second-order equations with the resistive pressure-airflow relationship of non-flow limited inspirations

	Linear	Second order
PIG 1	0.91 ± 0.04	0.92 ± 0.04
PIG 2	0.92 ± 0.04	0.93 ± 0.04

Values are means ± SD, n = 12.

large standard deviations. For the rectangular hyperbolic equation, α and β were similar for both pigs.

The mean correlation coefficients for Eqs. 2 and 3 are shown in Table 2. The rectangular hyperbolic equation had a very high mean correlation coefficient for both pigs which were significantly greater than the correlation coefficients for the second order equation.

The correlation between the measures of resistance (R_{avg} and R_{mid}) and the parameter estimates of the linear equation for the non-FL inspirations are shown in Table 3, and the correlation between resistance and the rectangular hyperbolic equation parameter estimates for the FL inspirations are shown in Table 4. For the linear equation, the slope parameter, m, had a positive correlation with R_{avg} for both pigs, but correlated significantly with R_{mid} for Pig 2 only. For the rectangular hyperbolic equation, α had a significant nega-

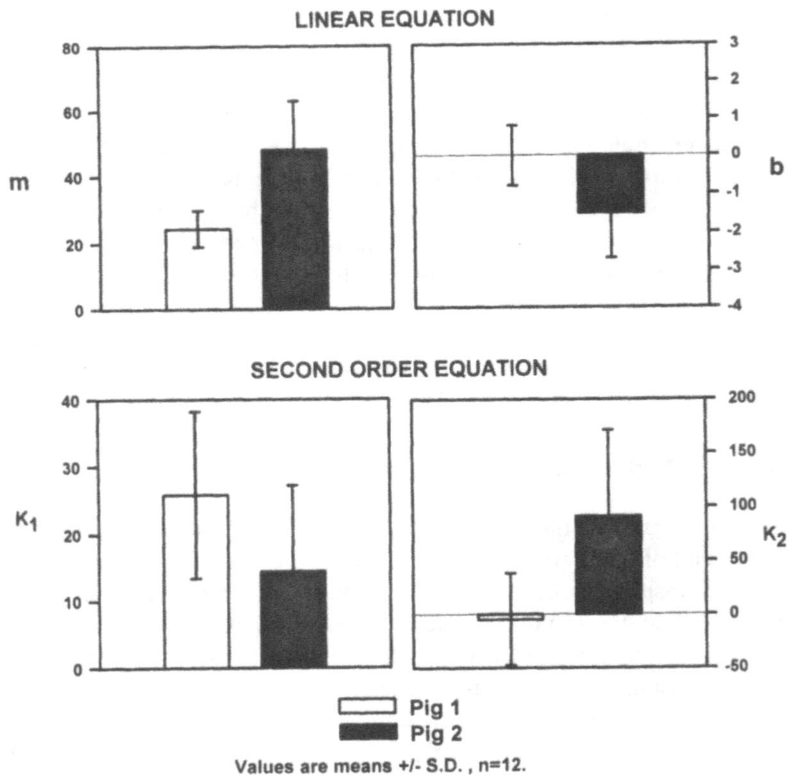

Figure 2. Mean parameter estimates of the linear and second order equations for non-flow limited inspirations.

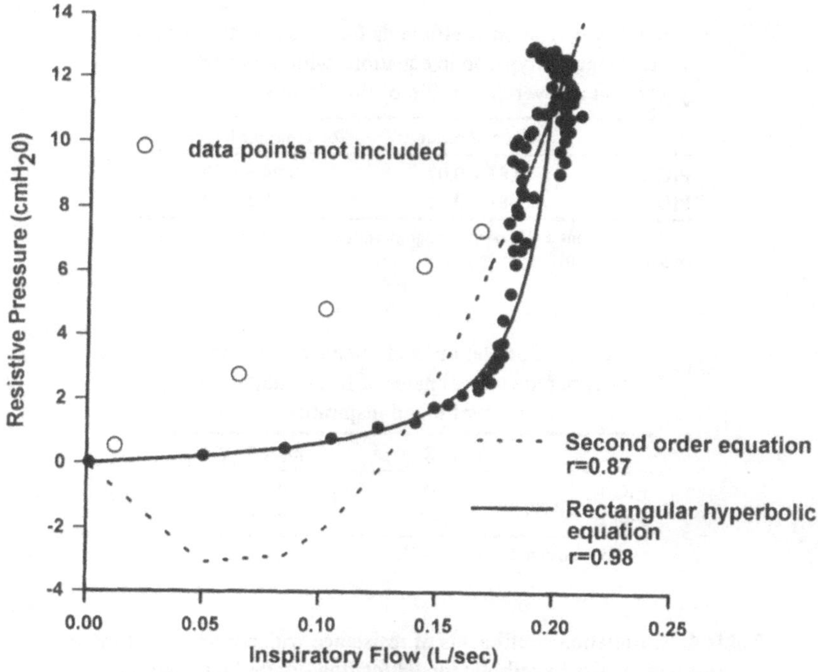

Figure 3. The application of the second order and rectangular hyperbolic equations to a typical flow-limited inspiration.

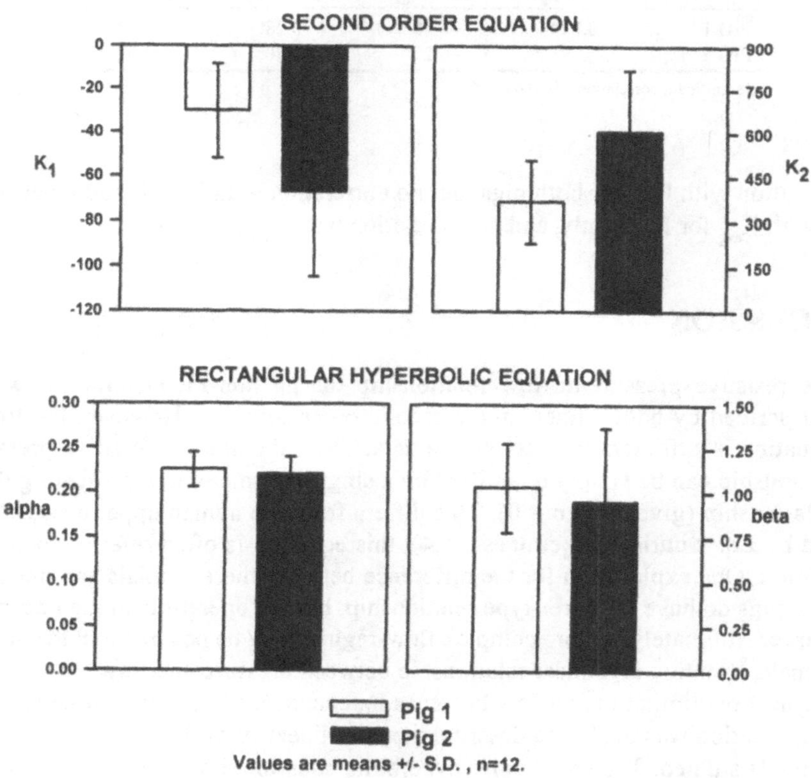

Figure 4. Mean parameter estimates of the second order and rectangular hyperbolic equations for flow-limited inspirations.

Table 2. Correlation coefficients for the second-order and rectangular hyperbolic equations with the resistive pressure-airflow relationship of flow limited inspirations

	Second order	Rectangular hyperbolic
PIG 1	0.83 ± 0.07	0.94 ± 0.02*
PIG 2	0.87 ± 0.03	0.95 ± 0.03*

Values are means ± SD, n = 12. *significantly different from second-order equation, $p < 0.01$.

Table 3. Correlation coefficients of resistance with parameter estimates of linear model for non-flow limited inspirations

	m vs R_{avg}	m vs R_{mid}
PIG 1	0.77*	0.42
PIG 2	0.94*	0.92*

*significant correlation, $p < 0.05$.

Table 4. Correlation coefficients of resistance with parameter estimates of rectangular hyperbolic model for flow-limited inspirations

	α vs R_{avg}	α vs R_{mid}	β vs R_{avg}	β vs R_{mid}
PIG 1	−0.81*	−0.45	0.28	0.12
PIG 2	−0.79*	−0.27	−0.70*	0.02

*significant correlation, $p < 0.05$.

tive correlation with R_{avg} for both pigs, but no correlation with R_{mid}. β had a negative correlation with R_{avg} for Pig 2 only, and no correlation with R_{mid}.

4. DISCUSSION

The resistive pressure-airflow relationship during non-FL inspirations were adequately described by both a linear and a second order equation. However, the first order linear equation is sufficient to describe the data. Thus, the non-FL resistive pressure-airflow relationship can be simply quantified by a single parameter m representing the slope of the relationship (given that b = 0). This differs from the human upper airway which is described by Eq. 2 during wakefulness (1,4); this equation is often referred to as the Rohrer equation. One explanation for the difference between these animals and humans may be that the pigs do have a Rohrer-type relationship, but are operating on the linear portion of this curve. Alternately, a more complex flow-regime may be occurring in the airways of these animals, resulting in a linear relationship between pressure and flow.

During flow limitation, airflow becomes independent of resistive pressure. The second order equation was unable to describe the curvilinearity of this relationship observed in the animals studied. The rectangular hyperbolic equation, however, was adequate, and superior to the second order equation in describing the resistive pressure-airflow relationship during flow limitation. A similar finding was reported by Hudgel et al. (2) for humans, who concluded that the rectangular hyperbolic equation offered a better description than the Rohrer equation of the inspiratory pressure-flow relationship of the upper airway during inspiratory FL. The parameters of the rectangular hyperbolic equation, α and β, may therefore be useful in characterizing FL inspirations, with α describing the maximal flow, and β indicating how rapidly this maximal flow is attained.

Under conditions of flow limitation, resistance, in the conventional sense, may be an inappropriate measure as pressure becomes dissociated from airflow. However, this study shows that R_{avg}, the average resistance throughout inspiration, does correlate positively with the slope of the linear relationship for non-FL inspirations, and negatively with maximal flow for FL inspirations. Therefore, R_{avg} may usefully describe the mechanical load on the respiratory system during both FL and non-FL inspirations, even though the mechanical significance of R_{avg} is uncertain.

Although resistance is traditionally reported at mid-inspiration (R_{mid}), we consider R_{avg} to be a better estimate of resistance as it incorporates information from the entire inspiration. This is obviously important in the FL inspirations, where resistance varies considerably within a breath as airflow is a non-linear function of resistive pressure. Furthermore, a measure which uses information from the entire inspiration will be less prone to noise in the data than one which uses a single data point.

An interesting observation was the "hysteresis" in the resistive pressure-airflow relationship during FL inspirations. A similar observation was made in the human upper airway during flow limitation (2). Assuming that the upper airway acts as a Starling resistor, narrowing of the upper airway is expected during flow limitation as described by Isono et al. (3). When the driving pressure is suddenly reduced at the end of inspiration, we observed airflow to be lower than would be expected. This suggests a difference in the geometry of the upper airway at end-inspiration compared to early-inspiration, likely related to the narrowing which occurred during flow limitation. This may result from viscoelastic properties of the airway wall, mucosal/surface forces acting within the upper airway, or upper airway constrictor muscle activity.

In conclusion, the resistive pressure-airflow relationship of the obese Vietnamese pot-bellied pig was best described by a linear equation during non-FL inspirations, and a rectangular hyperbolic equation during FL inspirations. Average inspiratory resistance may reflect the mechanical load on the respiratory system during non-FL as well as FL conditions.

ACKNOWLEDGMENTS

This research was supported by the Respiratory Health Network of Centres of Excellence, Inspiraplex.

REFERENCES

1. Anch, A.M., J.E. Remmers, and H. Bruce III. Supraglottic airway resistance in normal subjects and patients with occlusive sleep apnea. *J. Appl. Physiol.* 53:1158–1163, 1982.
2. Hudgel, D.W., C. Hendricks, and H.B. Hamilton. Characteristics of the upper airway pressure-flow relationship during sleep. *J. Appl. Physiol.* 64(5):1930–1935, 1988.
3. Isono, S., T.R. Feroah, E.A. Hajduk, R. Brant, W.A. Whitelaw, and J.E. Remmers. Interaction of cross-sectional area, driving pressure, and airflow of passive velopharynx. *J. Appl. Physiol.* 83(3):851–859, 1997.

4. Rohrer, F. The resistance in the human airway and the influence of branching of bronchial systems on frequency of breathing at different lung volumes. *Pfluegers Arch. Physiol.* 162:255–299, 1915.
5. Ruckebusch, Y. The relevance of drowsiness in the circadian cycle of farm animals. *Anim. Behav.* 20:637–643, 1972.

HUMAN VENTILATORY RESPONSE TO IMMERSION OF THE FACE IN COOL WATER

Lauren M. Stewart, Abraham Guz, and Piers C. G. Nye

University Laboratory of Physiology
Parks Road, Oxford OX1 3PT
United Kingdom

1. INTRODUCTION

The cardiovascular response to facial cooling in man can be dramatic (3), and this is shown by one exceptional subject studied by us (Figure 1). Here there was an increase in cardiac interval from one second during the control period to seven seconds as the cool water reached the eyes. This response is very pronounced in diving animals (1) and is therefore commonly known as the diving reflex. The bradycardia is accompanied by vasoconstriction which diverts blood flow away from the robust periphery towards the hypoxically sensitive heart and brain. The reflex is most prominent during breath-holding, indeed it may be completely overridden by the act of breathing. An abstract of this work has been published (4).

1.1. Ventilatory Drive from Facial Cooling

While setting up an undergraduate practical class to study the cardiovascular responses to facial immersion in cold water one of us (PN) reported intense discomfort which started immediately the eyes were cooled. This discomfort seemed to him to be the same as that felt at the break point of breath-holding We therefore decided to investigate the nature of this apparent drive to breathe. Ethical permission for the experiments was obtained from the Central Oxford Research Ethics Committee.

2. METHODS

The experimental setup is shown in Figure 2. We studied the ventilatory and cardiovascular responses of six healthy male subjects, aged between 19 and 21 years. They lay

Advances in Modeling and Control of Ventilation, edited by Hughson *et al.*
Plenum Press, New York, 1998.

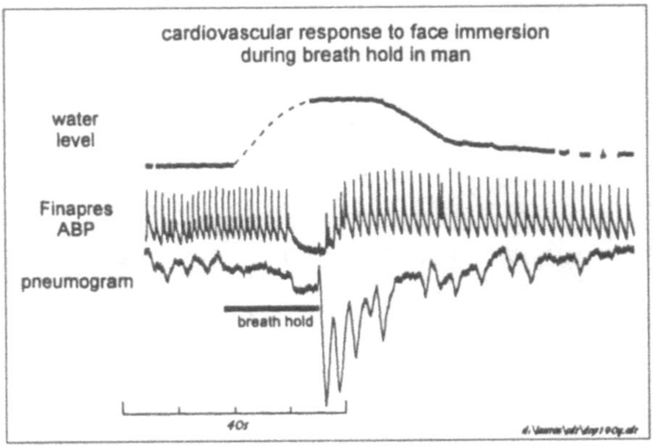

Figure 1. Slowing of heart rate by cooling of the face of a prone man with water at 17°C while breath-holding (see Figure 2 for setup). Top trace: water level in bucket rising to cover face (dashed line shows period when level-meter jammed). Middle trace: an index of blood pressure from a Finapres machine (uncalibrated). Bottom trace: pneumogram, inspiration down (uncalibrated). The long heart interval lasted for seven seconds.

prone on a bed, breathing through a mouthpiece in the bottom of a plastic bucket. The tip of a sample line to an infra-red meter, for the continuous measurement of CO_2, was positioned in the mouthpiece. A continuous index of arterial blood pressure was obtained non-invasively from a Finapres machine, and respiratory efforts were recorded from a differential transducer attached to a pneumograph bellows taped in place around the chest. Calibration of the latter's output against a spirometer trace showed that, in prone subjects, it gives a better representation of tidal volume than an inductance 'respitrace'. A float attached to an isotonic transducer was used to measure the level of the water in the bucket as it was poured through a wide-bore tube from another bucket.

The subject was asked to continue breathing on the mouthpiece as his face was wetted by filling and then draining the bucket. An attempt was made to hold end tidal CO_2 constant when ventilation increased, thereby eliminating possible dampening of the drive to breathe by the inevitable fall in arterial P_{CO_2}.

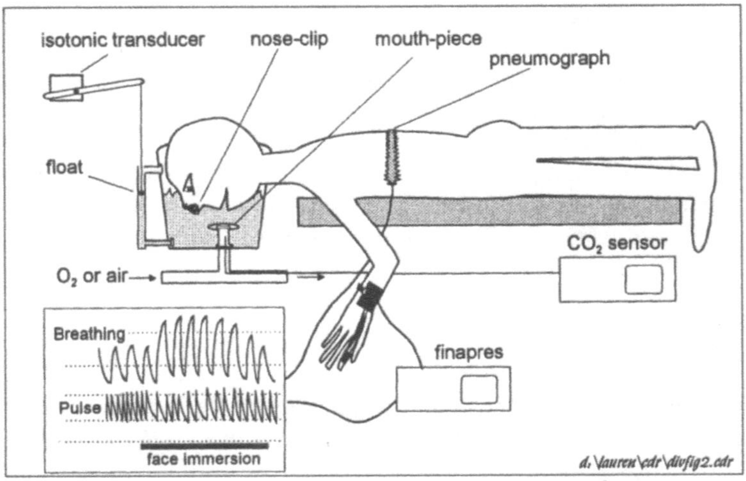

Figure 2. Experimental setup—see text for details.

3. RESULTS

All subjects hyperventilated when their faces were immersed in cool (17°C) water. When the water was warm (35–37°C) there was little if any ventilatory response. The raw data traces from one subject immersed in cool water is shown in Figure 3A. Figure 3B shows the ventilatory effect of facial cooling in a different subject in cool and warm water (5 immersions; 2 in cool, 3 in warm water). The subject in which face immersion in cool water elicited the greatest hyperventilation had a mean peak ventilation that was 1147% of control and the average increase in ventilation across all subjects was 457% of control values. This increase in ventilation was mediated almost entirely by an increase in tidal volume with frequency sometimes increasing at the point where ventilation was greatest. The maximal ventilatory increase during face immersion in cool water was significantly greater ($P < 0.001$) than the maximal increase during immersion in warm water. This indicated that coolness, rather than wetness was the pertinent stimulus.

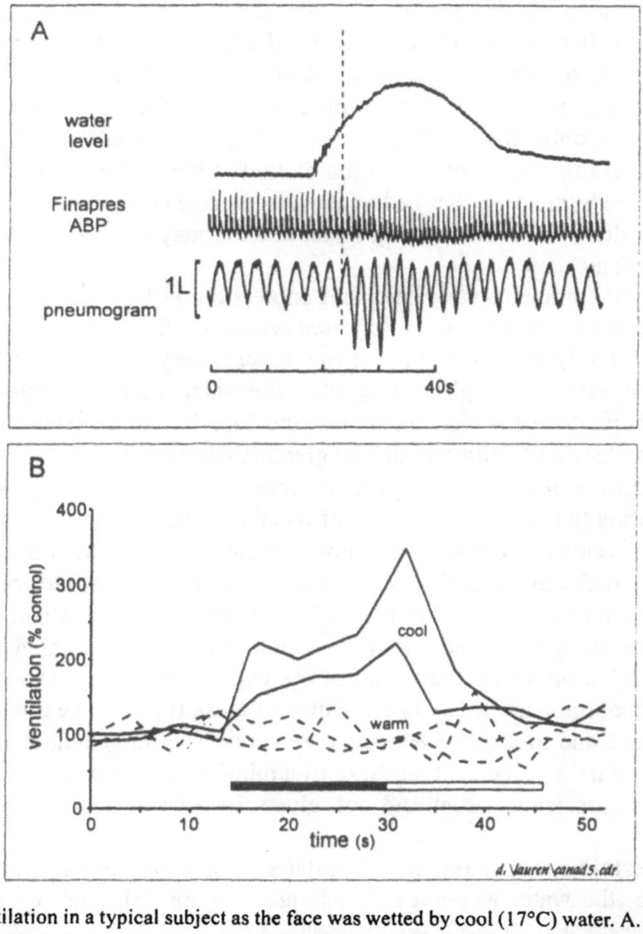

Figure 3. Hyperventilation in a typical subject as the face was wetted by cool (17°C) water. A. Traces as in Figure 1 with vertical dashed line showing approximate time of water reaching level of eyes. B. Interpolated (second-by-second) ventilation from five immersions in one subject—dashed lines: water temperature 35–37°C; solid lines: 17°C. The filled horizontal bar shows the period of rising water level, the open bar shows the period of falling water level.

4. DISCUSSION

The substantial hyperventilation and reported discomfort with face immersion in cool water suggests that it elicits a drive to breathe. If this is the case we might expect immersion in cool water to reduce breath-hold duration. In fact, breath-hold duration was reduced in four of our six subjects. It is possible that the lack of effect in the other two subjects arose because the drive from facial cooling is only transient, perhaps adapting with a time course similar to that reported by Hensel and Iggo (2) for cold receptors. If this is so, and adaptation is complete, a subject who can hold out through the initial period may be able to hold his breath for as long as he can under control conditions. This possibility is given some support by one subject who spoke of a sensation felt at the top of his chest upon immersion that was like "an explosion that dissipates" and that after this it "no longer felt unnatural not to be breathing". Hence in four subjects, the summed effects of an adapting trigeminal input and a rising chemoreceptor input may have been enough to make them to break their breath-hold whilst the other two may have been able to "hold out" until only the rising chemoreceptor input determined the drive.

Given the unnatural circumstances of breathing during complete immersion of the face in cool water, it would be understandable if a cortical input such as anxiety caused the subjects to hyperventilate. However, heart rate did not change from its control rate of 65 beats/min during immersion, suggesting that our subjects were not overly anxious. Moreover, although, collectively, their comments suggested that they found the experience moderately unpleasant, one of them said that it "didn't bother him at all" and another said that it was "wicked", implying that he had derived at least some enjoyment from the experience! Thus the drive to breathe during facial cooling may be comparable to that felt under hypercapnic conditions.

Our subjects' comments suggested that immersion of the eyes was especially important in mediating the drive to breathe. Reported sensations during breath-holding included: "can't breathe normally when water is above the nose; very, very hard when it reaches the eyes"; "worst at eyes"; "get a tight feeling when the water reaches the eyes".

The trigeminal nerve which innervates the face has three separate branches—the mandibular, maxillary and ophthalmic. The greatest increase in ventilation coincides with cooling of the ophthalmic branch, suggesting that input from this may be especially important in mediating the ventilatory effects of facial cooling.

If the hyperventilation observed is not cortical in origin, what anatomical basis is there for a reflex pathway involving the trigeminal nerve and respiratory centres? At the level of the tegmentum, trigeminal fibres divide into short ascending and long descending axons. The descending axons are small myelinated and unmyelinated fibres, many of which arise from the ophthalmic division of the trigeminal nerve and which convey primarily the senses of temperature and pain. Some of these fibres make direct monosynaptic contact in the pons and medulla where the respiratory rhythm generator resides (5). Thus whilst there is no direct evidence that these trigeminal fibres synapse onto respiratory nuclei, it would be surprising if they did not, given the area in which they make synaptic connections.

Hence, breathing can be reflexly stimulated simply by cooling the face. This does not involve pain; the water was not cold enough to elicit pain and our subjects did not report pain.

Whilst the benefits of a reduced cardiac output during diving are clear, namely to conserve oxygen for the heart and brain, an increased drive to breathe in response to facial cooling confers no obvious advantage. In fact, it might be expected to increase the likelihood of inspiring water in subjects threatened by drowning. The ventilatory response to facial cooling may however have clinical applications; for example an apneic patient might be revived by a sponge dipped in cold water.

ACKNOWLEDGMENTS

We wish to thank Chris Hirst, David O'Connor and Tim Pragnell for their technical and computing assistance in this project. Funding was generously provided by the Breathlessness Research Charitable Trust and by Balliol College Oxford.

REFERENCES

1. Daly, M.DeB. Breath-hold diving: mechanisms of cardiovascular adjustments in the mammal. In Recent advances in Physiology 10 ed. P.F. Baker: 201–245, 1984.
2. Hensel, H., and A. Iggo. Analysis of cutaneous warm and cold fibres in primates. Pflugers Arch. 329: 1–8, 1971.
3. Kawakami, Y., B.H. Natelson, and A.B. Dubois. Cardiovascular effects of face immersion and factors affecting diving reflex in man. J Appl Physiol. 23: 964–970, 1967.
4. Stewart, I.M., A. Guz, and P.C.G. Nye. Stimulation of human ventilation by face immersion in cold water. J Physiol. 501: 58P, 1997.
5. Truex, R.C., and M.B. Carpenter. Strong and Elwyn's Human Neuroanatomy (5th ed.). Williams & Wilkins: Baltimore, 1964.

VENTILATORY RESPONSE TO PASSIVE HEAD UP TILT

J. M. Serrador,[1] R. L. Bondar,[1] and R. L. Hughson[2]

[1]Cerebral Blood Flow Lab
School of Kinesiology
University of Western Ontario
London, Ontario N6A 3K7, Canada
[2]Cardiorespiratory and Vascular Dynamics Lab
Department of Kinesiology
University of Waterloo
Waterloo, Ontario N2L 3G1, Canada

1. INTRODUCTION

Humans have several adaptive mechanisms to deal with the effect of gravity during upright postures. The effect of passive upright tilt (HUT) on respiration has been shown to reduce the end tidal partial pressure of carbon dioxide ($P_{ET}CO_2$) and increase the end tidal partial pressure of oxygen ($P_{ET}O_2$) (1,4,5,9,15–19). The typical 4 mmHg drop in $P_{ET}CO_2$ has been shown to correspond with a decrease in arterial CO_2 (P_aCO_2) of approximately 2 mmHg (1,5). McHenry et al. (17) found that P_aCO_2 decreased by ~1 mmHg during 30° HUT. Boutellier and coworkers (6) demonstrated that as subjects went from +1 G_z force to +2 and +3, P_aCO_2 continued to decrease.

Three possible mechanisms for this decrease have been proposed. The first suggests that active hyperventilation is the cause. For this to occur, there would have to be an increase in alveolar ventilation (V_A) compared to CO_2 production ($V\text{CO}_2$). However, $V\text{CO}_2$ has been found to remain unchanged or slightly increase (1,5,15,16,18). This in combination with the fact that dead space (V_D) is known to increase in the upright posture (5,20) while tidal volume (V_T) and breathing frequency (F_B) remain relatively unchanged (4,18,20) suggests that it is unlikely that V_A increases. Hughes found that there was no change in V_A (12). Bjurstedt and colleagues found that V_A increased in 2 of 5 subjects, however $P_{ET}CO_2$ decreased in 4 of 5 subjects, suggesting that increased V_A could not explain changes in $P_{ET}CO_2$ and P_ACO_2 for all subjects (5). Matalon and Farhi (16) found that estimated V_A from breath-by-breath measures increased during HUT. Since they did not report $V\text{CO}_2$, V_T or F_B, it is unclear how this increase occurred.

Advances in Modeling and Control of Ventilation, edited by Hughson *et al.*
Plenum Press, New York, 1998.

The second suggests that the decrease is due to a change in the Ventilation-Perfusion (V_A/Q) ratio. In the upright posture there is a change in the distribution of both the pulmonary ventilation and blood flow (2,3,7). The force of gravity results in greater blood flow in the lower lobes of the lung, while ventilation is reduced. In the upper lobes we see a reduction in blood flow and increased ventilation. This mismatch of ventilation to blood flow may affect gas exchange. Thus if overall, ventilation increased more than blood flow, we would expect that more CO_2 would be expired and we would see a decrease in $P_{ET}CO_2$ with a rise in Vco_2. While it is unlikely that V_A increases, studies have shown that decreases by approximately 30% upon entering HUT (8,16,18). Thus it is possible that the lungs are over ventilated, however as mentioned previously we see little or no increase in Vco_2.

The third mechanism examines the redistribution of gas stores throughout the body. The change in posture may result in a movement of CO_2 from various tissues within the body to other possible stores. A 70-kg man can store approximately 123 litres of CO_2 in their lungs, blood and tissues (10). Farhi and Rahn have found that the time constant for changes in CO_2 stores differs with the tissue, ranging from approximately 2 minutes for heart, brain and other tissues to 30 minutes for muscle tissue, to several hours to days for bone and adipose tissue (11). Since these time constants are also affected by the blood flow through the organ, changes in posture could result in a change in perfusion of various tissues stores, and thus result in a redistribution of CO_2 throughout those body tissues.

This study examined both the ventilatory response to HUT as well as the role that CO_2 stores may play by recording the breath-by-breath response to HUT with a short or long accommodation period for redistribution of gas stores within the body. It was hypothesized that both changes in ventilation and distribution of CO_2 stores contributed to the magnitude of the $P_{ET}CO_2$ decrease with HUT.

2. METHODS

Twenty-three subjects (13 Female and 10 Male) participated in this study. All were healthy with no history of cardiovascular or respiratory disease and all were non-smokers. Subjects were randomly assigned to one of two conditions.

2.1. Condition 1: 20 Minute Supine Accommodation Period

14 Subjects were placed supine for 20 minutes while being instrumented prior to the initiation of the HUT protocol. The HUT protocol involved 10 minutes of supine baseline collection followed by 10 minutes of an 85° Passive Head Up Tilt.

2.2. Condition 2: 60 Minute Supine Accommodation Period

8 Subjects were placed supine for 60 minutes prior to the initiation of the HUT protocol. Breath by Breath gas exchange and ventilatory data were obtained with a respiratory mass spectrometer (RAMS, Marquette Electronics Inc., Milwaukee, WI), and ultrasonic flow meter (Kou Consulting Inc., Redmond, WA), and a dedicated microcomputer. Ventilation was monitored while the subjects wore a facemask allowing both oral and nasal breathing. HR was obtained from a standard 3 lead ECG placement (model 7830-A, Hewlett Packard, Andover, MA). One subject was removed from the study due to an inability to complete the HUT protocol without becoming presyncopal.

3. RESULTS

3.1. 20 Minute Supine Accommodation Period

On going to the HUT position subjects demonstrated an increase in HR ($p < 0.0001$) and $P_{ET}O_2$ ($p < 0.005$), and decrease in $P_{ET}CO_2$ ($p < 0.05$) with no significant change in V_E (see Fig. 1). Vco_2 increased ($p < 0.005$) as did Vo_2 ($p < 0.005$) (See Fig. 2). Calculated V_A did not significantly change from supine to HUT (4.96 ± 0.04 to 5.14 ± 0.05 L/min). Fig. 3 shows the time course of the V_E/Vco_2 decreases that were all significant from 10–20 minutes except minute 11. V_E/Vo_2 also significantly decreased from 13–19 minutes. Subjects did not demonstrate any significant changes in either breathing frequency (F_B) (15.7 ± 1.3 to 14.8 ± 1.3 breaths/min) or tidal volume (V_T) (559 ± 60 to 614 ± 61 mL).

3.2. 60 Minute Supine Accommodation Period

Subjects demonstrated an increase in HR ($p < 0.0005$) and $P_{ET}O_2$ ($p < 0.01$), decrease in $P_{ET}CO_2$ ($p < 0.0001$) again with no significant change in V_E (see Fig. 1). Fig. 2 shows an initial decrease then an increase in Vo_2 that became significant at minute 18. Vco_2 did not change significantly. Calculated V_A did not significantly change from supine to HUT (4.68 ± 0.04 to 4.89 ± 0.05 L/min). Fig. 3 shows the time course of V_E/Vco_2 changes with a transient decrease at minute 10 followed by a return to supine levels. V_E/Vo_2 increased at minute 11 and then again returned to supine levels. Subjects also did

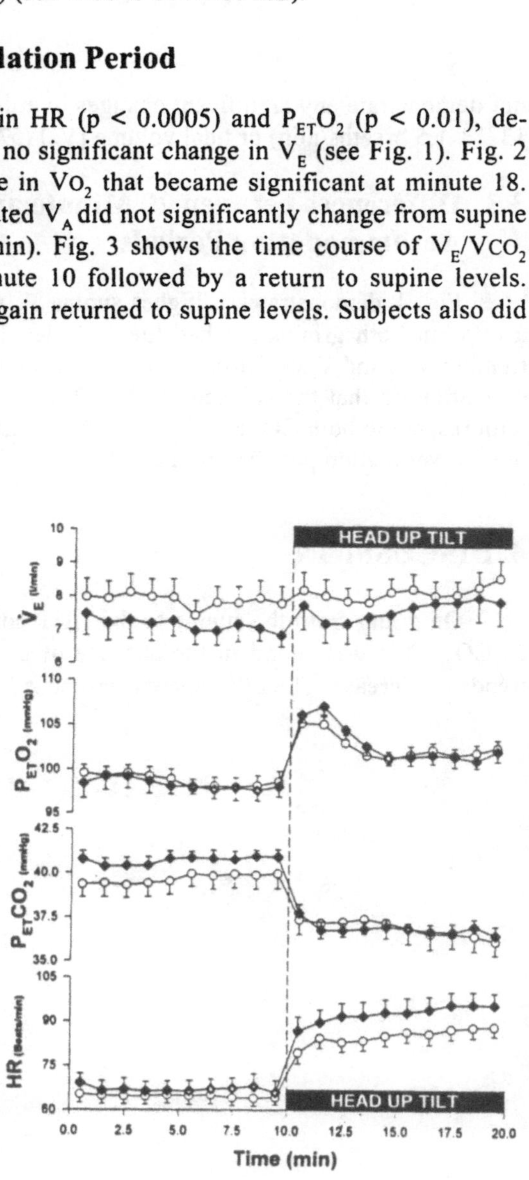

Figure 1. Ventilatory and cardiovascular response of subjects with previous 20 min supine accommodation (O) or 60 min supine accommodation period (◆) to 10 minute supine baseline followed by passive 85° head upright tilt. Data points represent one minute averages. Error bars are SE.

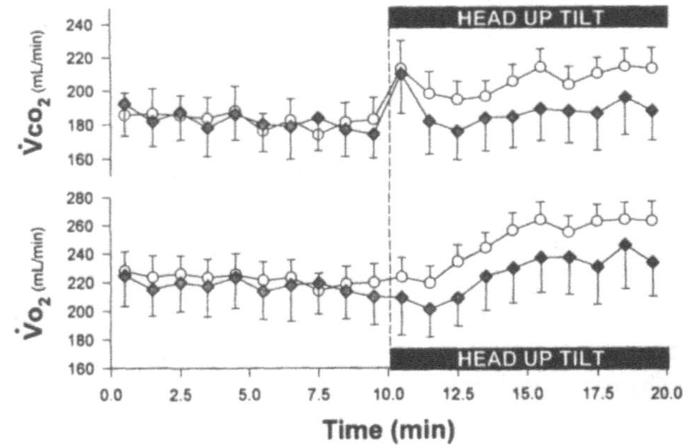

Figure 2. Changes in CO_2 production and O_2 consumption of subjects with previous 20 min supine accommodation (O) or 60 min supine accommodation period (◆) to 10 minute supine baseline followed by passive 85° head upright tilt. Data points represent one minute averages. Error bars are SE.

not demonstrate any significant changes in either breathing frequency (F_B) (13.9 ± 1.1 to 13.7 ± 1.5 breaths/min) or tidal volume (V_T) (529 ± 36 to 591 ± 59 mL).

3.3. Differences between 20 Minute and 60 Minute Supine Accommodation Periods

Fig. 1 demonstrates a higher supine $P_{ET}CO_2$ value in the 60 min accommodation group but both groups reached the same level during HUT. Fig. 2 shows an increasing trend in VO_2 and VCO_2 for the 20 min accommodation group. Fig. 3 shows the relative hypoventilation that the subjects in the 20 min accommodation condition show after HUT with respect to both CO_2 and O_2. The 60 min accommodation subjects remain at relatively normal ventilation post first minute of HUT.

4. DISCUSSION

On going from the supine to the HUT positions, there was a significant decrease in $P_{ET}CO_2$. This was found in the absence of a significant increase in V_A, but there were trends to increase. The difference in response between the 20 min accommodation and 60

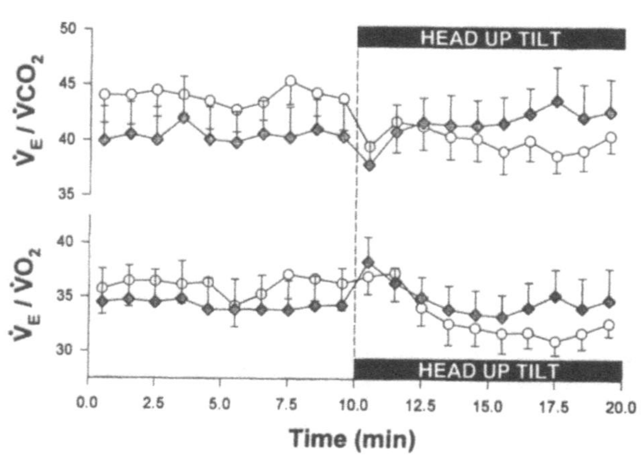

Figure 3. Changes in drive to breathe of subjects with previous 20 min supine accommodation (O) or 60 min supine accommodation period (◆) to 10 minute supine baseline followed by passive 85° head upright tilt. Data points represent one minute averages. Error bars are SE.

min accommodation groups suggests that redistribution of CO_2 stores may be directly involved in this $P_{ET}CO_2$ drop.

Active hyperventilation has been considered one possible mechanism to cause the decrease in $P_{ET}CO_2$. For this to occur the $V_A/V\text{CO}_2$ ratio must increase. The mean alveolar PCO_2 (21) was used to estimate V_A. We found that both V_E and estimated V_A did not increase significantly during HUT. There was an increase in $V\text{CO}_2$ for the 20 min supine accommodation group but $P_{ET}CO_2$ dropped to the same level in both groups (see Figs. 1 and 2). While these results are consistent with other research that also found no change in V_A while there was a decrease $P_{ET}CO_2$ (5,12), Matalon and Farhi (16) concluded that an increase in V_A accompanied the decrease in $P_{ET}CO_2$.

Our data (Fig. 3) show clearly that in the 20 min accommodation group, both the $V_E/V\text{CO}_2$ and $V_E/V\text{O}_2$ relationships decreased during HUT suggesting an apparent hypoventilation. That is, we have an indication of hyperventilation from the reduction in measured $P_{ET}CO_2$ and of hypoventilation from the decrease in $V_E/V\text{CO}_2$. This would suggest that there was an alteration in the V_D/V_T with a reduction in V_D on going to an upright position. While others have suggested that V_D should not decrease in the head up position (5,12), the supine position can be associated with increased air trapping (16). This can cause an increase in V_D so that on going to HUT, V_D/V_T is reduced.

A role for a change in the V_A/Q relationship is also a possible source of this $P_{ET}CO_2$ decrease. While our data did not include measures of pulmonary blood flow, we found no change in V_E or estimated V_A. This would suggest that the V_A/Q relationship did change during HUT. Jones and colleagues (13) demonstrated that a decrease in cardiac output can result in a decrease in $P_{ET}CO_2$. The decrease in Q with HUT thus could be another possible cause of this drop in $P_{ET}CO_2$; however, Q should remain relatively constant after the initial few minutes of HUT (8). However we noted a slow continual decrease in $P_{ET}CO_2$ over the 10 minutes of HUT (see Fig. 1). We would also expect that $P_{ET}O_2$ would remain elevated with the new V_A/Q relationship, however we saw a decrease in $P_{ET}O_2$ towards supine levels after the first 4 minutes of HUT. Thus while we cannot exclude V_A/Q as the possible cause of this decrease, it would appear unlikely to be the mechanism for the slow drop in $P_{ET}CO_2$ seen over the 10 minutes of HUT.

The final mechanism to be examined is the shifting of CO_2 between body tissues stores. Liner and co-workers (14) found that there was little change in CO_2 stores when subjects went from immersion to dry conditions. This would agree with the lack of change in $V\text{CO}_2$ seen in the 60 min accommodation group. The continued increase in $V\text{CO}_2$ in the 20 min accommodation group may indicate that CO_2 is evolving from tissue stores with faster time constants. Thus subjects in the 60 min accommodation group had sufficient time to store CO_2 in tissues with a time constant greater than the 10 min of HUT. It is possible that CO_2 does not begin to evolve out of these tissues within the short period of the HUT.

Our observations of a reduction in $P_{ET}CO_2$ probably coincide with a reduction in P_aCO_2. It has commonly been observed that the $P_{ET}CO_2$ to P_aCO_2 difference at rest is about 2 mmHg or less (1,5,16,21). On going from supine to HUT positions, this difference might be reduced (1,5), although others have found no change (16). Anthonisen and colleagues demonstrated that during HUT there was a drop in both $P_{ET}CO_2$ and P_aCO_2 suggesting that $P_{ET}CO_2$ does represent changes at the arterial level (1).

An unexpected finding was the decrease in both the $V_E/V\text{CO}_2$ and $V_E/V\text{O}_2$ ratios in the 20 min accommodation group but not the 60 min. This would suggest that since V_E did not change in either group that this decrease was primarily due to the increased $V\text{CO}_2$ and $V\text{O}_2$ in the 20 min group. The relative hypoventilation seen in the 20 min group may be a compensatory mechanism to prevent a further decline in $P_{ET}CO_2$. If V_E were to be increased to match the increasing VCO_2, we would expect there to be a further decrease in $P_{ET}CO_2$. Thus while more CO_2 is expired, there is no resulting change in ventilation. This raises the question as to why the peripheral and central chemoreceptors do not detect this decrease in $P_{ET}CO_2$ and modify ventilation accordingly to eliminate the drop.

One possible explanation for this unexpected ventilatory response is that the cerebrospinal fluid may contain a different level of CO_2 and H^+ than arterial blood. Thus our measurement of P_aCO_2 may not reflect what the central chemoreceptors are detecting. Examination of changes in cerebral blood flow may provide further information on this mechanism. Previously, Matalon and Farhi (16) speculated that a reduction in cerebral blood flow might account for the increased V_A.

This study demonstrated that the significant reduction in $P_{ET}CO_2$ was not accompanied by an increase in V_E. However, V_A must have increased and V_D must have decreased to allow for the reduction in $P_{ET}CO_2$. To what extent changes in V_A/Q ratio and CO_2 tissue stores contributed to this response is not known. Future investigations of the mechanisms responsible for the decrease in $P_{ET}CO_2$ must include direct measurement of P_aCO_2, and concurrent measures of cerebral blood flow and cerebrospinal fluid during postural changes.

ACKNOWLEDGMENTS

This research was supported by grants from the Natural Sciences and Engineering Research Council of Canada and the Canadian Space Agency.

REFERENCES

1. Anthonisen, N. R., J. R. Bartlett, and S. M. Tenney. Postural effect on ventilatory control. *J. Appl. Physiol.* 20: 191–196, 1965.
2. Anthonisen, N. R. and J. Milic-Emili. Distribution of pulmonary perfusion in erect man. *J. Appl. Physiol.* 21: 760–766, 1966.
3. Anthonisen, N. R., P. C. Robertson, and W. R. Ross. Gravity-depende sequential emptying of lung regions. *J. Appl. Physiol.* 28: 589–595, 1970.
4. Barrett, J., F. Cerny, J. A. Hirsch, and B. Bishop. Control of breathing patterns and abdominal muscles during graded loads and tilt. *J. Appl. Physiol.* 76: 2473–2480, 1994.
5. Bjurstedt, H., C. M. Hesser, G. Liljestrand, and G. Matell. Effects of Posture on Alveolar-Arterial CO_2 and O_2 Differences and on Alveolar Dead Space in Man. *Acta Physiol. Scand.* 54: 65–82, 1962.
6. Boutellier, U., R. Arieli, and L. E. Farhi. Ventilation and CO_2 response during +Gz acceleration. *Respir. Physiol.* 62: 141–151, 1985.
7. Bryan, A. C., J. Milic-Emili, and D. Pengelly. Effect of gravity on the distribution of pulmonary ventilation. *J. Appl. Physiol.* 21: 778–784, 1966.
8. Butler, G. C., Xing HC, Northey DR, and Hughson RL. Reduced orthostatic tolerance following 4 h head-down tilt. *Eur. J. Appl. Physiol.* 62: -30, 1991.
9. Cencetti, S., G. Bandinelli, and A. Lagi. Effect of PCO_2 changes induced by head-upright tilt on transcranial Doppler recordings. *Stroke* 28: 1195–1197, 1997.
10. Cherniack, N. S. and G. S. Longobardo. Oxygen and carbon dioxide gas stores of the body. [Review] [219 refs]. *Physiol. Rev.* 50: 196–243, 1970.
11. Farhi, L. E. and H. Rahn. Dynamics of Changes in Carbon Dioxide Stores. *Anesthesiology* 21: 604–614, 1960.
12. Hughes, J. M. Regional lung function: physiology and clinical applications. *Clin. Physiol.* 5: 19–31, 1985.
13. Jones, P. W., W. French, M. L. Weissman, and K. Wasserman. Ventilatory responses to cardiac output changes in patients with pacemakers. *Journal of Applied Physiology: Respiratory, Environmental & Exercise Physiology* 51: 1103–1107, 1981.

14. Liner, M. H. Tissue gas stores of the body and head-out immersion in humans. *J. Appl. Physiol.* 75: 1285–1293, 1993.
15. Loeppky, J. A. and U. C. Luft. Fluctuations in O_2 stores and gas exchange with passive changes in posture. *J. Appl. Physiol.* 39: 47–53, 1975.
16. Matalon, S. V. and L. E. Farhi. Cardiopulmonary readjustments in passive tilt. *J. Appl. Physiol.* 47: 503–507, 1979.
17. McHenry, L. C., J. F. Fazekas, and J. F. Sullivan. Cerebral hemodynamics of syncope. *Am. J. Med. Sci.* 241: 173–178, 1961.
18. Miyamoto, Y., T. Tamura, T. Hiura, T. Nakamura, J. Higuchi, and T. Mikami. The dynamic response of the cardiopulmonary parameters to passive head-up tilt. *Jpn. J. Physiol.* 32: 245–258, 1982.
19. Newberry, P. D., A. W. Hatch, and J. M. MacDonald. Cardio-respiratory events preceding syncope induced by a combnation of lower body negative pressure ad head-up tilt. *Aerosp. Med.* 41: 373–378, 1970.
20. Rea, H. H., S. J. Withy, E. R. Seelye, and E. A. Harris. The effects of posture on venous admixture and respiratory dead space in health. *Am. Rev. Respir. Dis.* 115: 571–580, 1977.
21. Whipp, B. J., N. Lamarra, S. A. Ward, J. A. Davis, and K. Wasserman. Estimating arterial PCO_2 from flow-weighted and time-average alveolar PCO_2 during exercise. In: *Respiratory Control—A Modeling Perspective*, edited by G. D. Swanson, F. S. Grodins, and R. L. Hughson. New York: Plenum Press, 1988.

23

DO SEX-RELATED DIFFERENCES EXIST IN THE RESPIRATORY PHARMACOLOGY OF OPIOIDS?

Elise Sarton, Albert Dahan, and Luc Teppema

Departments of Anesthesiology and Physiology
Leiden University Medical Center
2300 RC Leiden, The Netherlands

1. INTRODUCTION

There are strong indications from human and animal studies (especially from studies using inbred strains of mice), that strain and sex-related differences exist in the analgesic potency of endogenous and exogenous administered opioids, as well as in the neurochemical and genetic mechanisms activated to modulate pain.[1–6] These differences are not restricted to the analgesic properties of opioids. Opioid-induced lethality, changes in locomotor activity, opioid addiction and opioid discrimination also exhibit sex- and/or strain-related differences.[1,3,7] Studies on strain- or sex-related differences in the influence of opioids on ventilatory control are scarce. We retrieved one study from the literature. Muraki and Kato studied the influence of morphine on the occurrence of hypothermia and respiratory rate in six strains of male mice.[8] They observed significant strain differences in these two measures of morphine action.

In humans, we recently investigated if sex-related differences exist in the respiratory pharmacology of morphine.[9] Morphine is the prototype μ-opioid receptor agonist and is widely used for treatment of acute and chronic pain states. Since respiratory depression is a common and serious side effect of morphine, knowledge on the existence of sex-related differences is of evident clinical importance. With the use of the Dynamic End-Tidal Forcing technique, we determined the normoxic steady-state hypercapnic ventilatory response (HCVR) in 24 healthy young volunteers (12 men: mean age 25.6 ± 1.7 yr; and 12 women: mean age 24.8 ± 4.5 years [mean ± SD]) before and during administration of intravenous morphine (bolus = 100 μg/kg, followed by a continuous infusion of 30 μg/kg.h). The study had a double-blind, placebo controlled, randomized design.

Advances in Modeling and Control of Ventilation, edited by Hughson *et al.*
Plenum Press, New York, 1998.

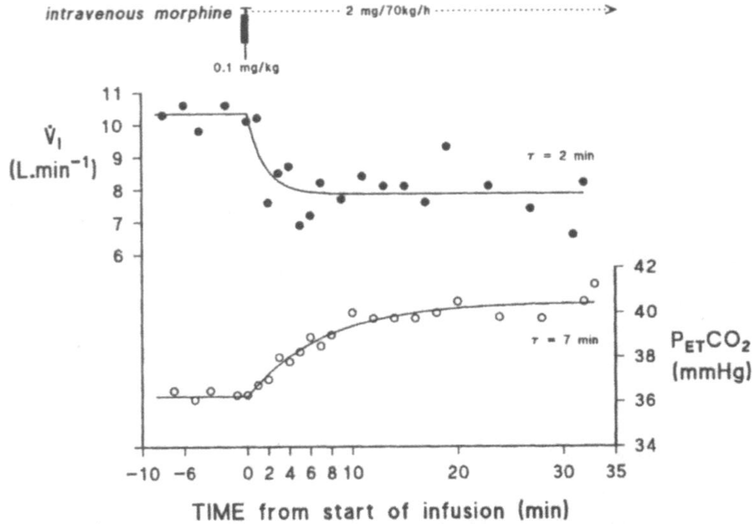

Figure 1. Influence of intravenous morphine on resting ventilation and resting end-tidal PCO_2 in a single subject. At time t = 0 min the intravenous adminsitration of morphine was started (bolus 100 μg/kg followed by a continuous infusion of 30 μg/kg·h). Data points are mean values of 10 breaths. A simple one component exponential was fitted to the data (separate analysis were performed on $P_{ET}CO_2$ and $\dot{V}_I$). The time constant (τ) for $\dot{V}_I$ was 2 min, and for $P_{ET}CO_2$ 7 min.

A summary of the results, relevant to the issue of gender, is given here:

1. Placebo had no effect on resting $\dot{V}_I$, resting $P_{ET}CO_2$, the position and slope (S) of the $\dot{V}_I$-$P_{ET}CO_2$ response in men and women. In addition, we observed no sex differences in these variables and parameters.
2. Morphine infusion caused a decrease in $\dot{V}_I$ and an increase in $P_{ET}CO_2$ of similar magnitude and similar temporal profile in men and women (see Figure 1).
3. In women, after morphine, the slope of the $\dot{V}_I$-$P_{ET}CO_2$ response was reduced by ~30% without affecting the extrapolated $P_{ET}CO_2$ at which $\dot{V}_I$ = 0 L/min (B).
4. In men, after morphine, the position of the $\dot{V}_I$-$P_{ET}CO_2$ response was shifted to higher $\dot{V}_I$-values without any significant change in slope.
5. The morphine-induced changes in B and S differed significantly between men and women.
6. A post-hoc study (n = 9 women) on the influence of phase of menstrual cycle in women revealed no differences in morphine-induced changes in B and S between the follicular and luteal phases (see Figure 2).

Our results indicate important sex-related differences in the respiratory pharmacology of the μ-receptor agonist morphine in young and healthy volunteers. We suggest that the mechanisms involved are not different from those causing sex-differences in the antinociceptive action of morphine, as observed in animal studies.

Possible mechanisms responsible for the observed sex differences include:

a. Sex differences in the blood and/or brain concentration of morphine and/or its metabolites;
b. Acute effects of sex steroids (i.e. their mere presence and/or absence);
c. Long-term developmental effects of sex steroids occurring in perinatal life;
d. Sex-independent factors.

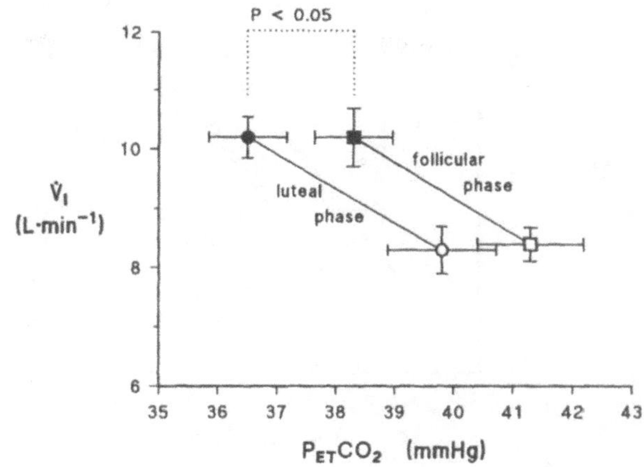

Figure 2. Influence of morphine on resting $P_{ET}CO_2$ and resting $\dot{V}_I$ in the follicular and luteal phases of the menstrual cycle. Closed symbols = control; Open symbols = morphine; Circles = luteal phase; squares = follicular phase.

2. SEX DIFFERENCES IN BLOOD AND BRAIN CONCENTRATION OF MORPHINE AND MORPHINE-6-GLUCURONIDE

We are unable to exclude sex differences in the concentration of morphine or its active metabolite, morphine-6-glucuronide (M6G), at the site of μ-opioid receptors involved in respiratory control. One of the causes of a difference in active agent concentration at the target receptors sites may be a sex-difference in the morphine or M6G blood-effect site equilibration times (defined by the time constant τ). For example, larger values of τ in men may have caused lower morphine or M6G concentration at target sites at the time the respiratory experiments. We determined the values of τ for changes in resting $P_{ET}CO_2$ and resting $\dot{V}_I$ (see Figure 1). For both variables, they ranged between 2 and 7 min and did not differ between men and women. Furthermore, since we performed the experiments 40 min after the start of morphine administration (i.e. at least 6 times the blood-effect site τ), an effect of differences in blood-effect site equilibration time constant is not of any importance.

Another cause for a difference in morphine or M6G concentration in the brain may be sex-differences in steady-state morphine pharmacokinetics. For example, this may cause higher active agent concentrations at target receptors sites, after a dose given on weight basis, in women compared to men, independent of the time of measurement. Evidence against this hypothesis comes from a study of Bourke and Warley[10] on the influence of intravenous morphine on the hyperoxic steady-state HCVR. This study was performed in men exclusively. They showed that the ventilatory response curve was shifted to higher $P_{ET}CO_2$ levels without changes in slope after 0.07 and 0.21 mg/kg morphine (see Figure 3). The latter dose was clearly higher than the dose used in our study. The results of Bourke and Warley indicate that, in men, the effect of morphine on the slope of the $\dot{V}_I$-$P_{ET}CO_2$ response is dose independent (at least in the dose range studied). We therefore conclude that the differences in the effect of morphine on S in men and women is not related to differences in active agent concentrations at target receptors sites.

3. ACUTE EFFECTS OF TESTOSTERONE, ESTROGEN AND PROGESTERONE

It is possible that an acute effect of testosterone is the cause for the observed sex differences. Testosterone increases metabolic rate together with an increase in hypoxic re-

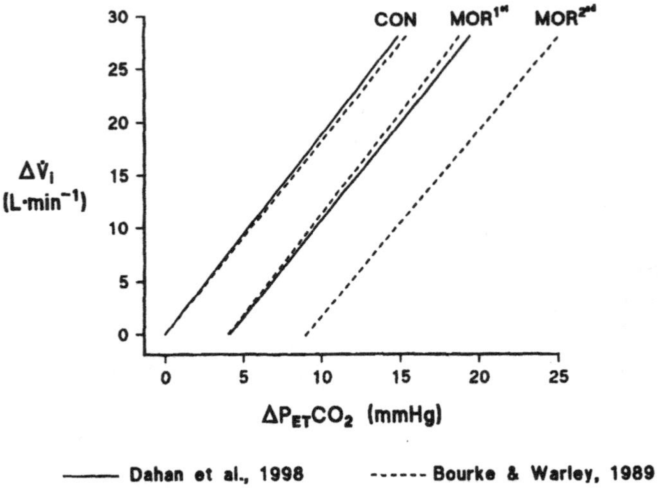

Figure 3. Comparison of the results from the study of Bourke and Warley[10] and this study. Mor1st = initial dose of morphine in the study of Bourke and Warley (0.07 mg/kg morphine); MOR2nd = the second dose (cumulative dose = 0.21 mg/kg). Note the absence of a dose-related effect on the slope of the $\dot{V}_I$-$P_{ET}CO_2$ response.

sponses.[11] We have to keep in mind, however, that control studies and placebo studies did not show sex-differences.

An acute effect of female sex hormones seems an unlikely mechanism, since we observed no differences in the respiratory effects of morphine in the follicular and luteal phases of the menstrual cycle.

4. LONG TERM DEVELOPMENTAL EFFECTS OF SEX HORMONES

A mechanism through which gonadal hormones could mediate sex differences is by their long-term developmental and/or organizational effects, occurring in perinatal life, and resulting in sexual dimorphism (i.e. differences in brain morphology and resultant neurobiology).[12,13] It is generally accepted that this mechanism is accountable for the findings that the analgesic potency of exogenous administered opioids and endogenous opioids is higher in male than in female animals, and the finding that the sexes modulate pain using neurochemically and genetically distinct mechanisms. An example of the organizational effects of sex hormones, relevant to our study, is the significant variation in the density and distribution of methionine-enkephalin-immunoreactive μ-opioid-receptors in the median preoptic area of the rat brain between males and females.[14] We hypothesize that long-term organizational effects triggered by sex steroids may well be responsible for the differences in the respiratory effects of morphine in men and women, as observed in our study. It then follows that sex differences exist in the density, distribution and/or affinity of μ-opioid receptors in brain areas directly or indirectly involved in respiratory control. An other possibility is that sex differences exist in the central translation of information from stimulated μ-opioid receptors. Candidate areas where sex-related differences in the respiratory pharmacodynamics could find their origin include peripheral sites (the carotid bodies)[15] and μ-opioid receptors containing sites within the central nervous system (median preoptic area)[14].

REFERENCES

1. Mogil, J.S., S.P. Richards, L.A. O'Toole, M.L. Helms, S.R. Mitchell, and J.K. Belknap. Genetic sensitivity to hot-plate nociception in DBA/2J and C57BL/6J inbred mouse strains: possible sex-specific mediation by δ_2-opiod receptors. Pain 70: 267–277, 1997
2. Mogil J.S., W.F. Sternberg, P. Marek, B. Sadowski, J.K. Belknap, and J.C. Liebeskind. The genetics of pain and pain inhibition. Proc Natl Acad Sci USA 1996; 93: 3048–55, 1996.
3. Frischknecht H.R., B. Siegfried, and P.G. Waser. Opioids and behavior: genetic aspects. Experientia 44: 473–81, 1988.
4. Cicero, H.J., B. Nock, and E.R. Meyer. Gender-related differences in the antinociceptive properties of morphine. J. Pharmacol. Exp. Ther. 279: 767–773, 1996.
5. Kaveliers, M., and D.G.L. Innes. Stress-induced opioid analgesia and activity in deer mice: sex and population differences. Brain Res. 425: 49–56, 1987.
6. Gear, R.W., C. Miaskowski, N.C. Gordon, S.M. Paul, P.H. Heller, and J.D. Levine. Kappa-opioids produce significantly greater analgesia in women than in men. Nature Med. 2: 1248–50, 1996.
7. Moskowitz, A.S., G.W. Terman, K.R. Carter, M.J. Morgan, and J.C. Liebeskind. Analgesic, locomotor and lethal effects of morphine in the mouse: strain comparisons. Brain Res. 361: 46–51, 1985.
8. Muraki, T., and R. Kato. Strain differences in the effects of morphine on the rectal temperature and respiratory rate in male mics. Psychopharm 89: 60–64, 1986.
9. Dahan, A., E. Sarton, L. Teppema, and C. Olievier. Sex-related differences in the influence of morphine on ventilatory control in humans. Anesthesiology, in press.
10. Bourke, D.L., and A. Warley. The steady-state and rebreathing methods compared during morphine administration in humans. J. Physiol. Lond. 419: 509–517, 1989.
11. White, D.P., B.K. Schneider, R.J. Santen, M. McDermott, C.K. Pickett, C.W. Zwillich, and J.V. Weil. Influence of testosterone on ventilation and chemosensitivity in male subjects. J. Appl. Physiol. 59: 1452–1457, 1985.
12. Arnold, A.P., and S.M. Breedlove. Organizational and activational effects of sex steroids on brain behavior: a reanalysis. Horm. Behav. 19: 469–498, 1985.
13. Breedlove, S.M. Sexual differentiation of the human nervous system. Ann. Rev. Psychol. 45: 389–418, 1994.
14. Hammer, R.P. The sexually dimorphic region of the pre-optic area in rats contains denser opiate receptor binding sites in females. Brain Res. 308: 172–176, 1984.
15. McQueen, J.S., and J.A. Ribeiro. Inhibitory actions of methionine-enkephalin and morphine an the cat carotid chemoreceptor. Br. J. Pharmacol. 71: 297–305, 1980.

ARE THE RESPIRATORY RESPONSES TO CHANGES IN VENTILATORY ASSIST OPTIMIZED?

Yoshitaka Oku and Shigeo Muro

Department of Clinical Physiology
Chest Disease Research Institute
Kyoto University
53 Kawahara-cho, Shogoin, Sakyo-ku, Kyoto, 606-8397, Japan

1. INTRODUCTION

The input-output relationship of the CO_2 respiratory controller can be described in two different ways. The first method is to describe the output as a function of the inputs; this type of controller model may be called a reflex controller model. The other method is to describe the relationship by an operating principle. The operating principle is often the minimization of a certain criterion or parameter; in this case, the controller is called an optimal controller. Several investigators have proposed that the respiratory controller responds to various stimuli in order to minimize certain criteria, which represent the energetic cost of breathing (3,4,9,13). In the concept proposed by Poon (9), the maintenance of arterial blood gas tensions and mechanical work are competing priorities for the respiratory controller, and the controller functions to minimize the net operating cost of both work and deviations of the blood gas tensions from given set points.

One of the advantages of an optimal controller over a reflex controller is that it can predict a wide range of behaviors without additional parameters or hypotheses, because an optimal controller is described by its operating principle. In contrast, a reflex controller requires both new parameters and new hypotheses to describe different behaviors. For example, some optimal controllers can predict exercise hyperpnea (9), but the reflex controller requires an additional parameter associated with the metabolic rate or the intensity of the exercise plus an additional hypothesis characterizing the relationship between the output and these new parameters. Due to its unique feature of predictability, an optimal controller can be tested by applying different inputs and then comparing the predicted versus actual responses. This is an important procedure in order to ascertain the extent to which a proposed operating principle could be applied under various circumstances.

Advances in Modeling and Control of Ventilation, edited by Hughson *et al.*
Plenum Press, New York, 1998.

Recently, a new mode of artificial ventilation, termed proportional assist ventilation has been developed by Younes (14). In this mode, the ventilator simply amplifies the patients' instantaneous effort throughout the inspiration while giving the patient complete control over all aspects of his or her breathing pattern. Since this new mode of ventilation has a number of potential advantages (greater comfort, a reduction in peak airway pressure required to maintain adequate ventilation, etc.) over other ventilatory modes, it may be particularly beneficial for those patients requiring sustained ventilatory support. However, since this approach preserves the patients' own respiratory control system, it is crucial to understand how the respiratory control system responds to different levels of ventilatory assist at different $PaCO_2$ levels. In the present study, we compared the responses to changes in ventilatory assist at different inspired CO_2 levels between predicted and experimentally obtained data, and then determined whether the proposed controllers were adequate for describing these responses.

2. MODEL PREDICTIONS

Optimal controllers operate in order to minimize certain criteria. In Eqs. 1 and 2, criteria J_1 and J_2 are represented as mathematical functions, called energy function or cost function:

$$J_1 = k_1 \cdot C_{aut}{}^2 + k_2 \cdot (PaCO_2 - P_0)^2 \quad (1)$$

$$J_2 = k_1 \cdot \ln\left(\frac{C_{aut}}{1 - C_{aut} / C_{max}}\right) + k_2 \cdot (PaCO_2 - P_0)^2 \quad (2)$$

where C_{aut} and C_{max} are the automatic respiratory motor command and its maximum respectively, P_0 is the arterial CO_2 partial pressure ($PaCO_2$) set point, and k_1 and k_2 are constants. The first term represents the work of breathing and the second term represents the deviation from the $PaCO_2$ set point, although the work of breathing is expressed differently in these two functions. In both equations, the optimal controller functions to minimize the conflicting challenges between less work for breathing and smaller deviations in the blood gas tensions. In the first equation, both terms have a quadratic expression, but in the second equation, the term for the work of breathing is logarithmic such that the controller produces a linear hypercapnic ventilatory response. Eq. 1 is the expression that has been proposed for the quantitative relationship between the chemical and mechanical inputs to the respiratory controller and breathing discomfort (6). Eq. 2 is a modified version of the cost function proposed by Poon (9). In the original equation, the logarithmic term is a function of ventilation, and how the controller determines the level of ventilation is not clearly defined. The automatic motor command is related to the minute ventilation ($\dot{V}$) and the maximal minute ventilation ($\dot{V}_{max}$) by the linear coefficient R.

$$\dot{V} = R \cdot C_{aut} \quad (3)$$

$$\dot{V}_{max} = R \cdot C_{max} \quad (4)$$

Substituting Eqs. 3 and 4 into Eq. 2 yields Poon's original cost function. Furthermore, we can relate the ventilation input to the arterial CO_2 output according to Eq. 5 below:

$$PaCO_2 = \frac{A \cdot \dot{V}CO_2}{\dot{V} - \dot{V}_D} + 713 \cdot FICO_2 \tag{5}$$

where A is a constant, $\dot{V}CO_2$ is the minute CO_2 production (assumed to be constant), $\dot{V}_D$ is the minute dead space ventilation (assumed to be constant), and $FICO_2$ is the inspired CO_2 fraction. Substituting Eq. 3 into Eq. 5 yields

$$PaCO_2 = \frac{A \cdot \dot{V}CO_2}{\left(R \cdot C_{aut} - \dot{V}_D\right)} + 713 \cdot FICO_2 \tag{6}$$

In the present study, we examined the responses of the controller when the coefficient R is changed. This corresponds to the situation where the ventilatory assist is changed in an artificially ventilated subject with a proportional assist ventilator (14) or a phrenic nerve driven servo respirator (10, see Experiments).

First, we will consider the quadratic function. At a constant inspired CO_2 level, when R or the ventilatory assist decreases, then the $PaCO_2$ increases. Therefore, the second term of Eq. 1 increases, and the optimal controller attempts to suppress this increase by increasing the automatic motor command, C_{aut}. However, this action increases the first term, and therefore there must be an optimal combination of $PaCO_2$ and C_{aut} to minimize J_1. The optimal solution is derived by a numerical method. The parameter values used are as follows: $k_1 = 0.0055$, $k_2 = 0.01$, $P_0 = 35.0$ (Torr), $A \cdot \dot{V}CO_2 = 246.57$ (Torr·L/min), $\dot{V}_D = 3.0$ (L/min), and $C_{max} = 100$. These values have been chosen arbitrarily so that the controller produces reasonable ventilatory responses. Fig. 1 shows the optimal values of these parameters at different R values at $FICO_2 = 0$.

We next examined the output of the optimal controller in response to changes in R at the two different inspired CO_2 levels, 0% and 5% (Fig. 2). At a constant $PaCO_2$, the optimal controller predicts a reduction in respiratory output at a higher $FICO_2$. As it is depicted in Fig. 2B, a higher level of ventilatory assist is required at a higher $FICO_2$ to maintain a given level of $PaCO_2$.

These characteristics hold true of the energy function expressed in Eq. 2. Fig. 3 shows the numerically derived responses of the optimal controller described by Eq. 2. The parameter values used are: $k_1 = 6.0$, $k_2 = 0.01$, and $P_0 = 30.0$ (Torr). The values for the

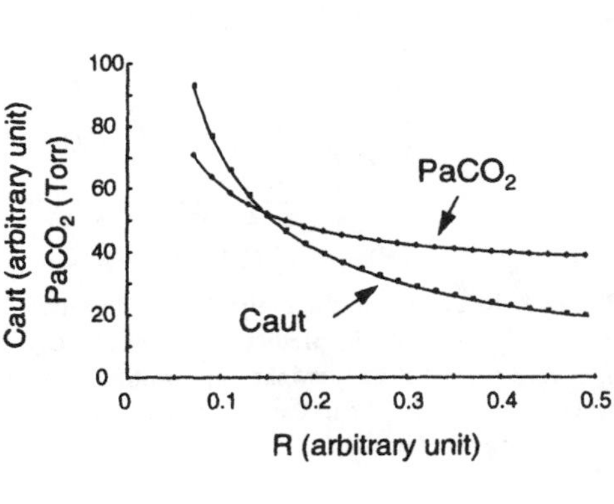

Figure 1. Predicted responses of the optimal controller described by Eq. 1 to changes in ventilatory assist. The responses represent the values of two parameters, C_{aut} and $PaCO_2$, that yield a minimum value for J_1 at a given level of ventilatory assist, R.

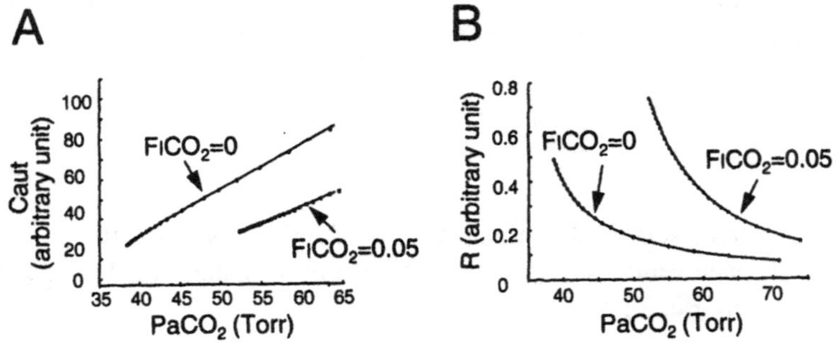

Figure 2. Predicted responses of the optimal controller described by Eq. 1 to changes in ventilatory assist at different FICO$_2$ levels. A: the hypercapnic Caut response was almost linear. The response curve shifts downward (or rightward) when 5% CO_2 is added to the inhaled gas. B: the relationship between R (ventilatory assist) and the PaCO$_2$. When 5% CO_2 is inhaled, a higher level of assistance is necessary to maintain a given PaCO$_2$.

other parameters are the same as those used for the controller described by Eq. 1. Again, the values have been chosen arbitrarily so that the controller produces reasonable ventilatory responses. At a constant PaCO$_2$, the controller predicts a reduction in respiratory output at a higher FICO$_2$ and consequently a higher ventilatory assist. The relationship between PaCO$_2$ and Caut predicted by Eq. 2 was concave downward.

3. EXPERIMENTAL PROCEDURES

We recently examined the responses to changes in ventilatory assist at different inspired CO_2 levels in cats (5). The experiments were performed on decerebrate, paralyzed, and artificially ventilated cats at three different FICO$_2$ levels (0, 0.03, 0.05). As illustrated in Fig. 4, the airway pressure (Paw) delivered by the ventilator is proportional to the moving time-averaged phrenic activity (Phr), with the use of the phrenic-driven servo respirator described by Remmers et al. (10). This is similar to the situation where a subject is ventilated with a proportional assist ventilator. We defined the assist gain as the average ratio of Paw to Phr in one breath (Fig. 4). The assist gain corresponds to R in our model.

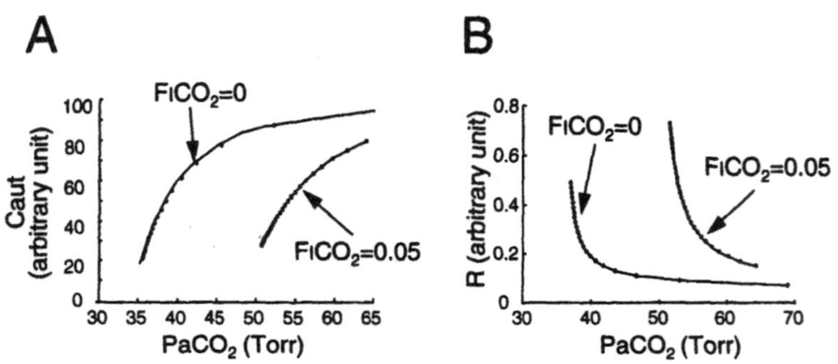

Figure 3. Responses of the optimal controller described by Eq. 2 to changes in ventilatory assist at different inspired CO_2 levels. A: the predicted hypercapnic C_{aut} response was concave downwards. The response curve shifts rightward when 5% CO_2 is inhaled, resulting in a reduction in the C_{aut} at a given PaCO$_2$. B: the relationship between R and PaCO$_2$. As in the case of Eq. 1, when 5% CO_2 is inhaled, a higher level of assistance is necessary to maintain a given PaCO$_2$.

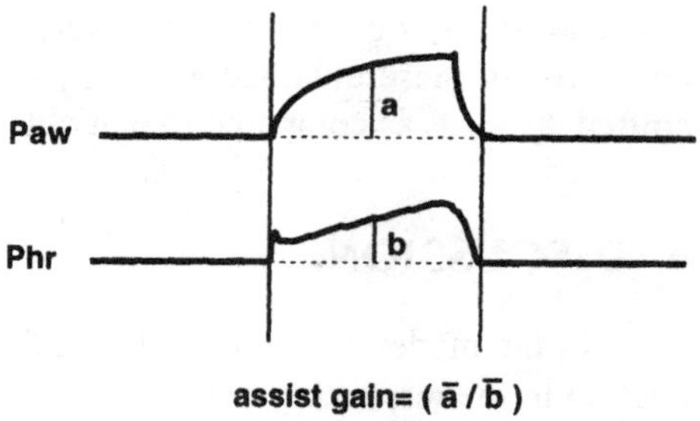

Figure 4. Schematic illustration of airway pressure (Paw) and moving time-averaged phrenic activity (Phr) during phrenic-driven ventilation. Using a phrenic-driven respirator, the tracheal pressure is proportional to the moving time-averaged phrenic activity. The assist gain is defined as the average ratio of Paw to Phr in one breath.

Furthermore, we defined the minute phrenic activity as the total moving time-averaged phrenic activity during a one minute period, and regarded this as the respiratory drive. This corresponds to C_{aut} in our model.

We changed the assist gain at a given $FICO_2$ within the range where $PaCO_2$ did not exceed 80 Torr by adjusting the flowrate control. This procedure corresponds to the present model analysis. Fig. 5 shows a representative response obtained from one cat.

Fig. 5A illustrates the relationship between the assist gain and $PaCO_2$ at different $FICO_2$ levels. When the assist gain decreases, $PaCO_2$ increases. To maintain the same $PaCO_2$, a higher level of ventilatory assistance was required for a higher $FICO_2$; this agrees with the model analysis (Figs. 2B and 3B). Figure 5B shows the relationship between $PaCO_2$ and the minute phrenic activity. The relationship between $PaCO_2$ and the minute phrenic activity was curvilinear, and similar between different inspired CO_2 levels. To compare the minute phrenic activities at different $FICO_2$ levels, the relationship between $\log_{10}PaCO_2$ and the minute phrenic activity was determined and fitted to the linear regression line with the least squares method. The regression lines fitted well to the experimental data ($n = 26$, $r > 0.82$, $p < 0.05$). The regression lines at different $FvCO_2$ levels were then compared by the F-test. In nine out of ten cats, the regression lines at $FICO_2 = 0.03$ were not significantly different from those obtained at $FICO_2 = 0$. In eight out of ten cats, the regression lines at $FICO_2 = 0.05$ were not significantly different from those ob-

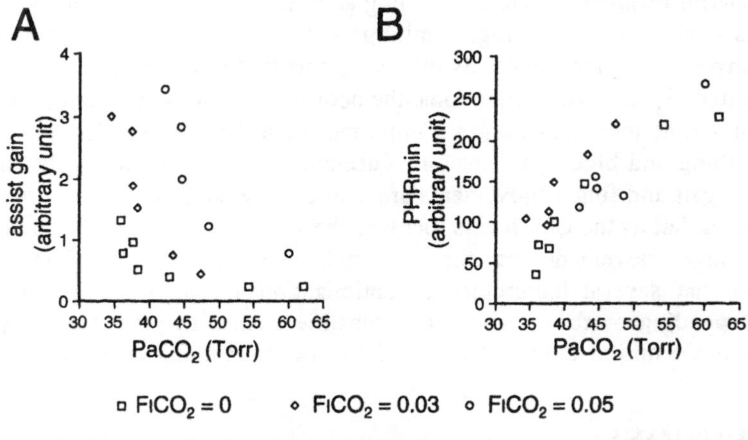

Figure 5. Changes in the minute phrenic activity and $PaCO_2$ in response to changes in the assist gain. A: the relationship between $PaCO_2$ and the assist gain is hyperbolic. CO_2 inhalation shifts the curve rightward. B: the hypercapnic response of the minute phrenic activity (PHRmin) is concave downwards. The slopes of the hypercapnic responses are similar between different inhaled CO_2 levels.

tained at $F_{I}CO_2 = 0$. In two cats, the regression lines at $F_{I}CO_2 = 0.05$ shifted downwards as compared to those obtained at $F_{I}CO_2 = 0$; in one cat, the regression line at $F_{I}CO_2 = 0.03$ shifted upwards as compared to that obtained at $F_{I}CO_2 = 0$.

4. DISCUSSION

In the model analysis, we found that the proposed optimal controllers predicted a reduction in the respiratory output (C_{aut}) at a given $PaCO_2$ during 5% CO_2 inhalation as compared to without CO_2 inhalation. Since at a given $PaCO_2$ the level of assist ventilation was higher with increasing $F_{I}CO_2$ levels, these results also suggest that a higher level of ventilatory assistance reduces the respiratory output at a given $PaCO_2$. In contrast, in the cat experiments, the respiratory output was reduced in only two out of ten cats. Therefore, the difference in the $F_{I}CO_2$ levels is not likely to influence the respiratory output. The comparison between the predicted and experimental data indicates that the respiratory controller does not behave in an optimal manner under the present experimental conditions with respect to the minimization of the proposed energy functions. This discrepancy could be explained in several ways. First, it has been reported that the responses to an airway CO_2 load via "slug" CO_2 breathing or an added dead space are also discrepant between the experimental data and the data predicted by the optimal controller model (11,12). There is a possibility that the airway CO_2 load attenuated the optimizing process. Alternatively, the discrepancy between the predicted and experimental data may indicate that some structures or pathways important for the optimization process were missed by our experimental procedure. Since the animals were decerebrate and paralyzed, the forebrain and midbrain were missing and thus the reflex pathway mediated by the muscle spindles was not functional; these structures and pathway may have been important for the optimization process.

A more fundamental philosophical question is whether a single operating principle or optimal controller theory must explain everything. There is evidence suggesting that the respiratory control system does not obey a single optimization principle associated with the energetic cost of breathing. First, it has been shown that the spontaneously-adopted level and pattern of breathing coincide with those which produce the least sensation of breathlessness (1), suggesting that the optimization of breathing may involve the alleviation of unpleasant or uncomfortable breathing sensations (6). Second, many motor functions such as swallowing, coughing, vomiting, and vocalization share common muscles, neuronal pathways, and neuronal networks with respiration (8), and these functions are well-coordinated (2). In certain situations, the coordination between respiration and these non-respiratory functions may be more important than the optimization in terms of the work of breathing and blood gas tensions. Furthermore, non-respiratory rhythmic movements such as gait and finger movements are also closely coordinated with respiration. It has been shown that as the CO_2 levels increase the respiratory rhythm becomes more entrained with finger movements (see the chapters by Ebert and Rassler). These considerations suggest that several hierarchies of optimization are involved in the control of breathing. Depending on the situation, different hierarchies may have priority, and thus compromises to the optimization in terms of the work of breathing may occur.

5. CONCLUSIONS

It is concluded that the brainstem respiratory controller does not respond to changes in the level of ventilatory assistance in order to minimize the previously proposed energy functions. Such an optimization process may require higher brain structures and/or other peripheral feedback pathways. Alternatively, the brainstem respiratory center may operate according to different optimization criteria under different conditions.

ACKNOWLEDGMENTS

This work was supported by the Suzuken Memorial Foundation.

REFERENCES

1. Chonan, T., M.B. Mulholland, M.D. Altose, and N.S. Cherniack. Effects of changes in level and pattern of breathing on the sensation of dyspnea. *J. Appl. Physiol.* 69: 1290–1295, 1990.
2. Dick, T.E., Y. Oku, J.R. Romaniuk, and N.S. Cherniack. Interaction between central pattern generators for breathing and swallowing in the cat. *J. Physiol. Lond.* 465: 715–730, 1993.
3. Longobardo, G.S., N.S. Cherniack, and A. Damokosh-Giordano. Possible optimization of respiratory controller sensitivity. *Ann. Biomed. Eng.* 8: 143–158, 1980.
4. Luijendijk, S.C.M., and J. Milic-Emili. Breathing patterns in anesthetized cats and concept of minimum respiratory effort. *J. Appl. Physiol.* 64: 31–41, 1988.
5. Muro, S., Y. Oku, K. Chin, M. Mishima, M. Ohi, and K. Kuno. The effect of ventilatory assist on the level of respiratory drive in decerebrate cats. *Resp. Physiol.* 109: 205–217, 1997.
6. Oku, Y., G.M. Saidel, T. Chonan, M.D. Altose, and N.S. Cherniack. Sensation and optimization of breathing: A dynamic model. *Ann. Biomed. Eng.* 19: 251–272, 1991.
7. Oku, Y., G.M. Saidel, M.D. Altose, and N.S. Cherniack. Perceptual contributions to optimization of breathing. *Ann. Biomed. Eng.* 21: 509–515, 1993.
8. Oku, Y., I. Tanaka, and K. Ezure. Activity of bulbar respiratory neurons during fictive coughing and swallowing in the decerebrate cat. *J. Physiol. Lond.* 480: 309–324, 1994.
9. Poon, C.S. Ventilatory control in hypercapnia and exercise: Optimization hypothesis. *J. Appl. Physiol.* 62: 2447–2459, 1987.
10. Remmers, J.E., and H. Gautier. Servo respirator constructed from a positive-pressure ventilator. *J. Appl. Physiol.* 41: 252–255, 1976.
11. Swanson, G.D. The exercise hyperpnea dilemma. *Chest* 73: 270–272, 1978.
12. Ward, S.A., and B.J. Whipp. Ventilatory control during exercise with increased external dead space. *J. Appl. Physiol.* 48: 225–231, 1980.
13. Yamashiro, S.M., J.A. Daubenspeck, T.N. Lauritsen, and F.S. Grodins. Total work rate of breathing optimization in CO_2 inhalation and exercise. *J. Appl. Physiol.* 38: 702–709; 1975.
14. Younes, M. Proportional assist ventilation, a new approach to ventilatory support. Theory. *Am. Rev. Respir. Dis.* 145: 114–120, 1992.

SELECTIVE DEPRESSION OF PERIPHERAL CHEMOREFLEX LOOP BY SEVOFLURANE IN LIGHTLY ANESTHETIZED CATS

Luc Teppema,[1] Elise Sarton,[2] Albert Dahan,[2] and Kees Olievier[1]

[1]Department of Physiology
[2]Department of Anesthesiology
Leiden University Medical Center
PO Box 9604, 2300 RC Leiden, The Netherlands

1. INTRODUCTION

In animal studies it has been shown that volatile anesthetics depress the ventilatory response to hypoxia and hypercapnia compared to the awake state.[1,2] The findings that 0.5–1% halothane reduces activity in afferent nerve fibres of the carotid bodies,[3] and that halothane, enflurane and isoflurane inhibit CO_2–O_2 interaction[1,2], indicate that these inhalational anesthetics may directly act on the peripheral chemoreceptors. This, however, was not confirmed by Berkenbosch and coworkers,[4,5] who found that in cats lightly anesthetized with chloralose-urethane, overall anesthesia with halothane, or selective administration of this agent to the peripheral chemoreceptors reduced the gains of the peripheral and central chemoreflex loops to the same extent.

In men, subanesthetic doses of volatile anesthetics can cause a considerable reduction of the hypoxic ventilatory response,[6,7] but the extent to which this occurs depends on the arousal state of the subjects,[8] explaining the limited depressant effects of 0.1 MAC isoflurane reported by Temp et al.[8,9]

In the present study we measured the effects of 0.5% and 1% sevoflurane on the peripheral and central chemoreflex loops in cats under light chloralose-urethane anesthesia. To this aim, we used the Dynamic End-tidal Forcing (DEF) technique to determine the ventilatory response to changes in end-tidal PCO_2, and analyzed the data with a two-compartment model. In this way we were able to separate the effects of sevoflurane into actions on the peripheral and central chemoreflex loops, respectively.[10]

Advances in Modeling and Control of Ventilation, edited by Hughson *et al.*
Plenum Press, New York, 1998.

2. METHODS

Eight cats were lightly anesthetized with a bolus infusion of 20 $mg \cdot kg^{-1}$ chloralose and 100 $mg \cdot kg^{-1}$ urethane (i.v.), followed by a continuous infusion of, respectively, 1 $mg \cdot kg^{-1} \cdot h^{-1}$ and 5 $mg \cdot kg^{-1} \cdot h^{-1}$. Arterial blood pressure and rectal temperature were monitored continuously.

In all conditions (control, 0.5% sevoflurane, 1.0% sevoflurane), at least three DEF-runs were performed. After a steady state period of about two minutes, end-tidal PCO_2 was elevated by about 1–1.5 kPa within a few breaths, maintained at constant level for about 7 minutes and then lowered step-wise to the previous value and kept constant for another 7 minutes.

3. DATA ANALYSIS

In the steady state, minute ventilation can be described by:

$$\dot{V}_I = (G_p + G_c)(P_{ET}CO_2 - B) \tag{1}$$

G_p = sensitivity of the peripheral chemoreflex loop, G_c = sensitivity of the central chemoreflex loop, B = apnoeic threshold (X-intercept of the CO_2 response curve).

For the analysis of the dynamic ventilatory response, we used a two-compartment model:[10]

$$\tau_c \, d/dt \, \dot{V}_c(t) = G_c \, [P_{ET}CO_2(t - T_c) - B] - \dot{V}_c(t) \tag{2}$$

$$\tau_p \, d/dt \, \dot{V}_p(t) = G_p \, [P_{ET}CO_2(t - T_p) - B] - \dot{V}_p(t) \tag{3}$$

$$\tau_c = \tau_{ON} \, x + (1 - x) \, \tau_{OFF} \tag{4}$$

$$\dot{V}_I(t) = \dot{V}_c(t) + \dot{V}_p(t) + C \cdot t \tag{5}$$

$\dot{V}_c$ and $\dot{V}_p$ are the contributions of the central and peripheral chemoreflex loops to total ventilation ($\dot{V}_I$). τ_c: time constant of the central chemoreflex loop; τ_{ON}: central time constant of the on-transient; τ_{OFF}: central time constant of the ventilatory off-transient. *C* in Eq. 5 represents a drift-term.

Results are presented as means ± S.D. Different treatments (i.e. control, 0.5 and 1% sevoflurane) were compared using ANOVA. This study was approved by the Animal Ethics Board of the Leiden University.

4. RESULTS

The mean (± S.D) values for the apnoeic threshold B, and for G_p and G_c in the control situation were respectively: 4.27 ± 0.55 kPa, 0.14 ± 0.06 $l \cdot min^{-1} \cdot kPa^{-1}$, and 0.45 ± 0.28 $l \cdot min^{-1} \cdot kPa^{-1}$. The ratio G_p over G_c was 0.38 ± 0.12.

Increasing the end-tidal sevoflurane concentration from zero to 0.5% resulted in a small decrease of B to 3.71 ± 0.90 kPa ($P < 0.01$); G_p and G_c decreased to 0.05 ± 0.02 and 0.15 ± 0.05 $l \cdot min^{-1} \cdot kPa^{-1}$, respectively ($P < 0.01$ in both cases). The ratio G_p over G_c was 0.34 ± 0.11 and did not differ from that during the control situation.

When the end-tidal sevoflurane concentration was 1%, the value of B was no longer significantly different from control: 3.94 ± 1.55 kPa. G_p further reduced to 0.02 ± 0.03 $l \cdot min^{-1} \cdot kPa^{-1}$ ($P < 0.01$); G_c, however, did not further reduce and was 0.13 ± 0.06 $l \cdot min^{-1} \cdot kPa^{-1}$. The ratio G_p over G_c reduced to 0.15 ± 0.10.

5. DISCUSSION

By measuring the effects of sevoflurane in lightly anesthetized cats we were able to study its effects on metabolic control (i.e. chemical control by blood gases) of ventilation, not disturbed or overruled by cortical influences, as may occur in awake subjects.

Our finding of no consistent effect of sevoflurane on the apnoeic threshold B is in good agreement with previous findings on the effect of halothane and other anesthetics on the control of breathing in the anesthetized cat.

In contrast to G_c, G_p—and consequently also the ratio G_p/G_c—was affected in a dose dependent way by sevoflurane. This may indicate that the agent directly interferes with peripheral chemoreception, confirming previous data on the effect of 0.5–1% halothane on afferent carotid body activity.[3] Berkenbosch and coworkers[4,5] could not demonstrate a selective effect of 0.8–1.2% halothane on the peripheral chemoreflex loop, since in their study the ratio G_p/G_c remained constant. Apart from possible differences in the effects of halothane and sevoflurane (for example different dose-response curves), we have no explanation for the discrepancy between their and our present results.

Our finding that upon increasing the end-tidal sevoflurane concentration from 0.5 to 1.0% G_c did not further decrease is very interesting, and may indicate a very steep dose-response curve in the lower range of end-tidal sevoflurane concentrations, even under conditions of a background baseline anesthesia. If, on the other hand, under our experimental conditions an increase in the end-tidal sevoflurane concentration from 0.5 to 1% resulted in a deepening of anesthesia, then this would not be accompanied by a decrease in central respiratory drive.

The present data do not allow us to draw firm conclusions as to a central site of action of sevoflurane. Apart from the peripheral chemoreceptors or sites within the central nervous system (c.q. brainstem) where processing of afferent chemoreceptor takes places, the respiratory integrating centers could also be affected (cf. the unchanged ratio of G_p over G_p at 0.5% sevoflurane). We can also not exclude a direct action on the central chemoreceptors. Further studies involving simultaneous EEG and ventilatory measurements could yield relevant information as to the site(s) of action of inhalational anesthetics under different experimental paradigms (e.g. awake or baseline anesthesia).

REFERENCES

1. Weiskopf, R.B., L.W. Raymond, and J.W. Severinghaus. Effects of halothane on canine respiratory responses to hypoxia with and without hypercarbia. *Anesthesiology* 41, 350–360, 1974.
2. Hirshman, C.A., R.E. McCullough, P.J. Cohen, and J.V. Weil. Depression of hypoxic ventilatory response by halothane, enflurane and isoflurane in dogs. *Br. J. Anaesth.* 49, 957–963, 1977.
3. Davies, R.O., M. Edwards, and S. Lahiri. Halothane depresses the response of carotid body chemoreceptors to hypoxia and hypercapnia in the cat. *Anesthsiology* 57, 153–159, 1982.
4. van Dissel, J.T., A. Berkenbosch, C.N. Olievier, J. DeGoede, and Ph. Quanjer. Effects of halothane on the ventilatory response to hypoxia and hypercapnia in cats. *Anesthesiology* 62, 448–456, 1985.
5. Berkenbosch, A., J. DeGoede, C.N. Olievier, and Ph. H. Quanjer. Site of action of halothane on respiratory pattern and ventilatory response to CO_2 in cats. *Anesthesiology* 57, 389–398, 1982.

6. Knill, R.L., H.T. Kieraszewicz, and B.G. Dodgson. Chemical regulation of ventilation during isoflurane sedation and anesthesia in humans. *Can. Anaesth. Soc. J.* 30, 607–614, 1983.
7. Dahan, A., M.J.L.J. van den Elsen, A. Berkenbosch, J. DeGoede, I.C.W. Olievier, J. van Kleef, and J.G. Bovill. Effects of subanaesthetic halothane on the ventilatory response to hypercapnia and acute hypoxia in healthy volunteers. *Anesthesiology* 80, 727–738, 1994.
8. van den Elsen, M.J.L.J., A. Dahan, A. Berkenbosch, J. DeGoede, J. van Kleef, and I. Olievier. Does subanaesthetic isoflurane affect the ventilatory response to acute isocapnic nhypoxia in healthy volunteers? *Anesthesiology* 81, 860–867, 1994.
9. Temp, J.A., L.C. Henson, and D.S. Ward. Does a subanesthetic concentration of isoflurane blunt the ventilatory response to hypoxia? *Anesthesiology* 77, 1116–1124, 1992.
10. DeGoede, J. A. Berkenbosch, D.S. Ward, J.W. Bellville, and C.N. Olievier. Comparison of chemoreflex gains obtained with two different methods in acts. *J. Appl. Physiol.* 59, 170–179, 1985.

PULMONARY RAPIDLY ADAPTING RECEPTORS AND AIRWAY CONSTRICTION

Jerry Yu

Pulmonary and Critical Care Medicine
Department of Medicine
University of Louisville
ACB-3, 530 S. Jackson St.
Louisville, Kentucky 40292

1. BACKGROUND AND HYPOTHESIS

In general, the airway and pulmonary receptors can be divided into three groups: 1) slowly adapting pulmonary stretch receptors (PSRs); 2) rapidly adapting receptors (RARs); 3) receptors innervated by C-fibers. PSRs and RARs are mechanoreceptors, although RARs are also activated by many biological and chemical substances (3, 16, 17, 29). On the other hand, C-fibers are believed to be chemosensitive endings (2).

Our understanding of RARs is largely from the early studies of Widdicombe and his co-workers in the nineteen sixties. RARs are believed to be very important because they are associated with cough and can be stimulated by mediators released during anaphylaxis, such as histamine and PGF_{2a}. It has been reported that histamine can provoke a vagally mediated bronchoconstriction (7, 11). Gold and co-workers (9) further demonstrated that unilateral challenge of one lung with antigen resulted in bilateral bronchoconstriction, which could be abolished by cooling only the vagus nerve innervating the challenged lung. This reflex bronchoconstriction could also be blocked by atropine, suggesting a vagal reflex response. The above authors believed that RARs were responsible for the vagally mediated bronchoconstriction. Furthermore, while histamine produces airway constriction, it also stimulates RARs (14). Increased RAR activity is often accompanied by bronchoconstriction under many conditions (14, 21–24). In addition, contraction of airway smooth muscle is believed to stimulate RARs (4, 14, 20). Based on this information, a neurogenic theory is postulated: during asthmatic attack the local release of mediators, such as histamine, can stimulate RARs. The increased RAR activity could induce bronchoconstriction, causing further stimulation of RARs. This process forms a positive feedback which facilitates airway constriction (24).

Advances in Modeling and Control of Ventilation, edited by Hughson *et al.*
Plenum Press, New York, 1998.

The hypothesis that RARs cause airway constriction was based on the observation that in several conditions increased RAR activity was often accompanied by bronchoconstriction. Actually there is no direct evidence to support that RARs can evoke a vagally mediated bronchoconstriction (18). It may be dangerous to come to such a conclusion by simply relying on a correlation, as pointed out by Coleridge and Coleridge (3). The lack of direct evidence to prove or to refute the hypothesis is probably due to the fact that there has been no reliable stimulus to selectively activate RARs. Recently, it has been found that a decrease in lung compliance is a strong stimulant to RARs (30). This stimulus activates RARs without significant influence on the other pulmonary afferents (32).

2. CHALLENGE TO THE HYPOTHESIS

Using decreased lung compliance as a stimulus, Yu and Mink (31) tested the hypothesis that activation of RARs can reflexly induce bronchoconstriction in lightly anesthetized dogs. The right and left lung were ventilated separately at a PEEP of 4 cm H_2O. RARs in one lung were stimulated by decreasing lung compliance by removing and then restoring PEEP. The airway pressure in the non-stimulated lung was monitored as an index of airway muscle tone. No increases in the pressure swing could be detected, while RARs in the other lung were stimulated. On the other hand, electrical stimulation of the distal end of the cervical vagus nerve increased the pressure swing bilaterally (ipsilateral dominant), suggesting that a reflex response could be detected in the preparation. Moreover, deflation (or inflation) of the lung increased (or suppressed) diaphragmatic activity, also suggesting intact vagal afferents and a central response. Thus, it is concluded that activation of pulmonary RARs does not appear to induce airway constriction. If any effect is present, it appears to be small.

It can be argued that in the above observation (31) the failure to detect a reflex is due to no crossover of the RAR signals. However, RAR afferents likely have a bilateral projection (6, 13), about 8% of PSRs can be recorded in contralateral vagus nerve (personal observation). More importantly, since stimulation of vagal efferents produced a bilateral increase in airway resistance (31), it seems likely that RARs do not reflexly produce bronchoconstriction. Even if the afferent signal would not crossover, a unilateral input should give a bilateral output, due to the bilateral efferent innervation.

RARs at a different location may have a different reflex function. In the above experiments (31), the tip of the bronchial tube passed the carina and was in the main stems of the right and left bronchi. Thus, a large portion of a special group of RARs which reflexly induce cough was excluded. These results cannot exclude that those RARs can produce bronchoconstriction. In a further study, activation of RARs including those at the carina was examined in anesthetized, open chest, and artificially ventilated rabbits (33), in which total lung resistance and dynamic lung compliance were measured. Once again, activation of RARs by decreased lung compliance did not increase the resistance.

3. FURTHER EVIDENCE TO REFUTE THE HYPOTHESIS

It is widely accepted that stimulation of RARs by mechanically tickling the trachea can cause cough and bronchoconstriction (21, 22). Experimentally, directly poking the receptor field evokes a surging discharge of the RARs. It could be argued that this surging discharge of RARs could be responsible for cough and airway constriction; the failure to observe both cough and airway constriction during activation of RARs by decreasing lung compliance (31, 33) could be due to the absence of a surging discharge of RARs. In the present study, the association between cough and airway constriction during tickling the trachea was examined.

Rabbits were anesthetized with a mixture of 1% α chloralose and 10% urethane intravenously and were artificially ventilated. Airflow was monitored with a Fleisch pneumotachograph. The flow signal was integrated to give tidal volume. Airway pressure was also measured, and resistance of the total respiratory system and thoracic compliance were calculated by a Pulmonary Function Analyzer (Buxco, LS-20). Mucosa in the upper portion of the trachea was stimulated by strokes of a cotton tip applicator (tickling).

In the artificially ventilated rabbits, tickling the airway mucosa evoked cough, which was manifested by active respiratory movements. Airway pressure swing increased. It became negative during the deflation phase of the ventilator (inspiratory movement of the rabbit) and more positive during the inflation phase of the ventilator (expiratory movement), suggesting that both inspiratory and expiratory breathing mechanisms were activated (Fig. 1). This increased airway pressure swing is not due to an increase in resistance, because both inspiratory and expiratory flows increased concomitantly with the increase in fluctuation of airway pressure (Fig. 1). Direct measurements of the resistance of the total respiratory system and thoracic compliance cannot be obtained during cough because of the vigorous movement and the bizarre signals in airflow and airway pressure (Fig. 1, see the breaths between arrows). Immediately after the signals returned to a discernible state (for example the breath after the second arrow in Fig. 1), the results did not show an increase in resistance. To make sure that there was no increase in the resistance, we paralyzed the rabbit and reexamined the cough reflex. After succinylcholine (1 mg, i.v.), tickling the airway did not increase airflow or airway pressure (Table 1), suggesting that during cough there was no reflex increase in the resistance. It is clear that any increase in airway pressure swing during tickling is due to cough action of respiratory muscle, not due to constriction of airway smooth muscle. Fig. 2 summarizes the effect of tickling on peak inspiratory airflows, and airway pressure swing in the experiments. Fig. 3 demonstrates that vagotomy abolished cough due to tickling. Thus, as expected, the cough in response to tracheal tickling was a reflex action mediated by the vagus nerve. In this preparation, stimulation of vagal efferent can increase airway resistance and intravenous injection of histamine can cause a vagally mediated airway constriction (data not shown), suggesting that a reflex increase in airway resistance could be detected in this preparation if present.

Table 1. Effects of tracheal tickling on lung mechanics in close chest, artificially ventilated rabbits (n = 8) after paralysis

	Control	Tickling
F_{pi} (ml/s)	25.5 ± 1.4	25.4 ± 1.4
F_{pe} (ml/s)	60.8 ± 3.1	61.2 ± 3.2
P_t (cm H_2O)	10.7 ± 0.4	10.9 ± 0.5
Compliance (ml/cm H_2O)	2.42 ± 0.11	2.36 ± 0.12
Resistance (cm H_2O·s/ml)	0.042 ± 0.003	0.040 ± 0.003

F_{pi}, peak inspiratory flow; F_{pe}, peak expiratory flow; P_t, tracheal pressure. Note that there is no change in all the measured variables.

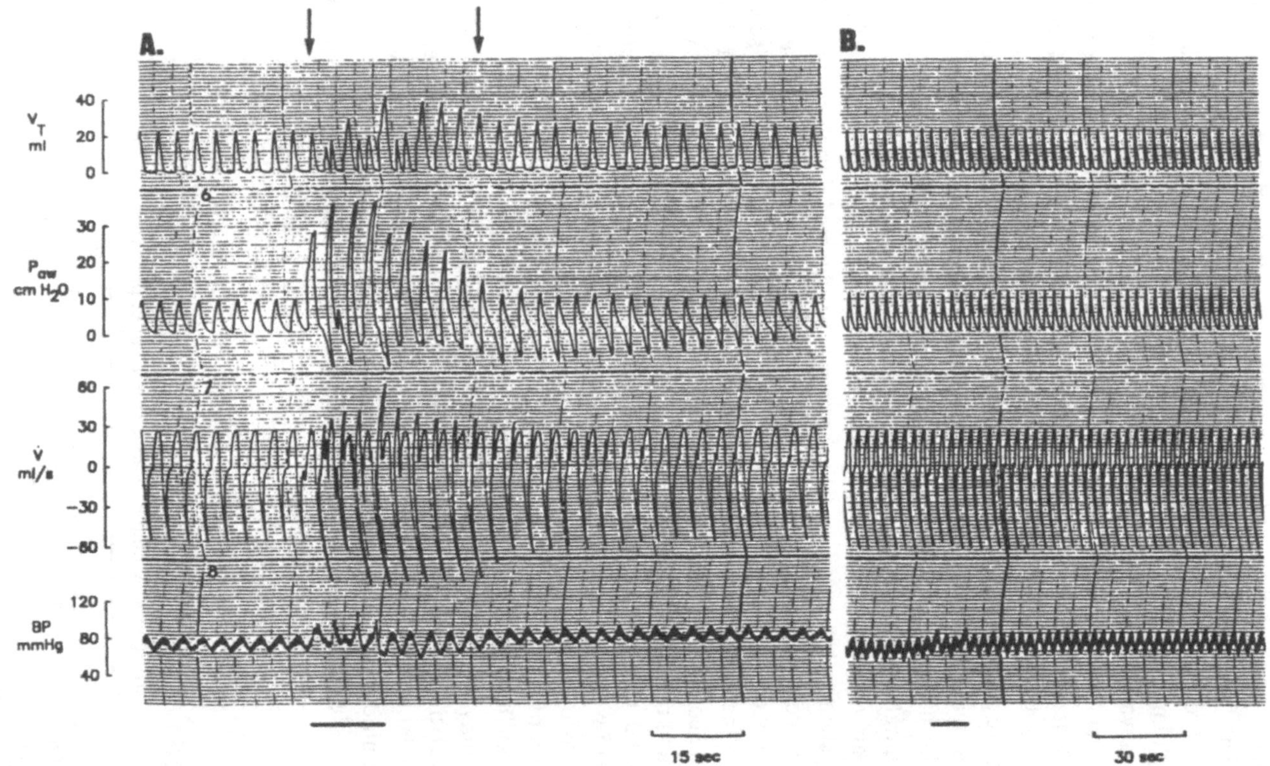

Figure 1. Effect of tickling tracheal mucosa on lung mechanics in an anesthetized, artificially ventilated rabbit before (A) and after (B) paralysis by intravenous injection of succinylcholine (1 mg). From top to bottom, the traces are tidal volume, V_T; airway pressure, P_{aw}; airflow, V; arterial blood pressure, BP. The horizontal bar at the bottom denotes the duration of tickling. Note that after paralysis, tickling did not increase airway pressure swing. In panel A, after second arrow the animal has an active inspiratory effort, demonstrated by a negative airway pressure during deflation phase of the ventilator. Note also that in B the paper speed was reduced to half of that in A.

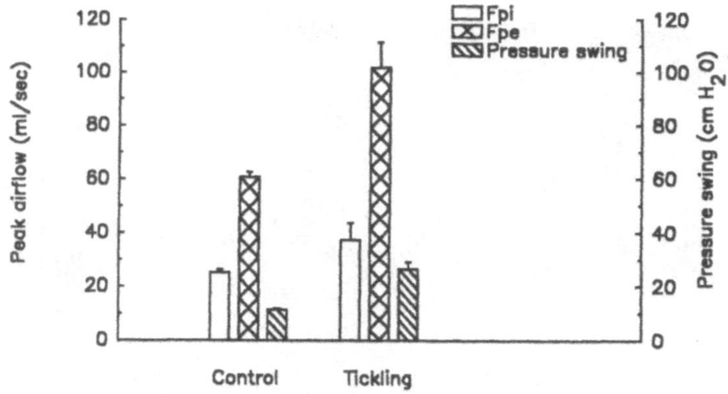

Figure 2. Effect of mechanical stimulation of trachea (tickling) on lung mechanics in artificially ventilated rabbits ($n = 17$). Fpi, peak inspiratory airflow; Fpe, peak expiratory airflow. Note that during tracheal tickling, peak inspiratory and expiratory airflows and airway pressure swing increased. The differences were significant for all variables ($p < 0.001$). Note that the increase in airway pressure swing is out proportion of the increase in airflow. This is due to the animal's inspiratory effort (see Fig. 1, negative airway pressure during the deflation phase).

4. DISCUSSION

The major finding in the present study is that mechanically probing the tracheal mucosa, which elicited cough, did not increase airway resistance. This lends further support to the notion that activation of airway RARs does not cause bronchoconstriction (31, 33).

Pulmonary congestion stimulates RARs and increases tracheal tone, which is blocked by cooling the vagus nerve to 8°C (RARs are blocked at this temperature) or by atropine (10). Thus, RARs are believed to be responsible for the increased tracheal tone. However, the reflex effects blocked at 8°C can only provide evidence for not rejecting the hypothesis that stimulation of RARs can produce bronchoconstriction in that experimental setting. At this temperature, the transmission of C-fiber activity is also greatly reduced. On the other hand, under these conditions C-fibers are also stimulated (1, 17), and C-fibers are known to cause bronchoconstriction (2, 4).

Although there is evidence to suggest that tickling airways may induce airway constriction, the evidence is not strong. In an early study, an overflow method was used to measure the volume as an indicator for airway resistance in the cat (28). This method of measuring resistance is indirect, as commented on by Karczewski and Widdicombe (11). In addition, the cats in those experiments were not paralyzed; any respiratory movement could have interfered with the results.

It is reported that tickling the trachea induced bronchoconstriction in anesthetized and paralyzed cats (27). The investigators assessed the bronchoconstrictive response by measuring total lung resistance determined from the slope of transpulmonary pressure vs airflow, in which the pressure signal related to lung volume change had been subtracted. They tickled the airway with a nylon fiber through a tracheal opening and observed an increase in the slope, which represented an increase in resistance. The authors concluded that the airways constricted during tickling. However, I am reluctant to interpret this finding as an increase in airway resistance, because the increased slope was due to a decrease in inspiratory airflow during tickling (fig. 9 in reference 27). When a paralyzed animal is ventilated with a volume cycled ventilator at a fixed inspiratory time and at a fixed tidal

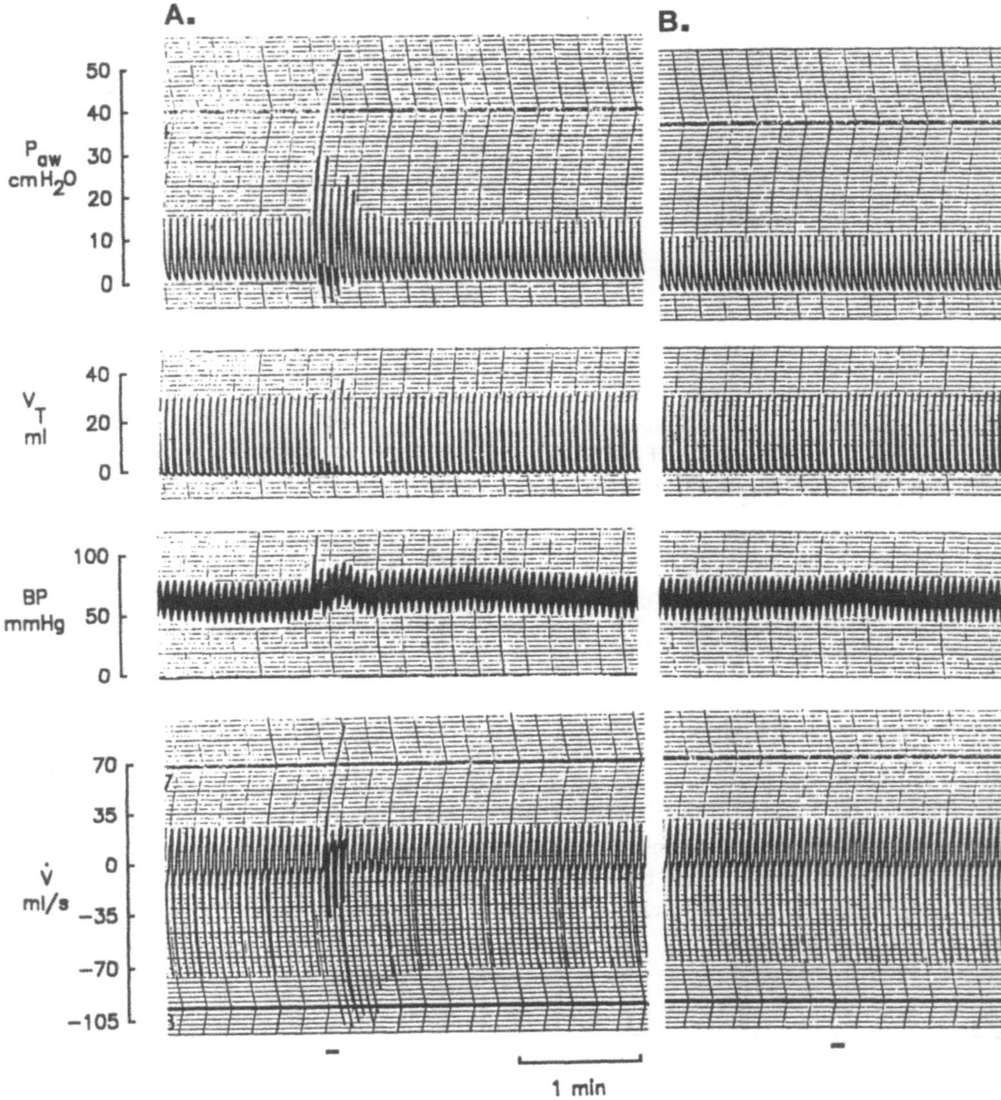

Figure 3. Bilateral vagotomy abolished the cough response to tracheal tickling in an artificially ventilated rabbit. A, vagal nerve intact; B, after vagotomy. See Fig. 1 legend for abbreviations.

volume, then inspiratory airflow should remain constant. It is not clear why airflow should decrease in such a ventilated preparation. Another frequently cited evidence is also from the paralyzed and artificially ventilated cat (15). It is reported that laryngeal stimulation increased airway resistance. However, during the stimulation the cat had a respiratory movement, exhibited by a negative transpulmonary pressure (fig. 6 in reference 15). As stated before, airway resistance cannot be well assessed with a respiratory effort by these techniques. Therefore, up to now there is no convincing evidence to show airway constriction by activation of RARs. Our data indicate that tickling the tracheobronchial tree or activating RARs by decreased lung compliance does not result in lower airway constriction.

While bronchoconstriction is often associated with cough, they are not necessarily mediated by the same afferents (5, 12). These two responses can be dissociated by aerosol of hypotonic saline (8) and distilled water (25). Thus, a question is raised whether cough and bronchoconstriction, are mediated through two different sets of receptors. In the present study, we have further shown that these two responses dissociate during the stimulation of tracheal mucosa by mechanical means. We believe that mechanical stimulation activates RARs, which elicits cough. C-fibers could also be activated by tickling in our experiment. Activation of C-fibers may also cause cough (2), although some investigators oppose this view (26, 29). Nevertheless, the activation would not be massive. If it had been massively activated, we would have observed a typical triad response: apnea followed by rapid shallow breathing, accompanied by bradycardia and hypotension.

In summary, activation of RARs by tickling the tracheal mucosa produced cough, but did not increase airway resistance. Together with previous reports (31, 33) where stimulation of RARs by decreased compliance did not increase airway resistance, it seems that airway constriction is not an important component of the reflexes evoked by RARs. On the other hand, it has been reported that C-fibers cause bronchoconstriction (2, 19). Thus, C-fibers may be the afferents responsible for the evoked reflex airway constriction from lower airways. If this is true, it solves the puzzle that there is a separation in cough and bronchoconstriction (12), and many other puzzles during the interpretation of results.

ACKNOWLEDGMENTS

This work was supported by grants from University of Louisville and Jewish Hospital Foundation.

REFERENCES

1. Armstrong, D.J., J.C. Luck and V.M. Martin. The effect of emboli upon intrapulmonary receptors in the cat. Respir. Physiol. 26:41–54, 1976.
2. Coleridge, J.C.G. and H.M. Coleridge. Afferent vagal C fibre innervation of the lungs and airways and its functional significance. Rev. Physiol. Biochem. Pharmacol. 99:1–110, 1984.
3. Coleridge, H.M. and J.C.G. Coleridge. Reflexes evoked from tracheobronchial tree and lungs. In: Handbook of Physiology, Section 3: The Respiratory System, Vol. II: Control of Breathing, part 1, edited by N.S. Cheniack and J.G. Widdicombe. Washington, D.C.: American Physiological Society, pp. 395–429, 1986.
4. Coleridge, H.M., J.C.G. Coleridge, and H.D. Schultz. Afferent pathways involved in reflex regulation of airway smooth muscle. Pharmac. Ther. 42:1–63, 1989.
5. Coleridge, H.M. and J.C.G. Coleridge. Pulmonary reflexes: neural mechanisms of pulmonary defense. Annu. Rev. Physiol. 56:69–91, 1994.
6. Davies, R.O. and L. Kubin. Projection of pulmonary rapidly adapting receptors to the medulla of the cat: An antidromic mapping study. J. Physiol. 373:63–86, 1986.
7. DeKock, M.A., J.A. Nadel, S. Zwi, H.J.H. Colebatch, and C.R. Olsen. New method for perfusing bronchial arteries: histamine bronchoconstriction and apnea. J. Appl. Physiol. 21:185–194, 1966.
8. Fuller, R.W. and J.G. Collier. Sodium cromoglycate and atropine block the fall in FEV_1 but not the cough induced by hypotonic mist. Thorax 39:766–770, 1984.
9. Gold, W.M., G.-F. Kessler, and D.Y.C. Yu. Role of vagus nerves in experimental asthma in allergic dogs. J. Appl. Phsyiol. 33:719–725, 1972.
10. Kappagoda, C.T., G.C. Man, K. Ravi, and K.K. Teo. Reflex tracheal contraction during pulmonary venous congestion in the dog. J. Physiol. Lond. 402:335–346, 1988.
11. Karczewski, W., and J.G. Widdicombe. The role of the vagus nerves in the respiratory and circulatory responses to intravenous histamine and phenyl diguanide in rabbits. J. Physiol. (London) 201:271–291, 1969.

12. Karlsson, J.A., G. Sant'Ambrogio, and J. Widdicombe. Afferent neural pathways in cough and reflex bronchoconstriction. J. Appl. Physiol. 65:1007–1023, 1988.
13. Lipski J., K. Ezure and R.B. Wong She. Identification of neurons receiving input from pulmonary rapidly adapting receptors in the cat. J. Physiol. (London) 443:55–77, 1991.
14. Mills, J.E., H. Sellick and J.G. Widdicombe. Activity of lung irritant receptors in pulmonary microembolism, anaphylaxis and drug-induced bronchoconstrictions. J. Physiol. (London) 203:337–357, 1969.
15. Nadel, J.A. and J.G. Widdicombe. Reflex effects of upper airway irritation on total lung resistance and blood pressure. J. Appl. Physiol. 17:861–865, 1962.
16. Pack, A.I. Sensory inputs to the medulla. Ann. Rev. Physiol. 43:73–90, 1981.
17. Paintal, A.S. Vagal sensory receptors and their reflex effects. Physiol. Rev. 53:159–226, 1973.
18. Paintal, A.S. The nature and effects of sensory inputs into the respiratory centers. Federation Proc. 36:2428–2432, 1977.
19. Roberts, A.M., M.P. Kaufman, D.G. Baker, J.K. Brown, H.M. Coleridge, and J.C.G. Coleridge. Reflex tracheal contraction induced by stimulation of bronchial C-fibers in dogs. J. Appl. Physiol. 51:485–493, 1981.
20. Sampson, S.R. and E.H. Vidruk. Chemical stimulation of rapidly adapting receptors in the airways. Adv. Exp. Med. Biol. 99:281–290, 1978.
21. Sant'Ambrogio, G. Afferent pathways for the cough reflex. Bull. Eur. Physiopathol. Respir. 23:19s-23s, 1987.
22. Sant'Ambrogio, G. Afferent nerves in reflex bronchoconstriction. Bull. Eur. Physiopathol. Respir. 23:81s-88s, 1987.
23. Sellick, H., and J.G. Widdicombe. The activity of lung irritant receptors during pneumothorax, hyperpnoea and pulmonary vascular congestion. J. Physiol. (London) 203:359–381, 1969.
24. Sellick, H. and J.G. Widdicombe. Stimulation of lung irritant receptors by cigarette smoke, carbon dust, and histamine aerosol. J. Appl. Physiol. 31:15–19, 1971.
25. Sheppard, D., N.W. Rizk, H.A. Boushey and R.A. Bethel. Mechanism of cough and bronchoconstriction induced by distilled water aerosol. Am. Rev. Respir. Dis. 127:691–694, 1983.
26. Tatar, M., S.E. Webber and J.G. Widdicombe. Lung C-fiber receptor activation and defensive reflexes in anaesthetized cats. J. Physiol. (London) 402:411–420, 1988.
27. Tomori, Z., J.G. Widdicombe. Muscular, bronchomotor and cardiovascular reflexes elicited by mechanical stimulation of the respiratory tract. J. Physiol. (London) 200:25–49, 1969.
28. Widdicombe, J.G. Respiratory reflexes from the trachea and bronchi of the cat. J. Physiol. (London) 123:55–70, 1954.
29. Widdicombe, J.G. Neurophysiology of the cough reflex. Eur. Respir. J. 8:1193–1202, 1995.
30. Yu, J., J.C.G. Coleridge and H.M. Coleridge. Influence of lung stiffness on rapidly adapting receptors in rabbits and cats. Respir. Physiol. 68:161–176, 1987.
31. Yu, J., and S. Mink. Activation of pulmonary rapidly adapting receptors does not induce bronchoconstriction in dogs. J. Appl. Physiol. 80:233–239, 1996.
32. Yu, J., H. Schultz, J. Goodman, J.C.G. Coleridge, H.M. Coleridge and B. Davis. Pulmonary rapidly adapting receptors reflexly increase airway secretion in dogs. J. Appl. Physiol. 67:682–687, 1989.
33. Yu, J., J.F. Zhang, A.M. Roberts, L. Collins and E.C. Fletcher. Effects of activation of airway rapidly adapting receptors (RARs) on airway resistance in the rabbit. Am. J. Respir. Crit. Care Med. 153:A846, 1996.

THE EFFECT OF EUCAPNIC AND ISOCAPNIC VOLITIONAL HYPERVENTILATION UPON BREATHLESSNESS

Andrew Binks and James Reed

Department of Physiological Sciences
Medical School
Framlington Place
University of Newcastle upon Tyne
Newcastle upon Tyne, NE2 4HH, United Kingdom

1. INTRODUCTION

Under given circumstances breathlessness increases with increasing ventilation, and is usually estimated with respect to the prevailing level of ventilation. Under normal reflex (medullary) control ventilation is set at a metabolically appropriate level; volitional (cortical) breathing allows inappropriate levels of ventilation. The role of such inappropriate ventilation in the generation of breathlessness has been previously addressed, but with conflicting results.

An inappropriately high level of volitional ventilation induced during hypercapnia has been reported to result in either a decrease in breathlessness (Adams *et al.* (1)), no effect or an increase (Schwartstwein (2)) or an increase (Chonan (3)). Chonan concluded that this increase could be due to an increased sense of effort.

Chonan's data was incorporated into Oku's model of respiratory sensation in 1995 (4), where it was suggested that isocapnic, voluntary hyperventilation above a spontaneously adopted level would increase the sensation of respiratory discomfort. However, Oku acknowledged the confusion and conceded that more physiological and psychological studies were needed to adequately describe the relationships between volitional, inappropriate ventilation, sensations of respiratory discomfort and the control of breathing.

The aim of this study was to investigate the effect of an inappropriately high level of ventilation, using exercise as a respiratory stimulant, in an attempt to simplify the situation by maintaining one of the unknown factors, the cortical drive to breath.

Advances in Modeling and Control of Ventilation, edited by Hughson *et al.*
Plenum Press, New York, 1998.

2. METHODS

2.1. Subjects

Subjects (n = 9, 20–38 years, 6 males) were recruited from the University. All were familiar with the laboratory procedures. They were informed as to the general protocols but were naive to the experimental purpose. Approval was gained from the local ethical commitee.

2.2. Experimental Protocol

The subjects' breathing pattern was recorded during a submaximal progressive exercise test upon a cycle ergometer (Rodby, 990). The test consisted of 20W increases in workload at the end of each minute after an initial rest period of two minutes. The test was ended when 75–80% of estimated maximal heart-rate was reached. The breathing pattern was digitally recorded from the flow signal of a pneumotachograph.

During the test breathing frequency (fR), tidal volume (Vt), expired ventilation (Ve), End-Tidal CO_2 ($PetCO_2$) and heart-rate (fC) were recorded continuously (Benchmark Pulmonary Exercise System, PK Morgan). At the end of each minute the subjects were asked to score their breathlessness using a Visual Analogue Scale (VAS). To allow for the effects of repetitive testing (17), all had undergone a minimum of three submaximal exercise tests over a period of approximately one week and were familiar with the use of the VAS. They were asked to score "*the sensation of breathlessness, the feeling of forced or laboured breathing or an air hunger; and not to score the feeling of their breathing increasing when they start to exercise.*"

Within the following 24–48 hours the subjects repeated the exercise test. They were asked to 'track' the previously recorded breathing pattern which was displayed on a monitor placed in front of them. An onscreen cursor, manipulated by their breathing, was used to follow a scrolling display of the original flow pattern. Although reflex respiratory drive would presumably still be present, under these circumstances the level and pattern of breathing would be determined cortically. On this occasion breathing was identical to the recorded, reflexly-driven initial exercise—since the experiment was not to primarily determine the effects of volitional breathing *per se* this trial was treated as a control.

In order to induce an inappropriately high level of ventilation, the subjects tracked the control breathing pattern but with one of two different work-load protocols:

1. level 1: the onset of workload increase was delayed by one minute. This resulted in effectively reducing the work-load, at any given time, by 20W compared to the control experiment. As the ventilation was the same as the control it was therefore inappropriately high (hyperventilation) for the imposed workload.
2. level 2: the onset of workload increase was delayed by two minutes. This resulted in lowering the workload by 40W at any given time during the test. This again resulted in the adoption of an inappropriately high level of ventilation, the degree of hyperventilation being approximately double that of level 1.

Hyperventilation by definition is associated with hypocapnia; as CO_2 is known to influence the sensation of breathlessness and is a potential confounding factor, the workload protocols were performed twice by each subject. On one occasion the $PetCO_2$ was unrestrained (eucapnia), whilst on the other occasion $PetCO_2$ was maintained at the same levels seen in the control trial (isocapnia). Isocapnia was maintained by increasing the level of CO_2 of the inspirate.

The differing workload protocols were presented in a randomised order.

Although the level of ventilation was inappropriately high during these experiments the pattern of breathing was appropriate for that ventilation.

2.3. Analysis

The accuracy of tracking was assessed by comparison, on a minute-by-minute basis, of the recorded data between the hyperventilation trials and the control. The relative change in ventilation was assessed by comparing the ventilations seen at each work-load. Differences were expressed as a percentage of control values and assessed for statistical significance using a paired two-tailed, t-test.

Breathlessness was scored as the intercept and slope of the linear relationship between VAS and minute ventilation.

Percentage changes were calculated between the onset and slope of the hyperventilation trials and the those of the control. Mean changes were calculated for the group data and assessed for statistical significance with a paired, two-tailed, t-test.

3. RESULTS

Percentage changes in relative ventilation are shown in Figure 1. A 20W reduction in work-load (level 1 hyperventilation) induced a hyperventilation of 8.3 ± 1.25% (mean ± sem) in the eucapnic state and 9.5 ± 1.24% in the isocapnic state. Figure 1 also shows the percentage changes in $PetCO_2$ for the group. During the eucapnic level 1 hyperventilation there was a fall in $PetCO_2$ (5.86 ± 0.58%), but no significant change during the isocapnic level 1 hyperventilation ($PetCO_2$, −0.19 ± 0.54%).

With the 40W reduction in workload (level 2 hyperventilation) a 14.98 ± 1.51% and 15.41 ± 1.34% relative hyperventilation was observed during eucapnia and isocapnia respectively (Figure 1). This was associated with a fall in $PetCO_2$ during eucapnia (10.06 ± 1.02%) with no significant change (−0.15 ± 0.29%) during the isocapnic trial.

With both eucapnic and isocapnic level 1 hyperventilation there were no significant differences from control values for either Vt or fR (Figure 2). During level 2 hyperventilation there was no change in fR during either eucapnia or isocapnia. There was a signifi-

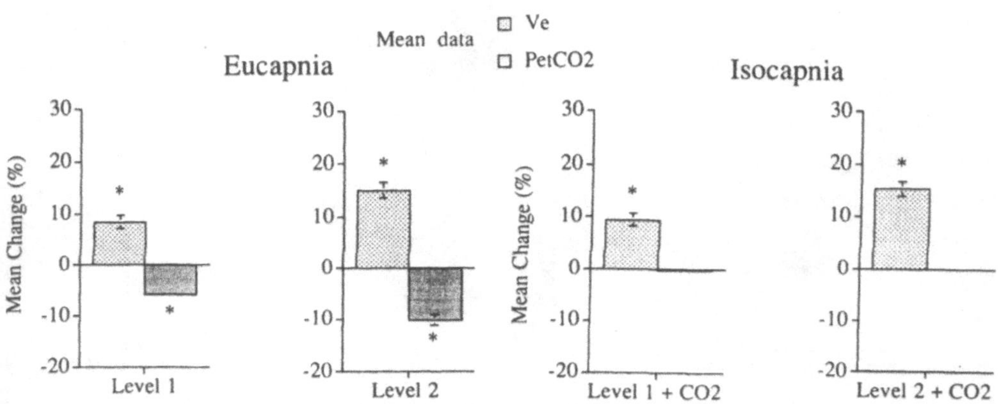

Figure 1. Mean percentage changes in relative Ventilation and End-tidal CO_2 ($PetCO_2$) during level 1 and level 2 eucapnic hyperventilations, and level 1 and level 2 isocapnic hyperventilations. n = 9. *$p < 0.05$.

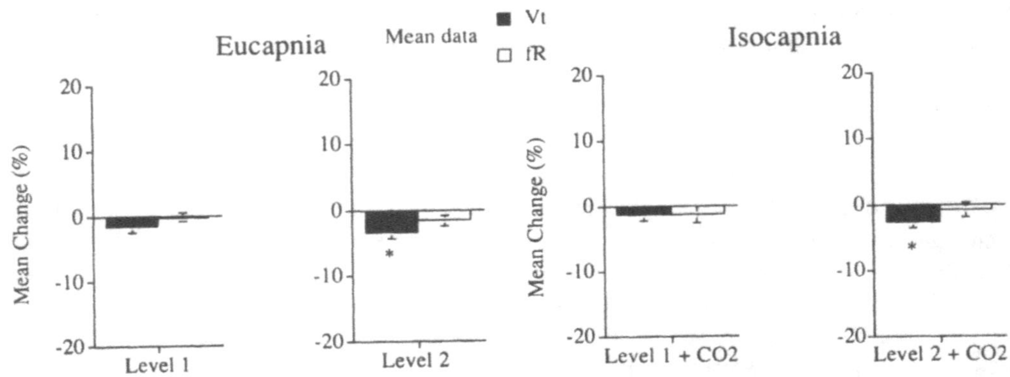

Figure 2. Mean percentage changes in Vt and fR during level 1 and level 2 eucapnic hyperventilations, and level 1 and level 2 isocapnic hyperventilations. n = 9. *$p < 0.05$.

cant change in Vt however, in both the eucapnic (−3.33 ± 0.88%) and isocapnic (−2.23 ± 0.89%) states (Figure 2).

There was no significant change in the VAS/Ve slope from the control values at either level of hyperventilation in either the eucapnic or isocapnic states. A significant delay in the onset of breathlessness was observed in both levels of eucapnic hyperventilation (Figure 3). During the level 1 hyperventilation the onset was delayed by 15.7 ± 3.2%, and in the level 2 hyperventilation by 20.8 ± 7.5%. Maintenance of $PetCO_2$ at the control levels resulted in no significant change in the onset of breathlessness at either level of hyperventilation (Figure 3).

4. DISCUSSION

The aim of the study was to investigate the effect of volitional hyperventilation upon breathlessness. To simplify a complex situation one potentially confounding factor was

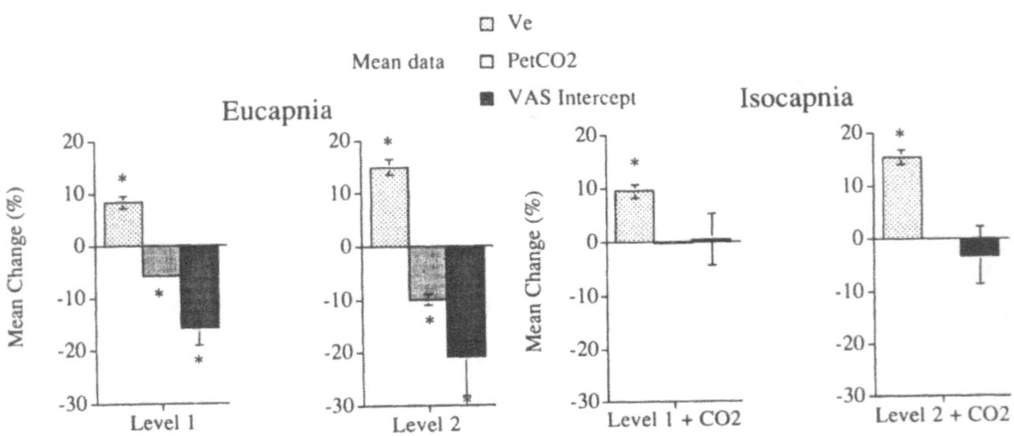

Figure 3. Mean percentage changes in relative Ventilation, $PetCO_2$ and the onset of breathlessness during two levels of hyperventilation in both eucapnia and isocapnia. n = 9. *$p < 0.05$.

controlled from the outset, ie the volitional drive to breath. At an initial trial the breathing pattern of the subjects was recorded during an incremental submaximal exercise test; this pattern of breathing was then reproduced at all subsequent experiments. Ventilation was under cortical control and an identical Ve/time profile was maintained throughout the study. It was assumed that the level of cortical drive would be the same during each of the individual trials.

The relationships between respiratory frequency, tidal volume and ventilation were kept constant throughout the trials. This reduced any effect that changes in breathing pattern might have upon breathlessness (3, 5, 6). However, at the higher levels of hyperventilation subjects did not achieve the desired level of Vt. This small, although statistically significant discrepancy appeared in both the eucapnic and isocapnic states to similar levels, it was not considered to have affected the interpretation of the data.

When subjects breathed at an inappropriately high level, and $PetCO_2$ was allowed to fall, a delay in the onset of breathlessness was seen. When $PetCO_2$ was maintained during the same level of hyperventilation then there was no change in breathlessness. There is some clinical evidence to suggest that the reflex and volitional drives are largely separate (7) However, during volitional breathing cortical efferents may inhibit the respiratory-related activity of the medulla (8) although it might be expected that some reflexly generated activity would remain (4). It is this residual activity which is of interest. When ventilation was raised above the metabolically appropriate level with no addition of CO_2, a fall in $PetCO_2$, (and presumably arterial CO_2), was observed. This hypocapnia would lead to a reduction in the excitatory afferent output of the chemoreceptors to the brainstem, and hence a further reduction in medullary respiratory activity. During this assumed reduction in medullary-derived respiratory output a reduction in breathlessness was observed. When the $PetCO_2$ was maintained at control levels no such reduction in chemoreceptive afferent activity would occur and reflex, medullary, activity would presumably remain at the same level. Under these circumstances, when medullary output was presumably maintained, there was no change in breathlessness. Given that cortical drive was constant, this is consistant with the hypothesis that breathlessness is a reflection of the reflex drive to breathe.

An alternative explanation would be that the reduction in breathlessness may have been caused by a direct effect of CO_2. This would assume however, that CO_2 is a specific dyspnogenic agent, a view which is still somewhat contentious (9–14).

The results in the present study are therefore in agreement with Adams; an inappropriately high, cortically-derived drive to breathe does not cause an increase in breathlessness but rather a decrease, perhaps through inhibition of medullary drive (15).

There are some differences in the experimental procedure between these studies and those of Schwartstein and Chonan (where an increase in breathlessness was observed). In the present study an inappropriately high ventilation was induced by a reduction in the respiratory stimulus whilst ventilation was held constant. In the studies of Schwartstein and Chonan the same effect was achieved by increasing ventilation and maintaining the respiratory stimulus. Whilst both methods resulted in inappropriately high levels of ventilation the differences in method might have contributed to the differences in outcome.

There were in addition differences in the stimulus, that is the level of hyperventilation. Both Chonan and Schwartstein induced hyperventilation in excess of 50% of the spontaneously adopted ventilation. The levels in the present study were much lower (between 5–15%). It could be argued that the hyperventilation in this study may have been insufficient to induce an increase in breathlessness but this would not explain the observed decrease. Isocapnic hyperventilation of 30% above spontaneous breathing produces similar results (16) to those presented here.

A significant contributing factor to all studies of respiratory sensation is the understanding by the subjects of what they are being asked to report. It has been demonstrated that various respiratory sensations can be distinguished and differentiated during studies such as these (10). In this and Adams' study the subjects were briefed as to which specific sensation they were expected to score. In Chonan's and Schwartstein's study there was apparently no such briefing, the assumption being that the subjects understood what sensation was to be scored. Chonan's conclusion that the sensation of respiratory effort contributed to the increase in discomfort may demonstrate that the studies have been investigating different aspects of respiratory discomfort.

Whilst there are differences in the experimental procedures used here and those used previously the results of this study would suggest that isocapnic volitional hyperventilation does not affect breathlessness. However, if $PetCO_2$ is allowed to decrease then the onset on breathlessness is delayed. It is not possible to determine whether the decrease in breathlessness is a direct result of the hypocapnia or as a consequence of a reduction in the reflex drive to breathe.

REFERENCES

1. Adams, L., *et al.*, Breathlessness during different forms of ventilatory stimulation: a study of mechanisms in normal subjects and respiratory patients. *Clinical Science*, 1985. 69: 663–672.
2. Schwartzstein, R.M., *et al.*, Breathlessness induced by dissociation between ventilation and chemical drive. *American Review of Respiatory* Disease, 1989. 139: 1231–1237.
3. Chonan, T., *et al.*, Effects of changes in level and pattern of breathing on the sensations of dyspnea. *Journal of Applied Physiology*, 1990. 69 (4): 1290–1295.
4. Oku, Y., *et al.*, Model of respiratory sensation and wilful control of ventilation. *Medical & Biological Engineering and Computing*, 1995. 33: 252–256.
5. Killian, K.J., *et al.*, Effect of increased lung volume on perception of breathlessness, effort and tension. *Journal of Applied Physiology*, 1984. 57(3): 686–691.
6. Sakurai, M., *et al.*, Effects of changes in breathing pattern on the sensation of dyspnea during loaded breathing. *American Review of Respiratory disease*, 1991. 143: A594.
7. Plum, F., Neurological integration of behavioural and metabolic control of breathing, in *Breathing, Hering-Breuer Centenary Symposium*, R. Porter, Editor. 1970, Churchill: London. 168–171.
8. Bassal, M. and A.L. Bianchi, Inspiratory onset or termination induced by electrical stimulation of the brain. *Respiration Physiology*, 1982. 50: 23–40.
9. Lane, R., L. Adams, and A. Guz, The effects of hypoxia and hypercapnia on perceived breathlessness during exercise in humans. *Journal of Physiology*, 1990. 428: 579–593.
10. Demediuk, B.H., *et al.*, Dissociation between Dyspnea and Respiratory effort. *American review of Respiratory Diseases*, 1992. 146:1222–1225.
11. Patterson, J.L., *et al.*, Carbon dioxide-induced dyspnea in a patient with respiratory muscle paralysis. *American Journal of Medicine*, 1962. 32: 811–816.
12. Plum, F. and R.J. Leigh, Abnormalities of Central Mechanisms. *Regulation of Breathing part 2*, ed. T.F. Hornbein. 1981, New York: Marcel Dekker. 989–1067.
13. Shea, S.A., *et al.*, Respiratory sensations in subjects who lack a ventilatory response to CO_2. *Respiration Physiology*, 1993. 93: 203–219.
14. Chonan, T., *et al.*, Sensation of dyspnea during hypercapnia, exercise, and voluntary hyperventilation. *Journal of Applied Physiology*, 1990. 68 (5): 2100–2106.
15. Adams, L., R. Lane, and A. Guz, Breathlessness during reflex and voluntary stimulation of breathing: a study of mechanisms in normals. *Clinical Science*, 1984. 66: 31P.
16. Oku, Y., *et al.*, Effects of changes in ventilation on respiratory discomfort during isocapnic exercise. *Respiration Physiology*, 1996. 104: 107–114.
17. Reed, J.W., M.M.F. Subhan, Effect of repetitive exercise on breathlessness. *Journal of Physiology*, 1994. 480p. 54p.

INFLUENCE OF LOW DOSE DOPAMINE ON THE HEART RATE AND VENTILATORY RESPONSES TO SUSTAINED ISOCAPNIC HYPOXIA

Albert Dahan[1] and Denham S. Ward[2]

[1]Departments of Anesthesiology and Physiology
Leiden University Medical Center
Leiden, The Netherlands
[2]Department of Anesthesiology and Electrical Engineering
School of Medicine and Dentistry
University of Rochester
Rochester, New York

1. INTRODUCTION

In awake humans, acute isocapnic hypoxia causes a rapid increase in both ventilation and heart rate. Although the acute ventilatory response is mediated primarily through a stimulatory drive from the carotid body, the heart rate response is more complex. There is apparently a heart rate depressing drive arising from the carotid bodies (CB), and a stimulation arising from a central effect of hypoxia, stimulation from vagal reflexes from the increased pulmonary ventilation and perhaps an increased drive from the aortic bodies (see ref. 1 for a review).

The ventilatory response to sustained isocapnic hypoxia is biphasic: an initial increase in ventilation ($\dot{V}_I$), also termed the acute hypoxic response (or AHR) is followed by a slow decline in $\dot{V}_I$ (hypoxic ventilatory decline or HVD) and steady-state ventilation, about 25 to 50% above prehypoxic resting values, is reached within 15 to 20 min. Dopamine, administered at a low dose (3 $\mu g\ kg^{-1}\ min^{-1}$), reduces the magnitude of the AHR.[2,3] Concomitantly, the magnitude of the HVD is reduced.[2,3] The AHR is affected due to inhibitory influences of dopamine at the site of the peripheral chemoreceptors (i.e., carotid bodies).[4,5] The absence of an intact carotid body drive is most probably the reason for the reduction or abolishment of the HVD.[2]

The heart rate response to acute hypoxia is also biphasic but the effects of variations in carotid body sensitivity is less well studied. In this study we investigated the influence of the reduction of the carotid body drive by dopamine on the heart rate (HR) response to sustained isocapnic hypoxia in healthy young volunteers.

Advances in Modeling and Control of Ventilation, edited by Hughson *et al.*
Plenum Press, New York, 1998.

2. METHODS

Nine healthy subjects (2 women, 9 men) participated in the protocol after University of Rochester IRB approval. In each subject one control hypoxic study preceded a hypoxic study during the intravenous infusion of dopamine (dose = 3 μg kg^{-1} min^{-1}). To study the heart rate and ventilatory responses to sustained isocapnic hypoxia we made use of a computer-driven dynamic end-tidal forcing system. The target end-tidal PO_2 ($P_{ET}O_2$) pattern was as follows: (1) 10 min at 105 Torr; (2) a rapid decrease to 45 Torr; (3) 20 min at 45 Torr. After each hypoxic exposure all subjects breathed a hyperoxic gas mixture (F_I O_2 > 0.8) for at least 10 min. The end-tidal PCO_2 was elevated 3 Torr above individual resting values and maintained at this level throughout the studies. For procedures and apparatus see also reference 2.

2.1. Data Analysis

Inspired minute ventilation ($\dot{V}_I$), $P_{ET}O_2$, $P_{ET}CO_2$, arterial Hb-O_2 saturation (derived from pulse oximetry; S_PO_2) and HR were determined and collected with the TIDAL software package.[6] The studies were evaluated by taking mean values of the data over identical time segments: period A: last two min of normoxia before sustained hypoxia; period B: min 3 and 4 of sustained hypoxia; period C: min 19 and 20 of sustained hypoxia. The HR and ventilatory responses to hypoxia were defined as changes in HR and $\dot{V}_I$ per % drop in S_PO_2.

The HR acute and sustained hypoxic sensitivities are defined as:

$$\frac{\Delta\ \text{HR [period B} - \text{period A]}}{\Delta\ S_pO_2\text{[period B} - \text{period A]}}$$

and

$$\frac{\Delta\ \text{HR [period C} - \text{period A]}}{\Delta\ S_pO_2\text{[period C} - \text{period A]}}.$$

The $\dot{V}_I$ acute and sustained hypoxic sensitivities are defined as:

$$\frac{\Delta\ \dot{V}_I\ \text{[period B} - \text{period A]}}{\Delta\ S_pO_2\text{[period B} - \text{period A]}}$$

and

$$\frac{\Delta\ \dot{V}_I\ \text{[period C} - \text{period A]}}{\Delta\ S_pO_2\text{[period C} - \text{period A]}}.$$

3. RESULTS

Mean values of end-tidal PCO_2, end-tidal PO_2, arterial Hb-O_2 saturation, tidal volume, respiratory frequency, inspired minute ventilation and heart rate for control and dopamine studies of periods A, B and C are collected in Table 1. In Figures 1A and 1B, we plotted the HR- and $\dot{V}_I$-responses as percentage of pre-hypoxic baseline values.

Table 1. Mean values of $P_{ET}CO_2$, $P_{ET}O_2$, S_PO_2, tidal volume (V_T), respiratory frequency (f), $\dot{V}_I$ and heart rate (HR) for control and dopamine experiments[a]

		Period		
		A (= pre-hypoxia)	B (= early hypoxia)	C (= late hypoxia)
$P_{ET}CO_2$ (Torr)	Control	41.6 ± 3.8	41.8 ± 3.8	42.1 ± 4.1
	Dopamine	42.2 ± 4.0	41.9 ± 4.1	42.1 ± 4.0
$P_{ET}O_2$ (Torr)	Control	102.5 ± 3.1	45.3 ± 0.5*	45.4 ± 0.6*
	Dopamine	100.6 ± 0.8	44.6 ± 0.4*	45.4 ± 0.6*
S_PO_2 (%)	Control	97.6 ± 1.1	81.7 ± 2.6*	81.2 ± 2.9*
	Dopamine	98.1 ± 0.9	80.1 ± 3.4*	78.6 ± 3.4*
V_T (L)	Control	0.75 ± 0.20	1.47 ± 0.61*	1.20 ± 0.60*
	Dopamine	0.66 ± 0.14	0.86 ± 0.24*	0.68 ± 0.13
f (breaths min^{-1})	Control	15.3 ± 3.5	17.1 ± 5.6	17.0 ± 4.4
	Dopamine	16.0 ± 4.5	16.0 ± 3.7	16.2 ± 3.2
$\dot{V}_I$ (L min^{-1})	Control	10.7 ± 3.1	24.0 ± 11.6*	18.5 ± 6.8*·#
	Dopamine	9.9 ± 2.7	12.8 ± 2.6*	10.5 ± 2.7
HR (beats min^{-1})	Control	65.0 ± 10.1	85.1 ± 11.9*	80.8 ± 12.2*·#
	Dopamine	66.5 ± 10.0	81.6 ± 10.7*	76.7 ± 10.6*·#

[a]Period A = last 2 min of normoxia before sustained hypoxia; Period B = min 3 and 4 of sustained hypoxia; Period C = min 19 and 20 of sustained hypoxia; values are mean ± SD; 2-way ANOVA (fixed model): *$P < 0.05$ *versus* Period A, and #$P < 0.05$ *versus* Period B.

3.1. $\dot{V}_I$-Responses

Both control and dopamine hypoxic responses had a biphasic nature (see Figure 2A). Dopamine reduced the acute hypoxic sensitivity from 0.8 ± 0.6 L min^{-1} $\%^{-1}$ to 0.2 ± 0.1 L min^{-1} $\%^{-1}$ (paired t test: $P < 0.05$) and the hypoxic ventilatory sensitivity from 0.5 ± 0.3 L min^{-1} $\%^{-1}$ to 0.03 ± 0.1 L min^{-1} $\%^{-1}$ ($P < 0.05$).

3.2. HR-Responses

The HR-responses were biphasic in control and dopamine studies (see Figure 2B). Dopamine reduced the acute hypoxic sensitivity from 1.3 ± 0.2 beats min^{-1} $\%^{-1}$ to 0.86 ±

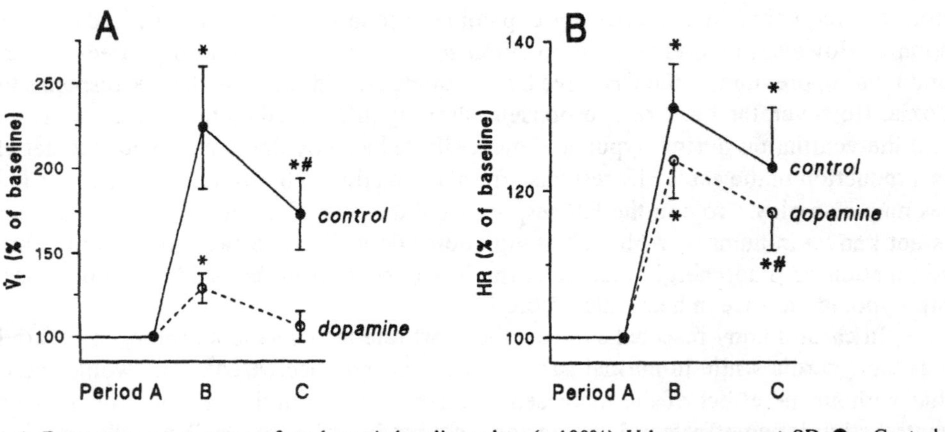

Figure 1. Responses as percentage of pre-hypoxic baseline values (= 100%). Values are mean ± SD. ● = Control; O = Dopamine. A: $\dot{V}_I$-data. B: HR-data.

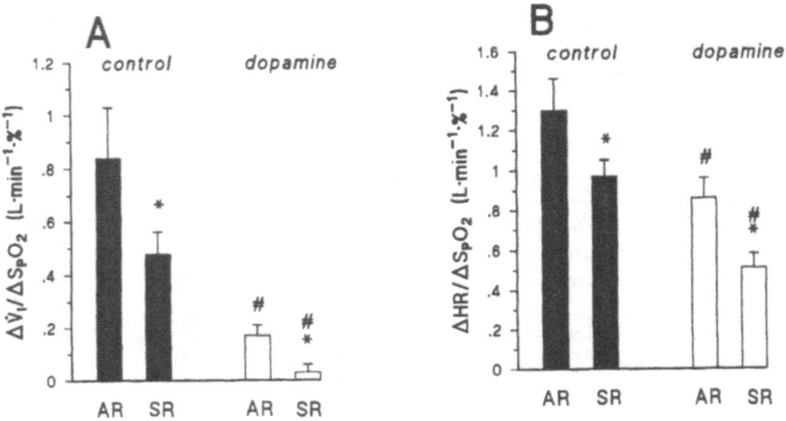

Figure 2. Acute (AR) and sustained (SR) hypoxic sensitivities for control (closed bars) and dopamine (open bars) studies. Hypoxic sensitivity is defined as the increase in $\dot{V}_I$ or HR per percentage drop in arterial Hb-O_2 saturation derived from pulse oximetry via a finger probe. ($\dot{V}_I/S_pO_2$ and HR/S_pO_2). Values are ± SD. A: $\dot{V}_I$-data. B: HR-data.

0.1 beats min^{-1} %$^{-1}$ ($P < 0.05$) and the sustained hypoxic sensitivity from 0.97 ± 0.1 beats min^{-1} %$^{-1}$ to 0.51 ± 0.1 beats min^{-1} %$^{-1}$ ($P < 0.05$).

3.3. HR-Responses *versus* $\dot{V}_I$-Responses

Dopamine reduced the HR-response to acute hypoxia by about 35%. In contrast the $\dot{V}_I$-response to acute hypoxia was reduced by 75%. The HR-response to sustained hypoxia was reduced by dopamine by about 45%, the equivalent $\dot{V}_I$-response was reduced by 95%.

4. DISCUSSION

While the effects of dopamine on the acute and sustained hypoxia on the ventilatory response and the heart rate response appear qualitatively similar, the mechanisms may be quite different. For example, while both hypercapnia and exercise increase the ventilatory response to hypoxia, neither increased the HR response[7]. Since the CB stimulates ventilation, the reduction in CB drive by dopamine accounts for the decreased ventilatory response. However, carotid body stimulation seems to have an inhibitory effect on the HR and thus suppression of the CB drive by dopamine should increase the HR response to hypoxia. However, the heart rate response is strongly influenced by the ventilatory response and the ventilation during hypoxia is markedly reduced by dopamine, Thus the net effect is a reduction in the acute HR response from a reduction in the ventilation. The aortic bodies may also play a role in the HR response and the role of dopamine in the aortic bodies is not known in humans. Although at high doses dopamine can cause tachycardia through stimulation of β adrenergic receptors, the low does used in this study did not cause any pre-hypoxia increase in heart rate (Table 1).

In carotid body resected subjects the heart rate response to a prolonged breath hold was tachycardia while in normal subjects the heart rate decreased[8]. This would indicated that without the effect of the increased ventilation the carotid body input overcomes the central stimulating effects of hypoxia and a net bradycardia results. When the carotid bodies are resected the unopposed central (or possibly aortic bodies) effect is tachycardia.

Since dopamine suppresses both the ventilation (Figure 1) and the carotid body drive during hypoxia, the resulting tachycardia comes from the unopposed central effect. The overall slight reduction in the response when both ventilation and carotid body drive is suppressed with dopamine would seem to indicate that the heart rate increasing effects of increased ventilation from hypoxia would seem to be slightly greater than the bradycardic effect of the carotid bodies.

The biphasic heart rate response to sustained hypoxia would seem to indicate that central hypoxia has a depressive effect similar to the effect on the ventilatory response. However, the central depressive effect of hypoxia on the ventilatory response seems to be mediated by modulating (i.e., reducing) the CB drive[2]. However, if central hypoxia worked to reduce the effects of the CB on the HR response then an increase in HR with sustained hypoxia should be seen. The protocol we used did not control for ventilatory changes and it is possible that the decrease in ventilation due to HVD was sufficient to overcome the stimulation that resulted by the central modulation of the CB input.

Further experiments with dopamine and fixed ventilation may allow for the roles of the CB drive and central hypoxia in the HR reduction with sustained hypoxia to be better defined.

5. CONCLUSIONS

1. The finding that the heart rate response to sustained isocapnic hypoxia is biphasic suggests that heart rate is affected by the central depressive effects of hypoxia in a similar fashion as minute ventilation.
2. Dopamine reduces the heart rate responses to acute and sustained isocapnic hypoxia. The reduction of HR responses may be related to the (dopamine-induced) reduction or absence of the ventilatory drive during acute and sustained hypoxia.

REFERENCES

1. De Burch Daly, M. Interactions between respiration and circulation. In: Handbook of Physiology, Section 3, Vol II Control of Breathing, Part 2. Alfred P. Fishman ed. American Physiological Society. Bethesda, 1986.
2. Dahan, A., D. Ward, M. van den Elsen, J. Temp, and A. Berkenbosch. Influence of reduced carotid body drive during sustained hypoxia on hypoxic depression of ventilation in humans. *J. Appl. Physiol.* 81: 565–572, 1996.
3. Ward, D.S., and M. Nino. The effects of dopamine on the ventilatory response to sustained hypoxia in humans. In: *Control of Breathing and its Modeling Perspective*, edited by Y. Honda, Y. Miyamoto, K. Konno, and J.G. Widdicombe. New York: Plenum Press, 1992, p. 291–298.
4. Llados, F., and P. Zapata. Effects of dopamine analogues and antagonists on carotid body chemosensors in situ. *J. Physiol. Lond.* 274: 487–499, 1993
5. Nishino, T., and S. Lahiri. Effects of dopamine on chemoreflexes in breathing. *J. Appl. Physiol.* 50: 892–897, 1981.
6. Jenkins, J.S., C.P. Valcke, and D.S. Ward. A programmable system for acquisition and reduction of respiratory physiological data. *Ann. Biomed. Eng.* 17: 93–108, 1989.
7. Sato, F., M. Nishimura, T. Igarashi, M. Yamamoto, K. Miyamoto, Y. Kawakami. Effects of exercise and CO_2 inhalation on intersubject variability in ventilatory and heart rate responses to progressive hypoxia. *Eur. Respir. J.* 9:960–967, 1996.
8. Gross, P.M., B.J. Whipp, J.T. Davidson, S.N. Koyal, and K. Wasserman. Role of the carotid bodies in the heart rate response to breath holding in man. *J. Appl. Physiol.* 41: 336–340, 1976.

ONDINE'S CURSE AND ITS INVERSE SYNDROME

Respiratory Failure in Autonomic vs. Voluntary Control

Fumihiko Yasuma,[1] Akiyoshi Okada,[2] Yoshiyuki Honda,[3] and Yoshitaka Oku[4]

[1]Department of Medicine
Suzuka National Sanatorium
3-2-1 Kasado, Suzuka, Japan 513
[2]Third Department of Internal Medicine
Nagoya City University School of Medicine
1 Kawasumi, Mizuho-cho, Mizuho-ku, Nagoya, Japan 457
[3]Department of Physiology
School of Medicine
Chiba University
1-8-1 Inoshishibana, Tyuou-ku, Chiba, Japan 260
[4]Department of Clinical Physiology
Chest Disease Research Institute
Kyoto University
53 Kawahara-cho, Seigoin, Sakyo-ku, Kyoto, Japan 606-01

1. INTRODUCTION

Breathing is controlled separately by the autonomic and voluntary pathways, which are, at least partially, anatomically different[1,2]. Rarely, a discrete lesion of the central nervous system may produce a selective paralysis of one type of respiration, but spare another. Recently, we encountered a patient with the paralysis of autonomic respiration (Ondine's curse) of unknown etiology, in whom the voluntary respiration remained intact. Then, we encountered another patient with its inverse clinical feature, in whom a localized, traumatic damage of the cerebral peduncle had induced a complete loss of voluntary respiration, while the autonomic respiration remained intact.

In this paper, we describe these two patients, which give us the insight into the dual-control system over ventilation with the autonomic and voluntary respiratory control.

Advances in Modeling and Control of Ventilation, edited by Hughson *et al.*
Plenum Press, New York, 1998.

2. CASE REPORT

2.1. CASE 1

A 53-year-old woman was referred to a city hospital due to dry cough, fever and restlessness. A chest X-ray showed cardiomegaly, pleural effusion and interstitial shadow of the right lung. Standard electrocardiography showed negative T waves in V1 to V4 chest leads. Arterial blood gas analysis during room air breathing showed hypercapnia and hypoxemia (pH 7.33, $PaCO_2$ 74.5 mmHg, PaO_2 32.4 mmHg). Three days later, she fell coma during oxygen inhalation, and mechanical ventilation was started. The etiology of the alveolar hypoventilation and heart failure had been investigated, however, the neurological, endocrine and orolaryngeal examinations, as well as the various imaging techniques, such as magnetic resonance imaging (MRI) of the brain and cervival spine, scintigraphy of the pulmonary blood flow and coronary angiography, were all normal. The mechanical ventilation could be weaned off in a month. Three months after the admission, she was discharged and back to the normal life.

Four months after the discharge, the patient was referred again due to the symptoms of common cold. Arterial blood gas analysis on admission during room air breathing showed hypercapnia and hypoxemia (pH 7.36, $PaCO_2$ 66.8 mmHg, PaO_2 42.2 mmHg). Three days later, she became comatous and mechanical ventilation was again initiated. Tracheostomy was subsequently performed. Two months later, her spontaneous ventilation

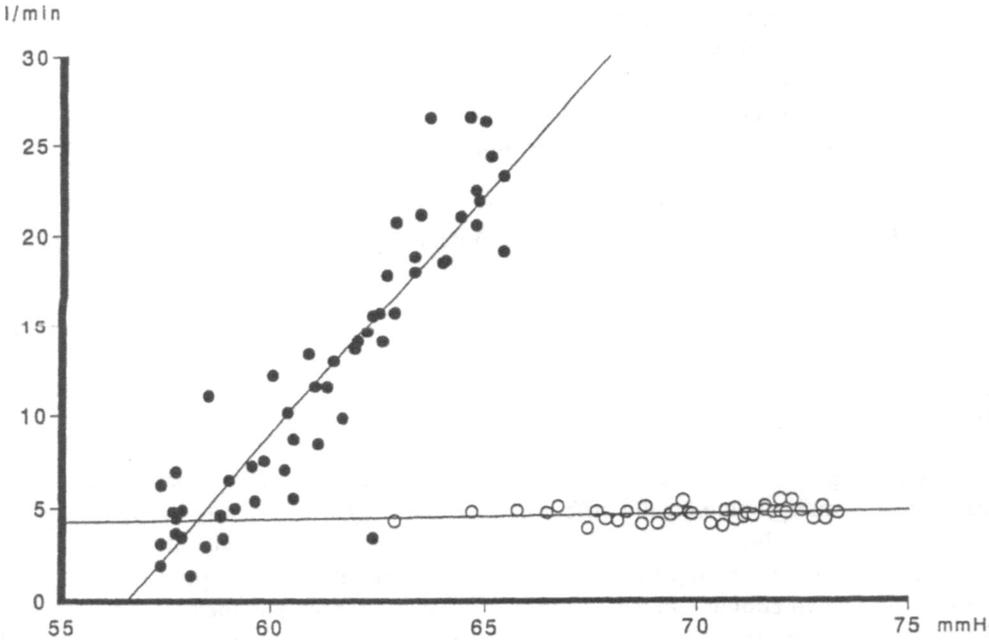

Figure 1. Ventilatory response to hyperoxic CO_2 rebreathing. Results of hyperoxic CO_2 rebreathing on breath-by-breath base are summarized for CASE 1 (○) and CASE 2 (●). There is no augmentation of ventilation in response to the progressive increase in $PetCO_2$ in CASE 1, in contrast, CO_2 rebreathing effectively augments ventilation in CASE 2. The horizontal axis represents PetCO2 (mmHg) and the vertical axis represents VI (l/min). Linear regression between $PetCO_2$ (=x) and VI (=y) is expressed as $y = 0.03x + 2.58$ ($p = 0.04$) in CASE 1, and $y = 2.63x - 148.8$ ($p = 0.83$) in CASE 2.

during wakefulness was normal, and, in contrast, it was significantlly suppressed during sleep. Therefore, she was treated with the nightly mechanical ventilation in the hospital.

Four months after the second admission, volitional hyperventilation and ventilatory response to hypercapnia were tested. She could hyperventilate following the oral instruction of the physician, and arterial blood gas level after hyperventilation lasting 3 min. was back to the normal range (pH 7.47, $PaCO_2$ 42.9 mmHg, PaO_2 90.6 mmHg). Hyperoxic CO_2 rebreathing by the Read method revealed no augmentation in ventilation or dyspnea sensation in response to the progressive increase in $PetCO_2$ (Figure 1, O)[3]. An overnight polygraphy showed hypoventilation without apnea. Accordingly, she was diagnosed as primary central alveolar hypoventilation syndrome (Ondine's curse)[4]. She was soon discharged and treated with the nocturnal mechanical ventilation at home.

2.2. CASE 2

A 51-year-old man was involved in the traffic accident while he was driving a car. He was transferred to the emergent intensive care unit due to coma and paralysis of the extremities. Computer tomography revealed the subarachnoid bleeding in the colliculus, and he was diagnosed as contusion of the brain and brainstem. Mechanical ventilation was started under orotracheal intubation and tracheostomy was subsequently performed. The level of his consciousness was improved to respond to the verval orders, however, the complete right-side hemiplegia and the incomplete paralysis of the left-side extremities were remained. He was unable to swallow or speak due to bulbar paralysis, therefore, the gastric tube feeding was maintained. He could be weaned off from the ventilator, three weeks after the admission, when he could communicate by an expression or by writing on the board with his barely functioning left hand. However, his respiration was completely irregular both in depth and frequency during wakefulness. He was unable to initiate breathing by voluntary effort, or to hyperventilate following the verbal instruction. MRI performed six weeks after the trauma revealed the localized, bilateral abnormal signals in the cerebral peduncle, lesion of which was predominant in the left side (Figure 2). This finding indicated dynfunction of the corticospinal tract (pyramidal tract), mediating the volitional respiration.

Eighteen months after the trauma, ventilatory response to hypercapnia was tested. Despite his irregular breathing, hyperoxic CO_2 rebreathing by the Read method effectively augmented his ventilation and dyspnea sensation[3] (Figure 1, ●). Arterial blood gas level during spontaneous room air breathing while awake was normal (pH 7.43, $PaCO_2$ 42.7 mmHg, PaO_2 74.5 mmHg). An overnight polygraphy showed more regular breathing than that during wakefulness, without apnea or oxygen desaturation. Above findings lead us to diagnosis him the selective paralysis of the voluntary respiration (inverse syndrome of Ondine's curse), while the autonomic respiration remained intact. He currently continued his rehabilitation program to improve the motor functions of the left-side extremities .

3. DISCUSSION

The center of the autonomic respiration is located in the pons and medulla oblongata, and that of the voluntary respiration is attributed to the cerebral cortex, The efferent pathway of the voluntary system is the corticospinal tract, passing through the cerebral peduncle and ventral basis pontis. The dual-control system over respiration with the autonomic and voluntary pathways is presented in Figure 3. Although the center (medullary respiratory networks and cerebral cortex) and the upper motoneuron originating from each

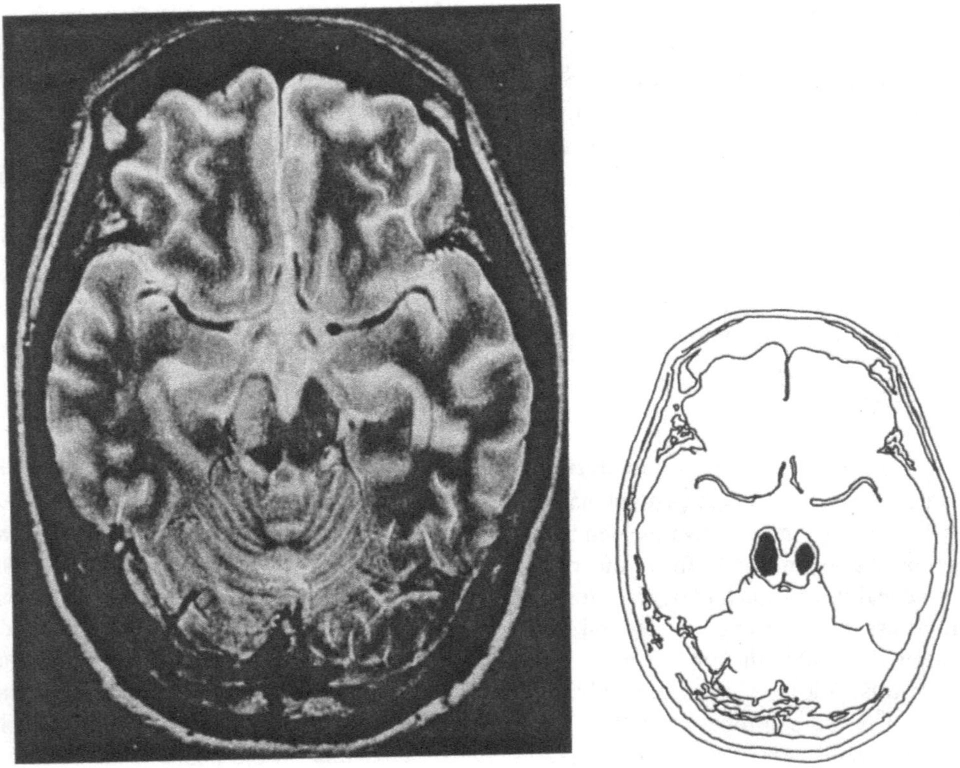

Figure 2. The T2-weighted magnetic resonance imaging in CASE 2. The MRI shows the abnormal signals in the bilateral cerebral peduncle, proximal to the basis pontis. This finding indicates that the lesions along the cortico-spinal tract (pyramidal tract) can selectively paralyze the voluntary respiration.

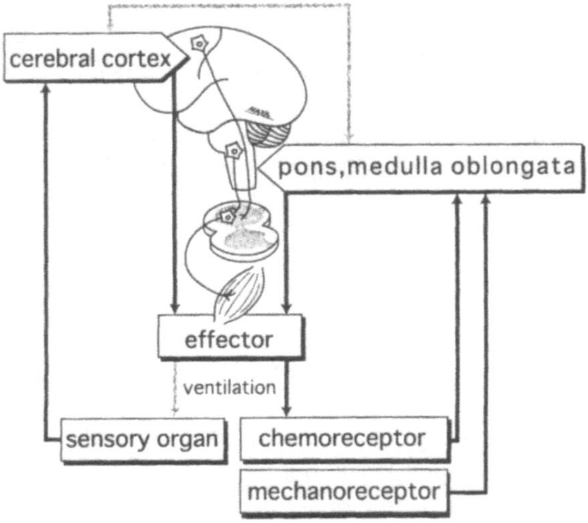

Figure 3. Conceptional framework of the respiratory control system. The voluntary respiratory control system is shown in the left, and the autonomicone is shown in the right side. Although the center of the voluntary and autonomic system, and the upper motoneuron originating from each center are anatomically different, the both systems have the lower motoneuron and effector organs in common.

center are anatomically different, the both systems have the lower motoneuron and effector organs in common.

A syndrome with the selective paralysis of the autonomic respiration was firstly named as Ondine's curse by Severinghaus and Mitchell[4] after German legend. Ondine of water nymph, having been jilted by Hans of her mortal husband, took all his autonomic functions, requiring him to remember to breathe. When Hans finally fell asleep, he died. The classical features of Ondine's curse were found in CASE 1, in whom the damage of the chemoreceptor or medullary respiratory networks was recognized with the blunted ventilatory response to hypercapnia[5]. Criteria to diagnose Ondine's curse are: 1. Alveolar hypoventilation corrected with voluntary hyperventilation or mechanical ventilation, 2. Blunted ventilatory response to hypercapnia, 3. Existence of sleep disordered breathing, and 4. Not having either apparent organic diseases or pharmacological effects, which induce alveolar hypoventiotion.

A syndrome with the selective paralysis of voluntary respiration, so called inverse syndrome of Ondine's curse, occurred mostly due to the brainstem infarction, in whom the volitional control of respiration was impossible, while the level of ventilation was maintained within normal range throughout a day.

The first case with the selective paralysis of the voluntary respiration, to our knowledge, was described by Meyer and Herndon in 1962[6]. The male patient with complete bilateral infarction of the pyramids and the adjacent ventromedial portion of the medulla exhibited tetraplegia and bulbar paralysis. He breathed spontaneously and regularly, however, he was unable to initiate any kind of voluntary movement below the neck, including breathing. Plum introduced a male patient with lacunar infarction of the cerebral hemispheres and basis pontis, who showed dysarthria and an uninterruptable autonomic breathing[7]. Although he retained the voluntary control over the extremities, he had completely lost it over the respiratory muscles. Feldman described a tetraplegic patient in 1971[8], who was chronically "locked-in" following a traumatic occlusion of the basilar artery. She was unable to carry out voluntary respiration, but was nevertheless able to laugh and cry despite her awareness of environment. Her resting blood gases were normal, and she showed a good ventilatory response to CO_2 stimulation with an increase in respiratory rate. The first of the four presented cases of "locked-in" syndrome by Nordgren and coworkers exhibited regular respiration at 14 per minute, over which he had no voluntary control[9]. The autopsy revealed bilateral infarction involved the basis pontis and both pyramical tract. Munschauer and his collegues reported that a male patient with medullary infarction exhibited tetraplegia and bulbar paralysis[10]. MRI demonstrated a well demarcated lesion restricted to the ventral basis pontis. The patient could not volitionally hold his breath, increase or decrease respiratory rate, or take large or small breath, despite obvious effort. However, his autonomic respiration was normal, showing normal ventilation during quiet wakefulness or sleep, as well as normal ventilatory response to hypercapnia. A couple of recent cases with the selective paralysis of voluntary respiration due to the brainstem infarction were described by Dawson and coworkers in 1994[11], or Heywood and coworkers in 1996[12]. The selective paralysis of voluntary respiration could occur due to the demyelinating lesion in the region of the cervicomedullary junction[13], multiple sclerosis[14], and the traumatic injury of the brainstem, as was shown in CASE 2 of the present investigation.

The above findings suggest that the voluntary breathing pathway may descend from the cerebrum to the lower bulbospinal level via the corticospinal pathways (pyramidal tract), lying along the cerebral peduncle and paramedian basis pontis. However, one cannot deduce whether the loss of voluntary respiratory control is resulted from interruption of the corticobulbar pathways to the medulla, or interruption of the corticospinal pathways to the spinal cord[1].

Therefore, diagnosis of the selective paralysis of voluntary respiration (inverse syndrome of Ondine's curse) should be based on the specific clinical features and the morphological changes along the pyramidal tract. Criteria to diagnose this syndrome are: 1. Inability of volitional control over respiration, 2. Normal ventilatory response to hypercapnia, 3. Not having either alveolar hypoventilation or sleep disordered breathing, and 4. Imaging or pathology showing the damaged corticospinal pathways (pyramidal tract).

4. CONCLUSION

We present a case with the selective paralysis of autonomic respiration and another with the selective paralysis of voluntary respiration, where we consider the conceptional framework of the respiratory control system.

ACKNOWLEDGMENTS

We thank for the kind comments by Dr. John W. Severinghaus, who first introduced the name of "Ondine's curse" in 1962[4], and attended to the Symposium on Advances in Modeling and Control of Ventilation in 1997, held in Huntsville, Ontario.

REFERENCES

1. Plum F., and R. J. Leigh. Abnormalities of central mechanisms. In T. F. Horbein., editor. Regulation of Breathing, Part 2. Lung Biology in Health and Disease Vol. 17. New York, Marcel Dekker Inc., 1981; 989–1067.
2. Newsom-Davis J., and F. Plum. Separation of descending spinal pathways to respiratory motor neurons. Exp. Neurol. 36:78–94, 1972.
3. Read D. J. C. A clinical method assessing the ventilatory response to carbon dioxide. Austral. Ann. Med. 16:20–32, 1971.
4. Severinghaus J. W., and R. A. Mitchell. Ondine's curse—Failure of respiratory center automaticity while awake. Clin. Res. 10:122, 1962.
5. Yasuma F., H. Nomura, I. Sotobata, H. Ishihara, H. Saito, K. Yasuura, H. Okamoto, S. Hirose, T. Abe, and A. Seki. Congenital central alvoelar hypoventilation (Ondine's curse); a case report and review of the literature. Eur. J. Pediatr. 143:81–83, 1987.
6. Meyer J., and R. M. Herndon. Bilateral infarction of the pyramidal tracts in man. Neurology. 12:637–643, 1962.
7. Plum F. Neurological integration of behavioral and metabolic control of breathing. In R. Porter., editor. Breathing: Hering-Breuer Centenary Symposium. (Ciba Foundation Symposium), London, England, J & A Churchill, 1970; 168–171.
8. Feldman M. H. Physiological observations in a chronic case of locked-in syndrome. Neurology. 21:459–478, 1971.
9. Nordgren R. E., W. R. Markesbery, K. Fukuda, and A. G. Reeves. Seven cases of cerebromedullospinal disconnection: The "locked-in" syndrome. Neurology. 21:1140–1148, 1971.
10. Munschauer F. E., M. J. Mador, A. Ahuja, and L. Jacobs. Selective paralysis of voluntary but not limbically influenced autonomic respiration. Arch. Neurol. 48:1190–1192, 1991.
11. Dawson K., M. D. Hourihan, C. M. Wiles, and J. C. Chawla. Separation of voluntary and limbic activation of facial and respiratory muscles in ventral pontine infarction. J. Neurol. Neurosurg. Psychiatry. 57:1281–1282, 1994.
12. Heywood P., K. Murphy, D. R. Corfield, M. J. Morrell, R. S. Howard, and A. Guz. Control of breathing in man: Insights from the "locked-in" syndrome. Respir. Physiol. 106:13–20, 1996.
13. Newsom Davis J. Autonomous breathing; report of a case. Arch. Neurol. 30:480–483, 1974.
14. Noda S., and H.Umezaki. Dysarthria due to loss of voluntary respiration. Arch. Neurol. 39:132, 1982.

CHEMOREFLEX MODEL PARAMETERS MEASUREMENT

R. M. Mohan,[1] C. E. Amara,[2] P. Vasiliou,[1] E. P. Corriveau,[1]
D. A. Cunningham,[2] and J. Duffin[1]

[1]Department of Physiology
University of Toronto
Toronto, Canada
[2]Department of Physiology
University of Western Ontario
London, Canada

1. INTRODUCTION

In 1966, Lloyd introduced a model of the control of breathing (12). In our current version this model, ventilation (V) is the sum of three components; central and peripheral chemoreflex drives to ventilation (Vc and Vp, respectively), and a basal ventilatory drive (Vb) dependent on state.

$$V = Vc + Vp + Vb$$

Vc is proportional (sensitivity Sc) to the central partial pressure of carbon dioxide ($PcCO_2$) when a threshold (Tc) was exceeded and 0 below it.

$$Vc = Sc(PcCO_2 - Tc);\ \{PcCO_2 > Tc\} \text{ and } Vc = 0;\ \{PcCO_2 < Tc\}$$

Similarly, Vp is proportional (sensitivity Sp) to the arterial partial pressure of carbon dioxide ($PaCO_2$) when a threshold (Tp) was exceeded and 0 below it.

$$Vp = Sp(PaCO_2 - Tp);\ \{PaCO_2 > Tp\} \text{ and } Vp = 0;\ \{PaCO_2 < Tp\}$$

Sp is not constant but varies rectangular hyperbolically with PaO_2. The rectangular hyperbolic relation is described by an area constant parameter (A), sensitivity asymptote (Spmin) and arterial partial pressure of oxygen asymptote PaO_2min.

$$Sp = Spmin + A/(PaO_2 - PaO_2min)$$

Advances in Modeling and Control of Ventilation, edited by Hughson *et al.*
Plenum Press, New York, 1998.

This model therefore describes the central and peripheral chemoreflexes in terms of their sensitivities and thresholds for the ventilatory response to carbon dioxide, and their variation with oxygen level.

These model parameters can be estimated using Read's (20) rebreathing technique modified by Duffin and McAvoy (6) as follows. First, rebreathing is started after a period of mild hyperventilation to lower body stores of carbon dioxide to a sub-threshold partial pressure of carbon dioxide. This modification allows detection of the central- and peripheral-chemoreflex thresholds and therefore determines separately both central- and peripheral-chemoreflex sensitivities as well as measuring the basal ventilation existing below the chemoreflex thresholds. The second modification is the utilisation of an oxygen feedback system to maintain iso-oxic conditions during rebreathing, so that variations of the chemoreflex parameters with oxygen can be determined by comparing several rebreathing tests at different iso-oxic levels.

The chemoreflex model sensitivity parameters can also be estimated using the dynamic end-tidal forcing technique. Computer driven alterations in inspired gases produce step changes in end-tidal partial pressures of carbon dioxide independent of ventilation at several levels of iso-oxia, and the resulting ventilatory responses are separated into central and peripheral components by exponential peeling.

Some investigators have found no difference in the central chemoreflex sensitivity to carbon dioxide measured using either steady-state techniques or rebreathing (4, 11, 14, 20, 21), whereas others have reported higher sensitivities to carbon dioxide with rebreathing compared with steady state (2, 10).

We compared central chemoreflex sensitivities measured using the modified rebreathing method, both with and without prior hyperventilation, and the dynamic end-tidal forcing technique. In addition, we used the modified rebreathing method to measure all of the model parameters.

2. METHODS

2.1. Rebreathing

The rebreathing apparatus and details of subject testing have been described in previous reports (6, 16). Two methods were used to analyze the resulting data. In the earlier studies we examined plots of ventilation vs. end-tidal partial pressure of carbon dioxide at each iso-oxic level to determine the breakpoints we interpreted as the peripheral- and central-chemoreflex thresholds (Fig. 3, upper). The segment below the first breakpoint defined the basal ventilation, and was fitted by a constant mean ventilation after discarding the initial equilibration data. A reduced major axis was fitted to the segment between the first and second breakpoints whose slope defined the peripheral-chemoreflex sensitivity. The segment above the second breakpoint was also fitted with a reduced major axis and its slope defined the sum of peripheral- and central-chemoreflex sensitivities. We varied the thresholds and minimized the sum of squares differences between measured and predicted values to best fit the line segments to the data.

More recently, we examined the plots of ventilation versus time and applied a model fit consisting of an exponential decay for points below the first breakpoint time and linear regression lines for the data points above the first breakpoint time (Figure 1, upper). The fitting process varied the time constant, starting value and ending value (Vb) for the exponential decay segment, and the breakpoints times and linear segment slopes to minimize

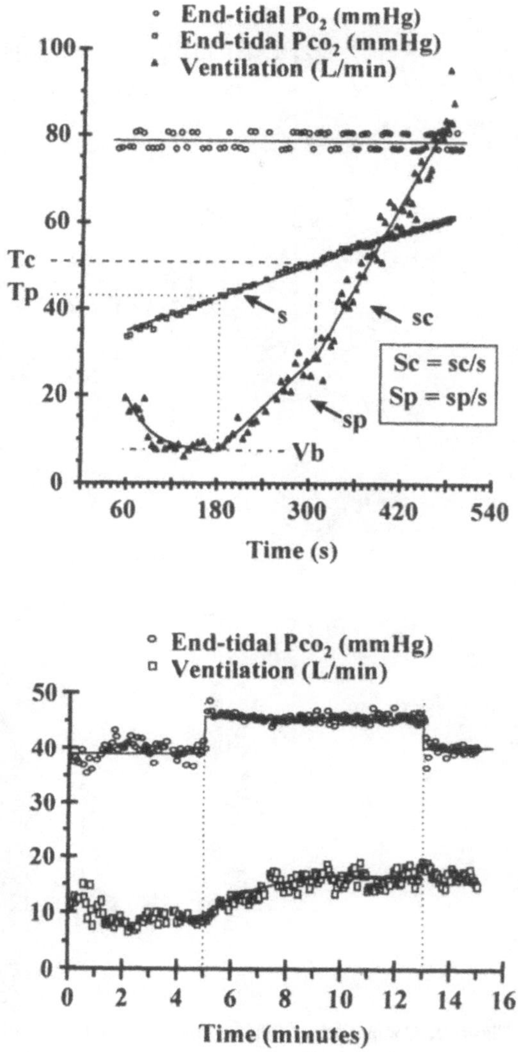

Figure 1. The variation of ventilation and the end-tidal partial pressures of carbon dioxide and oxygen with time during a modified rebreathing test and a dynamic end-tidal forcing test. Upper: A 3-segment model (see text) is fitted to the ventilation vs. time plot, and a single regression line is fitted to the end-tidal partial pressure of carbon dioxide vs. time plot. The chemoreflex sensitivities are calculated from the regression line slopes and their thresholds from the time breakpoints as shown. Lower: An exponential rise curve is fitted to the ventilation vs. time plot and a square wave to the end-tidal partial pressure of carbon dioxide vs. time plot. The final value of ventilation and the corresponding end-tidal partial pressure of carbon dioxide contribute one point to a plot of ventilation vs. end-tidal partial pressure of carbon dioxide.

the sum of squares differences between measured and predicted values. We plotted the times at which the breakpoints occurred on the graph of end-tidal partial pressure of carbon dioxide versus time to determine the chemoreflex thresholds (Tp and Tc), and divided the slopes of the regression lines fitted to the lines above the first breakpoint by the rate of rise of carbon dioxide to determine the chemoreflex sensitivities (Sp and Sc).

2.2. End-Tidal Forcing

For these experiments, we used a computer controlled fast gas mixing system similar to that described by Howson et al. (9) and Poulin et al. (18) to control end-tidal gases. In these experiments, the end-tidal partial pressure of carbon dioxide was held constant 1.5 mmHg above resting end-tidal partial pressure for 4 minutes, followed by a randomly selected square-wave increase of 3, 6, or 9 mmHg for 8 minutes, and a final 2 minutes recovery at the original end-tidal partial pressure. The end-tidal partial pressure of oxygen

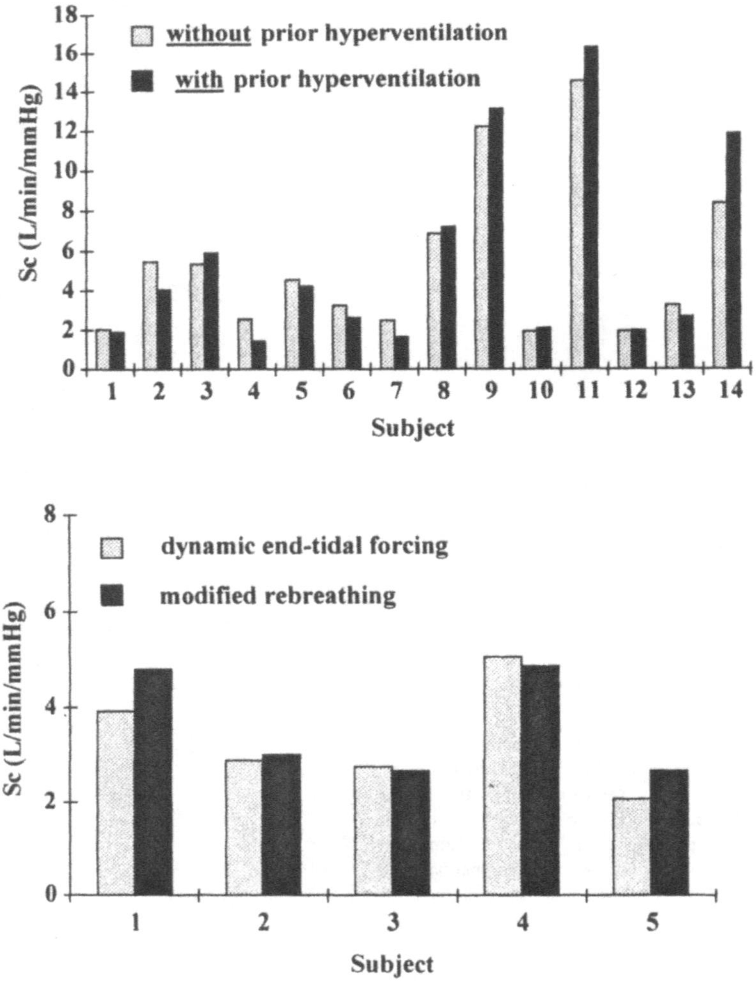

Figure 2. Comparisons of the central-chemoreflex sensitivity determined by different methods. Upper: Modified rebreathing method with a prior hyperventilation to that without. Lower: Modified rebreathing method to dynamic end-tidal forcing technique.

was held at 200 mmHg throughout the tests. Mean ventilation during the last minute of the initial 4 minute period and mean ventilation during the last 2 minutes of the 8 minute periods defined the central-chemoreflex sensitivity to carbon dioxide. Figure 1, (lower) shows an example of one ventilation response to a square wave of end-tidal partial pressure of carbon dioxide.

3. RESULTS AND DISCUSSION

3.1. Investigation 1

In this study we compared the central-chemoreflex sensitivity (Sc) measured with the modified rebreathing method, with or without a prior hyperventilation (randomly selected), at an iso-oxic level of 200 mmHg in 14 subjects. We found that the mean (SD) value of Sc, 5.5 (4.9) $L \cdot min^{-1} \cdot mmHg^{-1}$ during rebreathing with prior hyperventilation, was not significantly different from the mean value of 5.3 (3.9) $L \cdot min^{-1} \cdot mmHg^{-1}$ during rebreathing without prior hyperventilation (Figure 2, upper). These values are in agreement with previous measurements of 4.3 $L \cdot min^{-1} \cdot mmHg^{-1}$ made by this laboratory (6), and this finding supports the conclusion of other studies (3, 6, 20) that hyperventilation prior to rebreathing does not alter the central-chemoreflex sensitivity.

3.2. Investigation 2

In this study we compared the central-chemoreflex sensitivity (Sc) measured using the modified rebreathing technique with that measured using the dynamic end-tidal forcing method in 5 subjects. The subjects repeated each of the end-tidal forcing measurements 3 times on separate days to provide averaged values, and then performed one rebreathing measurement; all at iso-oxic end-tidal partial pressures of 200 mmHg.

For the end-tidal forcing experiments, Sc was determined from the slope of a line fitted to points seen to be above the threshold observed in the rebreathing experiments, often discarding the point 1.5 mmHg above the resting end-tidal partial pressure of carbon dioxide. The mean (SD) Sc, 3.3 (1.6) $L \cdot min^{-1} \cdot mmHg$, measured by the modified rebreathing method was not significantly different (t-test) from that of 3.4 (1.3) $L \cdot min^{-1} \cdot mmHg$ measured using the dynamic end-tidal forcing technique.

This finding agrees with Read's (20) original report which compared Sc measured using his rebreathing method to that obtained using the steady-state method as well as with similar findings reported by some other investigators (4, 11). However, some experimenters (2, 8, 10) found that Sc measured using rebreathing was greater than that measured using steady-state methods. We suggest that methodological differences can account for these different findings, based on the observations we made using the modified rebreathing method, and detail our arguments as follows.

None of the previous comparison studies measured the central-chemoreflex threshold. Although its presence may be inferred by extrapolating the ventilation vs. end-tidal partial pressure of carbon dioxide plots to zero ventilation, this end-tidal partial pressure intercept point is not the actual threshold. The actual threshold is the partial pressure at which ventilation equals basal ventilation. Nevertheless, a comparison of the intercepts between the two methods showed that the rebreathing method intercept was consistently greater than that for the steady-state method.

In making such a comparison it is important to realise that the rebreathing method considerably reduces the arterio-venous differences in blood gases, because of the initial equilibration at the onset of rebreathing, but in the steady-state method the normal existing arterio-venous difference is maintained. Therefore to compare the two responses, the rebreathing plot should be shifted to lower partial pressures of carbon dioxide (left) by about the arterio-venous difference existing at rest. When such a correction is applied to our data (e.g. 6–8 mmHg) the intercepts with the end-tidal partial pressure axis occur at 40–42 mmHg, similar to those observed using steady-state methods.

The difference in central-chemoreflex sensitivities between the two methods may be accounted for by a consideration of the hypocapnic region below the central-chemoreflex threshold, where basal ventilation is controlled primarily by neural inputs such as the "wakefulness" drive (7) and is unaffected by changes in end-tidal partial pressure of carbon dioxide. In this study we were able to observe the threshold directly during rebreathing, and so were able to avoid including points from the hypocapnic region in calculating the

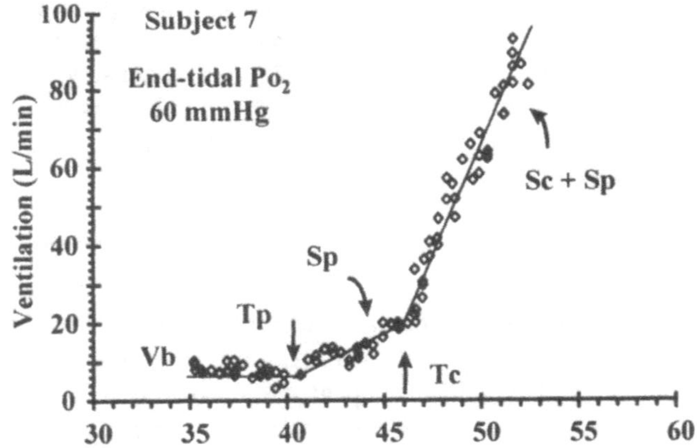

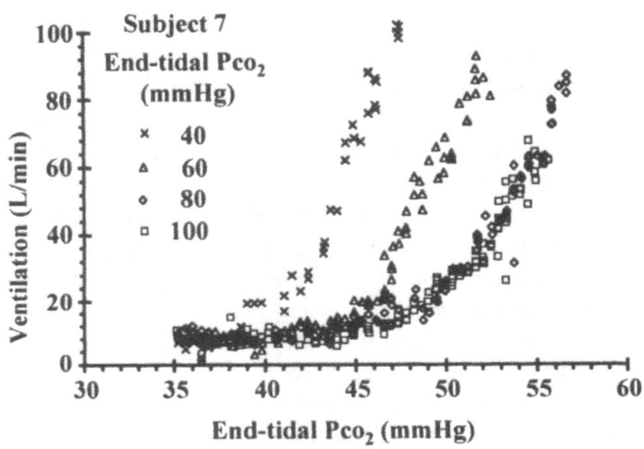

Figure 3. Ventilation vs end-tidal partial pressure of carbon dioxide during modified rebreathing. Upper: A single test at an iso-oxic end-tidal partial pressure of 60 mmHg showing the determination of the chemoreflex thresholds and sensitivities as well as the basal ventilation. Lower: Tests at 40, 60, 80 and 100 mmHg iso-oxic partial pressures in one subjects showing the variation with iso-oxic partial pressures.

central-chemoreflex sensitivity using the steady-state data. If these hypocapnic points were included in the calculation of central-chemoreflex sensitivity, the values were less than those measured using the rebreathing method, but if they were excluded as done in this study, the sensitivities were similar.

3.3. Investigation 3

We used the modified rebreathing method to measure the central- and peripheral-chemoreflex sensitivities (Sc and Sp) and thresholds (Tc and Tp) as well as Vb the ventilation in the hypocapnic region below the thresholds in 7 subjects (16). The rebreathing was carried out at iso-oxic end-tidal partial pressures of 40, 60, 80, 100 mmHg in order to examine the changes in chemoreflex parameters with hypoxia. Two other recent studies done for other purposes also used the modified rebreathing method at several iso-oxic end-tidal partial pressures to measures these same parameters.

Table 1. Parameter values for the chemoreflex control of breathing model

	Mean	SE	Number of subjects
Vb (L/min)	8.6	0.3	32
Tp (mmHg)	41	0.2	28
As (L.mmHg/min)	25	12	22
Sp min (L/min/mmHg)	0.7	0.2	22
PO_2 min (mmHg)	35	0.9	22
Tc (mmHg)	47	0.2	32
Sc (L/min/mmHg)	4.3	0.2	32

Figure 3 (upper) shows an example of a plot of ventilation vs. end-tidal partial pressure of carbon dioxide from an iso-oxic rebreathing test with the lines characterizing the response fitted to it. Single example plots for the same subject at each of the iso-oxic end-tidal partial pressures are also shown in Figure 3 (lower).

Basal ventilation (Vb) differed considerably between subjects in these studies, ranging from 0.9 to 28.4 $L \cdot min^{-1}$, with an overall mean (SD) of 8.6 (4.7) $L \cdot min^{-1}$ (Table 1); nevertheless for any particular subject mean Vb did not vary significantly with the iso-oxic end-tidal partial pressure (16). The latter finding is in agreement with some previous investigations but not others; similar disagreements occur between studies using steady-state methods (17) vs. (13) as well as between those using a progressive hypoxia technique (22) vs. (23). In some studies the actual thresholds were not known and hypocapnia was assumed to be sub-threshold, but may not have been, in which case hypoxia would affect ventilation. But in studies using steady-state methods where the coincident central and peripheral chemoreflex thresholds are discernible as a breakpoint ("dogleg" or "hockey stick"), some experimenters found a sub-threshold hypoxic ventilatory response (17) while others did not (13).

The present findings for basal ventilation agree with those of previous investigations carried out in our laboratory. Rapanos and Duffin (19) used rebreathing after hyperventilation to produce a progressive hypoxia at carbon dioxide end-tidal partial pressures below the chemoreflex thresholds and found no ventilatory response until carbon dioxide exceeded a threshold. Duffin and McAvoy (6) compared rebreathing tests at iso-oxic end-tidal partial pressures >150 mmHg with those at 75 mmHg and also showed that sub-threshold ventilation did not increase with hypoxia.

The peripheral-chemoreflex threshold (Tp) varied between subjects and also slightly between iso-oxic end-tidal partial pressures (16) ranging from 31 to 49 mmHg in these studies. The overall mean (SD) of 41 (3) mmHg (Table 1) is similar to a previous finding (39 mmHg) from this laboratory (6).

The peripheral-chemoreflex sensitivity (Sp) varied between subjects and between iso-oxic end-tidal partial pressures. Sp increased with hypoxia for all subjects, with most of the increase occurring at an iso-oxic end-tidal partial pressure of 40 mmHg. The model parameters describing the rectangular hyperbolic variation of Sp with PaO_2 (A, PaO_2min, Spmin) were determined by fitting the measurements of Sp made at different iso-oxic end-tidal partial pressures during the rebreathing tests for each subject, as shown in Figure 4. They are listed in Table 1.

The overall mean value of Sp for 7 subjects of 2.7 $L \cdot min^{-1} \cdot mmHg^{-1}$ at an iso-oxic end-tidal partial pressure of 80 mmHg (16) is similar to previous findings from this labo-

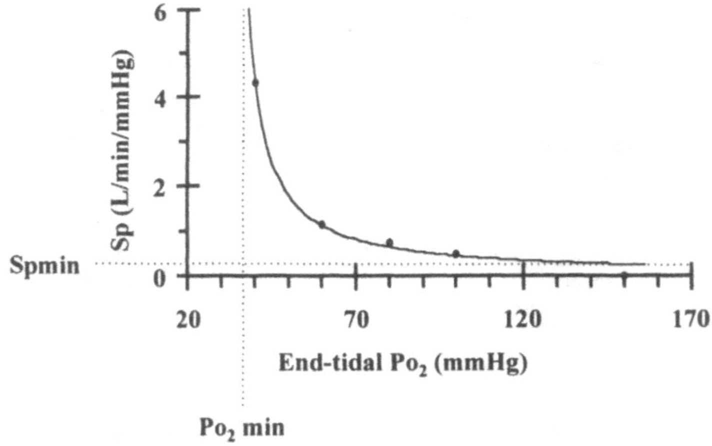

Figure 4. Peripheral-chemoreflex sensitivity vs. end-tidal iso-oxic partial pressure for a subject, showing the fitting of a rectangular hyperbola and the resulting model parameters (see text).

ratory of 3.5 $L{\cdot}min^{-1}{\cdot}mmHg^{-1}$ for 8 subjects at an iso-oxic end-tidal partial pressure of 75 mmHg (6). Using end-tidal forcing techniques, Poulin et al. (18) found a peripheral chemoreflex sensitivity of 2.16 $L{\cdot}min^{-1}{\cdot}mmHg^{-1}$ at an hypoxic end-tidal partial pressure of 50 mmHg in 7 subjects, comparable to our measurement of 3.0 $L{\cdot}min^{-1}{\cdot}mmHg^{-1}$ at 60 mmHg in 5 subjects (16). However, Dahan et al. (5) found a peripheral chemoreflex sensitivity in 9 subjects of 0.58 $L{\cdot}min^{-1}{\cdot}mmHg^{-1}$ in normoxia compared to ours of 2.4 $L{\cdot}min^{-1}{\cdot}mmHg^{-1}$ in 4 subjects; and 0.95 $L{\cdot}min^{-1}{\cdot}mmHg^{-1}$ at an hypoxic end-tidal partial pressure of 75 mmHg compared to ours of 2.7 $L{\cdot}min^{-1}{\cdot}mmHg^{-1}$. The latter differences may be due to differences in methodology, but likely represents variations between subjects.

The central-chemoreflex threshold (Tc) varied between subjects and slightly between iso-oxic end-tidal partial pressures (16), ranging from 34 to 58 mmHg. The overall mean (SD) of 47 (4) mmHg (Table 1) is similar to that (46 mmHg) found by Duffin and McAvoy (6).

The central-chemoreflex sensitivity (Sc) ranged from 0.2 to 20.4 $L{\cdot}min^{-1}{\cdot}mmHg^{-1}$ and while varying considerably between subjects, did not vary with iso-oxic end-tidal partial pressure (16). The overall mean (SD) of 4.3 (3.7) $L{\cdot}min^{-1}{\cdot}mmHg^{-1}$ is similar to our previous measurement of 4.3 $L{\cdot}min^{-1}{\cdot}mmHg^{-1}$ in 8 subjects (6) but greater than 1.8 $L{\cdot}min^{-1}{\cdot}mmHg^{-1}$ in 6 subjects (1). In general these measurements reflect the differences between individuals as found by other experimenters; for review see (15).

4. CONCLUSIONS

The rebreathing method described here, modified by including a prior hyperventilation and maintaining iso-oxia during rebreathing, allows rapid assessment of all the parameters necessary for describing a model of the chemoreflex control of breathing. Comparisons showed that the central-chemoreflex sensitivity is not altered by the inclusion of a prior hyperventilation. Other comparisons showed that the central-chemoreflex sensitivity measured using the dynamic end-tidal forcing technique and the modified rebreathing method were similar, providing that sub-threshold partial pressures of carbon dioxide are not included in the dynamic end-tidal forcing measurement.

REFERENCES

1. Baker, J. F., R. C. Goode, and J. Duffin. The effect of a rise in body temperature on the central-chemoreflex ventilatory response to carbon dioxide. *European Journal of Applied Physiology* 72: 537–541, 1996.
2. Berkenbosch, A., J. G. Bovill, A. Dahan, J. DeGoede, and I. C. W. Olievier. The ventilatory CO_2 sensitivities from Read's rebreathing method and the steady-state method are not equal in man. *Journal of Physiology* 411: 367–377, 1989.
3. Casey, K., J. Duffin, and G. V. McAvoy. The effect of exercise on the central-chemoreceptor threshold in man. *Journal of Physiology* 383: 9–18, 1987.
4. Clark, T. J. H. The ventilatory response to CO_2 in chronic airway obstruction measured by a rebreathing method. *Clinical Science* 34: 559–568, 1968.
5. Dahan, A., J. DeGoede, A. Berkenbosch, and I. C. Olievier. The influence of oxygen on the ventilatory response to carbon dioxide in man. *Journal of Physiology* 428: 485–499, 1990.
6. Duffin, J., and G. V. McAvoy. The peripheral-chemoreceptor threshold to carbon dioxide in man. *Journal of Physiology* 406: 15–26, 1988.
7. Fink, B. R. Influence of cerebral activity in wakefulness on regulation of breathing. *Journal of Applied Physiology* 16: 15–20, 1961.
8. Honda, Y., and M. Miyamura. Increased ventilatory response to CO_2 by rebreathing in consecutive daily trials. *Japanese Journal of Physiology* 22: 13–23, 1972.
9. Howson, M. G., S. Khamnei, M. E. McIntyre, D. F. O'Conner, and P. A. Robbins. A rapid computer controlled binary gas mixing system for studies in respiratory control. *Journal of Physiology* 394: 7P, 1987.
10. Jacobi, M. S., C. P. Patil, and K. B. Saunders. Transient, steady-state and rebreathing responses to carbon dioxide in man, at rest and during light exercise. *Journal of Physiology* 411: 85–96, 1989.
11. Linton, R. A., P. A. Poole-Wilson, R. J. Davies, and I. R. Cameron. A comparison of the ventilatory response to carbon dioxide by steady- state and rebreathing methods during metabolic acidosis and alkalosis. *Clinical Science & Molecular Medicine Supplement* 42: 239–49, 1973.
12. Lloyd, B. B. The interactions between hypoxia and other ventilatory stimuli. In: *Proc. Int. Symp. Cardiovasc. Resp. Effects of Hypoxia*, edited by J. D. Hatcher and D. B. Jennings. Basel: Karger, 1966, p. 146–165.
13. Lloyd, B. B., and D. J. C. Cunningham. A quantitative approach to the regulation of human respiration. In: *The Regulation of Human Respiration*, edited by D. J. C. Cunningham and B. B. Lloyd. Oxford: Blackwell, 1963, p. 331–349.
14. Lumb, A. B., and J. F. Nunn. Ribcage contribution to CO_2 response during rebreathing and steady state methods. *Respir Physiol* 85: 97–110, 1991.
15. McGurk, S. P., B. A. Blanksby, and M. J. Anderson. The relationship of hypercapnic ventilatory responses to age, gender and athleticism. *Sports Medicine* 19: 173–183, 1995.
16. Mohan, R., and J. Duffin. The effect of hypoxia on the ventilatory response to carbon dioxide in man. *Respiration Physiology* 108: 101–115, 1997.
17. Nielsen, M., and H. Smith. Studies on the regulation of respiration in acute hypoxia. *Acta Physiologica Scandinavica* 24: 293–313, 1952.
18. Poulin, M. J., D. A. Cunningham, D. H. Paterson, J. M. Kowalchuk, and W. D. F. Smith. Ventilatory sensitivity to CO_2 in hyperoxia and hypoxia in older aged humans. *Journal of Applied Physiology* 75: 2209–2216, 1993.
19. Rapanos, T., and J. Duffin. The ventilatory response to hypoxia below the carbon dioxide threshold. *Canadian Journal of Applied Physiology* 22: 23–36, 1997.
20. Read, D. J. C. A clinical method for assessing the ventilatory response to CO_2. *Australasian Annals of Medicine* 16: 20–32, 1967.
21. Soto Campos, J. G., S. Cano Gomez, J. Fernandez Guerra, M. Sanchez Armengol, F. Capote Gil, and J. Castillo Gomez. Hypercapnic stimulation and ventilation response in the syndrome of sleep obstructive apnea. Comparison of reinhalation and steady state. *Archive Bronconeumology* 32: 341–7, 1996.
22. Tenney, S. M., J. E. Remmers, and J. C. Mithoefer. Interaction of CO_2 and hypoxic stimuli on ventilation at high altitude. *Quarterly Journal of Experimental Physiology* 48: 192–202, 1963.
23. Weil, J. V., E. Byrne-Quinn, I. E. Sodal, W. O. Friesen, B. Underhill, G. F. Filley, and R. F. Grover. Hypoxic ventilatory drive in normal man. *Journal of Clinical Investigation* 49: 1061–1072, 1970.

VENTILATORY RESPONSE TO IMAGINATION OF EXERCISE AND ALTERED PERCEPTION OF EXERCISE LOAD UNDER HYPNOSIS

J. M. Thornton,[1] D. L. Pederson,[1] A. Kardos,[2] A. Guz,[1] B. Casadei,[2] and D. J. Paterson[1]

[1]University Laboratory of Physiology
Parks Road, Oxford, OX1 3PT
[2]University Department of Cardiovascular Medicine
John Radcliffe Hospital
Oxford, OX3 9DU

1. INTRODUCTION

Hypnotic suggestions have been used to assess the role of 'central command' in the ventilatory response to exercise. Some groups report an increase in ventilation (V_I) during imagined exercise under hypnosis (1) whereas others observe no significant ventilatory changes (2). The purpose of our study was to assess whether hypnosis can uncouple the role played by central command in exercise hyperpnoea. Some of these results have been presented in abstract form (3).

2. METHODS AND DISCUSSION

V_I and P_{ET} CO_2 were measured in seven subjects during imagination of exercise (protocol 1) and actual exercise on an electromagnetically-braked cycle ergometer (protocol 2), whilst under hypnosis. In protocol 1, resting hypnotised subjects were asked to imagine themselves exercising for 2 minutes at a heavy work rate. In protocol 2, hypnotised subjects exercised at a constant *moderate* work rate with the suggestion that the work rate became *harder*.

During imagined exercise (protocol 1) there was significant hyperventilation and hypocapnia (Figure 1, Table 1). During actual exercise (protocol 2), suggestion that the work rate had increased resulted in a significant increase in V_I and a fall in $P_{ET}CO_2$ (Figure 2, Table 1).

Advances in Modeling and Control of Ventilation, edited by Hughson *et al.*
Plenum Press, New York, 1998.

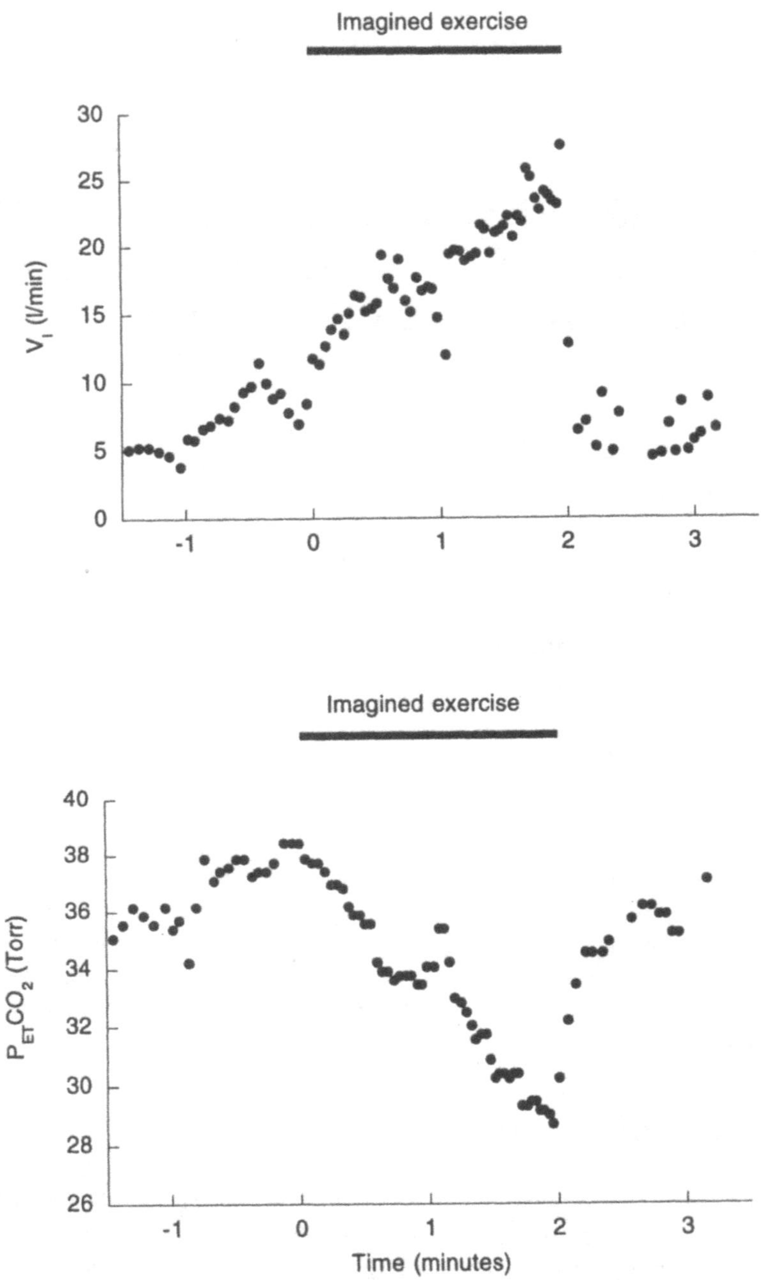

Figure 1. Breath-by-breath data from one subject. Imagination of heavy exercise for 2 minutes whilst under hypnosis (protocol 1) resulted in an increase in V_I (top panel) and a decrease in $P_{ET}CO_2$ (bottom panel).

In conclusion, our results suggest that 'central command' can drive ventilation during imagined exercise and actual exercise under hypnosis. Whether the neural sites activated during suggestion of exercise under hypnosis correlate with those previously identified as being important in the control of exercise hyperpnoea (cortical and sub-thalamic nuclei) remains to be determined.

Table 1. Mean (± SEM) data for protocols 1 and 2 (n = 7)[a]

	$P_{ET}CO_2$ (Torr)	V_T (litres)	f (bpm)	V_I (l/min)
Rest	39.5 ± 0.7	0.5 ± 0.1	14.6 ± 1.1	7.3 ± 0.6
Imagined exercise (protocol 1)	33.3 ± 1*	0.6 ± 0.1	27 ± 2.9*	15.1 ± 1.7*
Moderate exercise	43.9 ± 0.9	1.36 ± 0.2	18.2 ± 0.8	23.7 ± 2.9
Moderate, suggested heavy (protocol 2)	37.9 ± 1.7†	1.34 ± 0.1	26.9 ± 1.9†	34.3 ± 2.7†

[a]Protocol 1: * = significant difference between imagined exercise and rest. Protocol 2: † = significant difference between moderate (suggested heavy) and moderate exercise alone. ($p<0.05$, paired t-test).

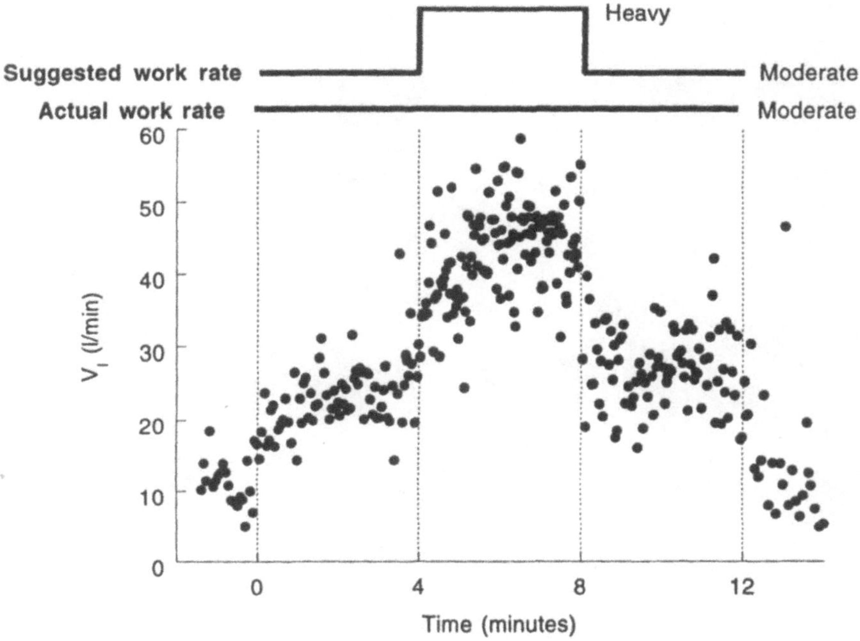

Figure 2. Breath-by-breath data from one subject showing the response to a suggested *increase* in exercise load (protocol 2) during 12 minutes of constant moderate intensity exercise.

ACKNOWLEDGMENT

This study was approved by the Central Oxford Research Ethics Committee.

REFERENCES

1. Arvidsson, T., Åström, H., Bevegård, S. & Jonsson, B. Circulatory effects of suggested leg exercise and fear induced under hypnotic state. Progr. Resp. Res. 5: 365–374. (1969).
2. Kraemer, W. J., Lewis, R. V., Triplett, N. T., Koziris, L.P., Hemyman, S. & Noble, B.J. Effect of hypnosis on plasma proenkephalin peptide F and perceptual and cardiovascular responses during submaximal exercise. Eur. J. Appl. Physiol. 65: 573–578. (1992).
3. Thornton, J. M., Pederson, D. L., Kardos, A., Guz, A., Casadei, B. and Paterson, D. J. Ventilatory response to the imagination of exercise and altered perception of exercise load under hypnosis. J Physiol. (1998) Abstract in press.

CARDIOLOCOMOTOR INTERACTIONS DURING DYNAMIC HANDGRIP AND KNEE EXTENSION EXERCISES

Phase-Locked Synchronization and Its Physiological Implications

Kyuichi Niizeki and Yoshimi Miyamoto

Laboratory of Biological-Informatics
Department of Electrical and Information Engineering
Faculty of Engineering
Yamagata University
Yonezawa, 992 Japan

1. INTRODUCTION

Coupling of locomotor and cardiac rhythms has been described during various locomotor activities in humans[1,5,6,7,9,10]. Although some prior work has attempted to clarify the advantage of the cardiac-locomotor coordination[1,6,9], direct evidence showing the functional significance of such locomotor modulation of heart beats has not yet been presented. It has generally been suggested that the synchronization phenomenon is a manifestation of nonlinear biological oscillators in which an inherent phase dependency to the periodically imposed input is involved[11]. Therefore, we thought that a phase dependency of cardiac rhythm with respect to the muscle contraction may also exist, and this would help us to infer the mechanism of coupling and its physiological significance. The purpose of this study was to investigate how the cardiac rhythm interacts with the muscle contraction rhythm during exercise, and to examine whether the coupling has such functional significance that ensures the exercising muscle blood flow.

2. METHODS

A total of 10 healthy male subjects (age 21–24 years) participated in this study. Dynamic handgrip (HG) and knee extension (KE) exercises were used to study changes in the

Advances in Modeling and Control of Ventilation, edited by Hughson *et al.*
Plenum Press, New York, 1998.

muscle blood flow. Both types of exercise consisted of 2 different protocols: one involved muscle contractions performed constant interval with an acoustic signal pacing, and the other exercise was synchronized to the heart beat. The data obtained from the heart beat-synchronized exercise was used for the phase response analysis as described below.

2.1. Handgrip Exercise

A subset consisting of 6 subjects participated in the HG experiment. Subjects performed rhythmic HG exercise at a load of 10% maximal voluntary isometric contraction (MVC) either with metronome-paced gripping or heartbeat-synchronized gripping. During the heartbeat-synchronized exercise, the subjects were asked to exercise so as to synchronize their exercise rhythm with their own heartbeats as closely as possible at a rate of one contraction to two heartbeats. To synchronize muscle contraction rhythm with heartbeats, the subjects listened a beep sound generated after a preset delay following the upstroke of the R wave of the ECG. The delay between the R wave onset and the beginning of the beep sound was changed stepwise to scan entire cardiac cycle.

2.2. Knee Extension Exercise

A subset consisting of 6 subjects participated in the KE experiment. The KE exercise was conducted in the seated position on a table with a back support in order to help relaxing of the subjects. The subject's right ankle was strapped to a supporting bar which was attached by a cable to a pulley system placed behind the subject. The rhythmic KE exercise were achieved by extending the right knee approximately from 90° to 120° against a 10 kg of weight (~30% MVC).

2.3. Data Acquisition

R-R interval (RRI), mean blood pressure (MBP), mean blood velocity (MBV) of brachial (for HG exercise) or femoral (for KE exercise) artery were measured beat by beat. The MBP was recorded continuously using the volume compensation method (Finapres, Ohmeda, 2300). The MBV responses were determined by pulsed Doppler ultrasound velocimetry (DVM 4200; Hadeco). A flat probe with an operating frequency of 5 MHz was used. For the HG exercise the probe was placed on the distal part of the upper arm over the brachial artery and for KE exercise it was fixed to the skin over the femoral artery ~5 cm distal to the inguinal ligament, with a beam angle of 45° with respect to the skin. The beat by beat MBV was determined on a computer by integrating the instantaneous blood velocity profile within one cardiac cycle. The signal of exercise onset was obtained from the onset of firing on the surface electromyography (EMG), monitored at the flexor carpi ulnaris muscle (for the HG exercise test) or the vastus lateralis muscle (for the KE exercise test) with bipolar electrodes. Those activities were amplified and R-C integrated. Respiratory flow was also monitored by a hot wire type flow meter (model RF-2, Minato) attached to the expiratory port of a breathing valve (model 7930, Hans Rudolph). The subjects were allowed to breath voluntarily. Analog waves of ECG, blood pressure, blood velocity and EMG signals were simultaneously recorded online on a FM DAT tape recorder (TEAC, RD-130TE).

2.4. Data Analysis

The recorded signals were digitized and sampled at 1 KHz on a PC-based system equipped with a 12-bit A/D converter. The time at which the R waves occurred, the MBP and MBV during one cardiac cycle were stored directly on a personal computer. For the qualitative evaluation of synchronization of the cardiac and exercise rhythms, relative phase transitions were calculated as previously described[9,10]. In brief, the period (Ts_i) at which the i-th onset of the integrated EMG wave appeared in one cardiac cycle was measured from the onset of the QRS complex. The phase difference ($\phi_{C,i}$) between the i-th MC and RRI (T) was then calculated as $\phi_{C,i} = Ts_i / T$. The synchronization implies that ϕ_C remains fixed at some particular value for a certain period. The chi-square (χ^2) values of the histogram of ϕ_C divided into 10 classes were calculated for every 20 consecutive point ϕ_C data sets, to determine whether the incidence of apparent coupling in the data differed significantly from that in the control condition. We defined synchronization when that the phase was distributed with a statistically significance χ^2 value and this condition lasted over three consecutive data set.

Changes in RRI, MBP, and MBV induced by one muscle contraction was expressed as a function of the timing of contraction in one cardiac cycle. Since the phase response is periodical, third order Fourier series regression curves were applied to the measured responses to characterize the phase dependency of the cardiovascular variables.

3. RESULTS

Figure 1 shows a representative tracing from the KE experiment in one subject, which includes RRI (A), MBP (B), MBV of femoral artery (C), and relative phase relationships between cardiac and exercise rhythms (ϕ_C, D). A temporal phase synchronization of the heartbeat and the knee extension rhythm was appreciable at a phase around 0.7 about 240 s after the experiment began, with the χ^2 value of 39.3 (Fig. 1D). In this period the fluctuation of MBV was decreased compared to those states of lost synchronization. Such significant coupling was also found in HG exercise. Figure 2 shows several characteristic phase patterns of synchronization during HG and KE exercises obtained from all the subjects. The identified coupling periods are indicated by closed circles. To examine whether there is a phase of cardiac cycle that favors spontaneous coupling to muscle contraction, we calculated the frequency histograms of the spontaneous locking phase which were obtained from all episodes of the coupling observed in HG ($n = 53$) and KE ($n = 55$) exercises, respectively (Figs. 3A and 3B). Although the coupling was identified at various phases during both exercises, the distributions were significantly different from a uniform one ($\chi^2 = 17.6$ for HG and 19.0 for KE, $p < 0.05$), i.e., the onset of the muscle contraction tended to cluster in the middle and the later half of the cardiac cycle.

We then analyzed how each cardiovascular variable varies depending on the relative phase between the onset of the muscle contraction and cardiac cycle. In this analysis, the respiratory effect on each measured variable were not be excluded, but we confirmed that the fluctuation of the RRI was well correlated with the muscle contraction rhythm independent of the respiratory rhythm by using power spectral analysis. Figures 4A–F show the group mean phase response curves (PRC) of RRI, MBP, and MBV. These are depicted as the averaged phase response by applying Fourier regression curve analysis including the standard deviation. All the variables were normalized by dividing by their mean value. When muscle contraction was given early in the cardiac cycle at around a ϕ_C of 0.2, this produces a shortening in the RRI, whereas muscle contraction given in the middle phase of the cardiac cycle or immediately before the R wave prolonged the RRI (Fig. 4A). The change in MBP appeared to reciprocate the change in RRI (Fig. 4B). The phase dependency of MBV was very clear; MBV was greatly restricted when muscle contraction occurs

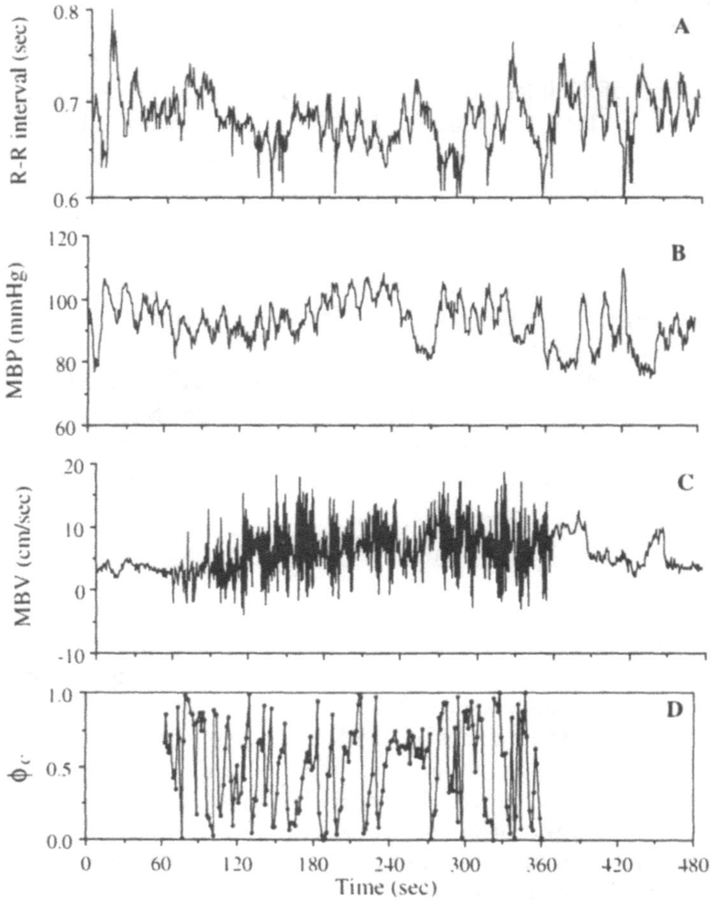

Figure 1. Typical examples of the R-R interval (RRI; A), mean blood pressure (MBP; B), mean blood velocity of the brachial artery (MBV; C), and the phase relationship between cardiac and muscle contraction rhythms (D) in a subject during knee extension exercise. Note that a phase locked synchronization occurred at periods of around 240 s after the experiment began.

in the systolic phase of the cardiac cycle and was increased when muscle contraction occurred in the later half of the cardiac cycle (Fig. 4C). The PRC properties for the KE exercise were similar as a whole to those obtained from HG exercise (Fig. 4D,F).

4. DISCUSSION

This study demonstrated that during HG and KE exercise spontaneous phase-locking could be induced when muscle contraction rhythm approached the cardiac rhythm. Examination of PRCs suggest that muscle contraction provides a phasic input for cardiac rhythm that is capable of entraining both rhythms.

The coupling phase can be estimated from the phase response of the RRI. From the mathematical analysis by Pavlidis[11], the necessary conditions for synchronization are that the gradient of the PRC ranges from 0 to 2, when the PRC is expressed in the normalized

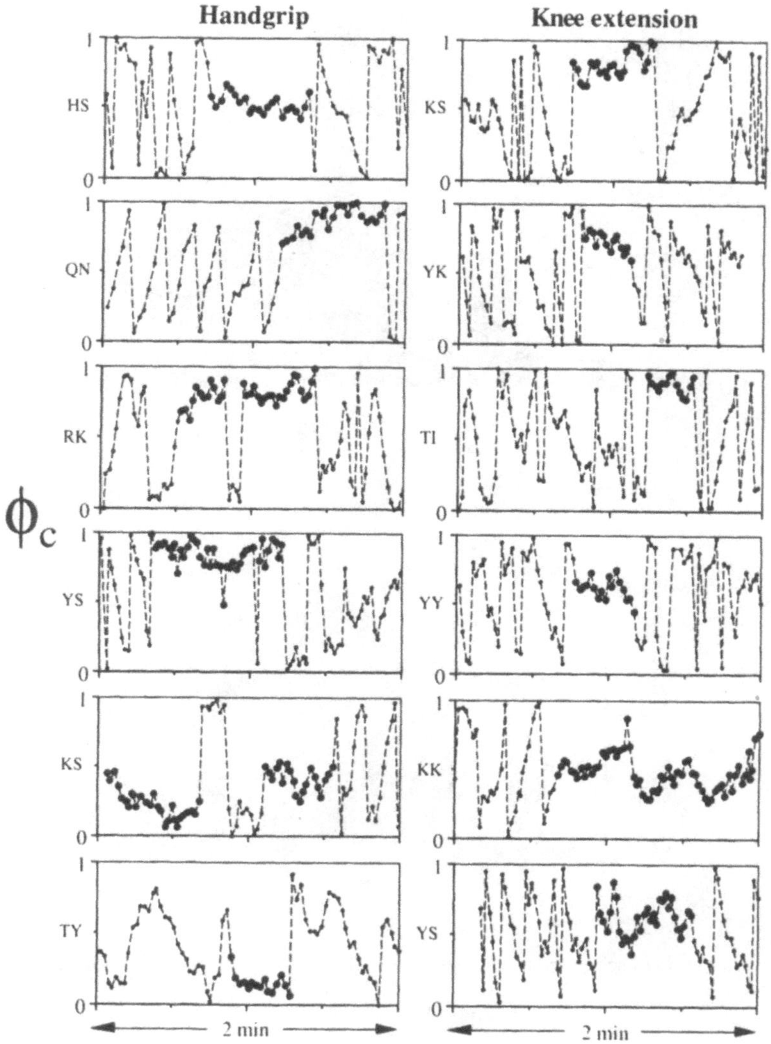

Figure 2. Examples of characteristics of phase difference patterns during HG and KE exercises while coupling was identified. The phases meeting the coupling criteria are indicated by closed circles.

form. Therefore the stable coupling between cardiac and muscle contraction rhythms would occur at a phase where the PRC of the RRI shows a positive gradient. As the estimated phase response curve showed a negative gradient at around 0 to 0.2 of ϕ_C (see Fig. 4), coupling would rarely occur at immediately after the R wave of ECG. This is supported by the observation that the spontaneous synchronized phase tended to gather mid or in the later half of the cardiac cycle (Fig. 3). As shown in the phase response of MBV, muscle contraction occurring immediately after the R wave greatly impedes MBV. It has been documented that muscle blood flow is occluded during the contraction phase of rhythmic exercise[14], this is probably because intramuscular pressure rises beyond the level of systolic blood pressure during contraction[13]. Therefore, synchronizing the heart beat and muscle contraction rhythm would be functionally favorable when it occurs in the later

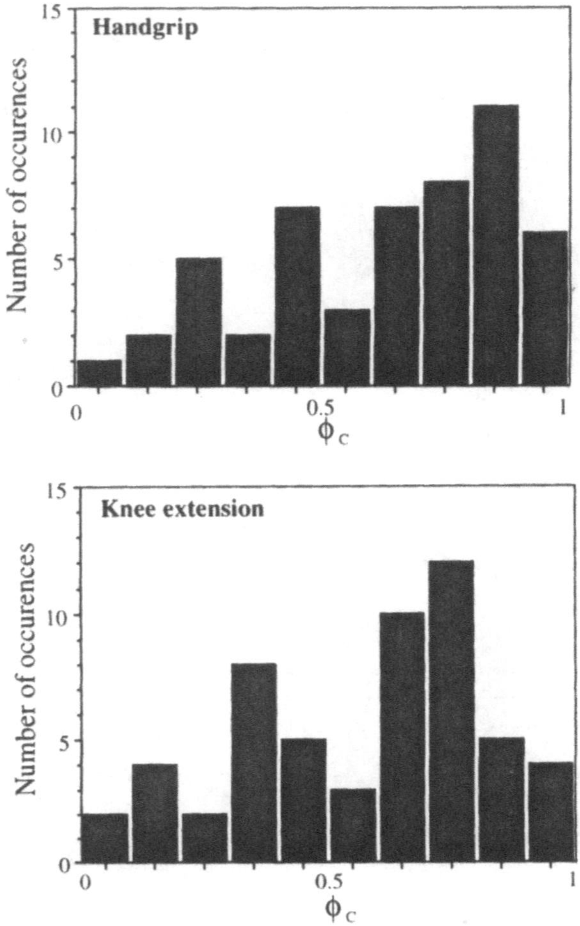

Figure 3. The distribution of spontaneous coupling phases obtained from all subjects for handgrip (A) and knee extension exercises (B).

half of the cardiac phase in these types of exercise. However, synchronization does not necessarily occur only in this region. It seems, therefore, that cardiac rhythm is so coordinated that the period of muscle contraction does not override the systolic phase of the cardiac cycle. Whether similar situations can be extrapolated to other natural locomotor activities such as walking, running, and cycling in humans is uncertain.

The mechanism(s) for the coupling and for the phase dependency of cardiac rhythm with the muscle contraction rhythm remains unknown. Several studies have demonstrated phase-resetting properties in isolated cardiac tissue or cells in response to brief electronic stimuli[3]. In isolated whole hearts, mechanical volume loading to the ventricle can induce ventricular excitation and it finally entrains the intrinsic heart rhythm[2]. While, it has been demonstrated that afferent fibers, classified as group III, are stimulated by dynamic exercise[11], and their reflex effects the autonomic nervous system[8]. Thus, the phase-dependent property may involve an intrinsic character such as in the case of the cardiac pacemaker (nonlinear oscillator) which interacts with afferent signals arising from the stimulation of mechano- and chemosensitive receptors in the contracting muscles, and/or it interacts with

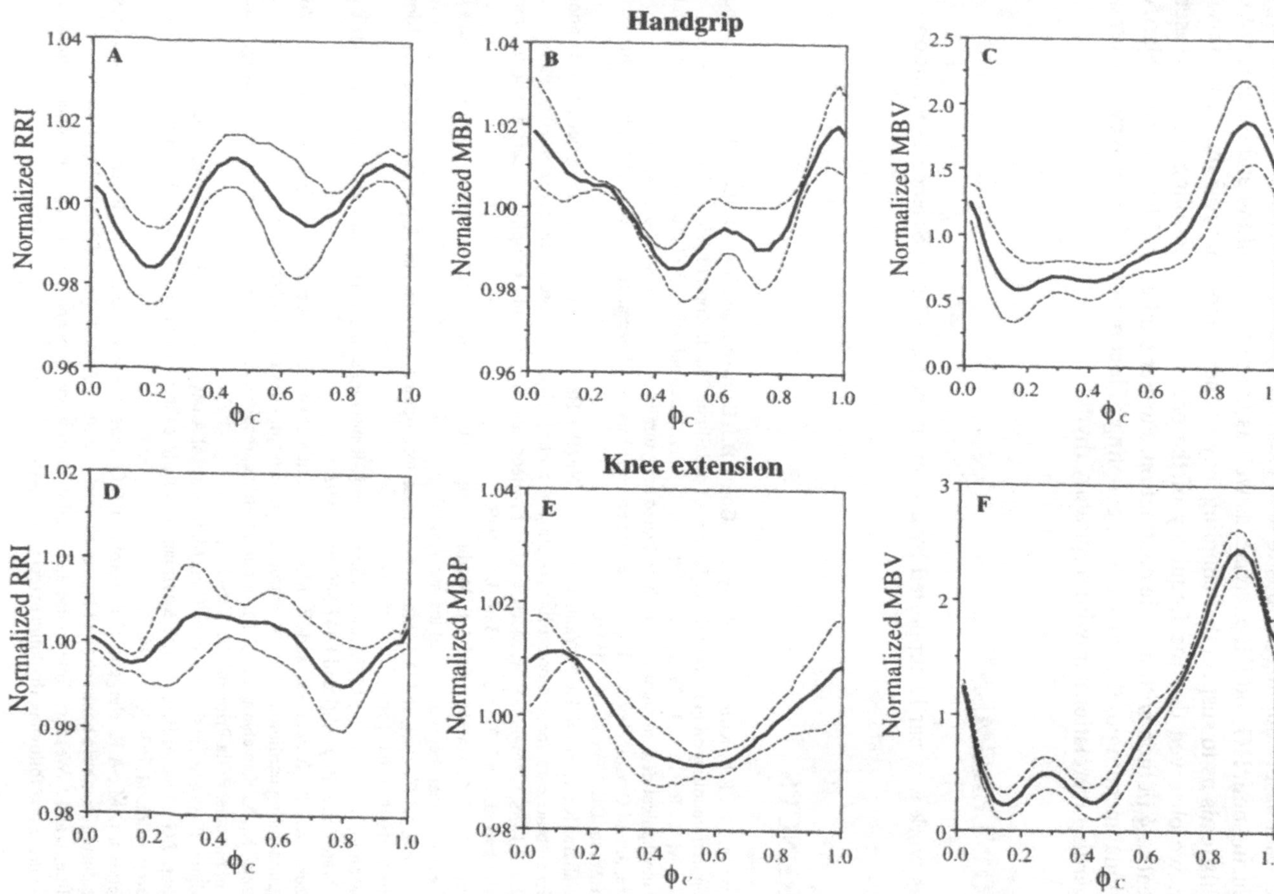

Figure 4. Influence of the phase difference on the RRI (A,D), MBP (B,E), and MBV (C,F) due to muscle contraction during handgrip (upper panels) and knee extension (lower panels) exercises. Solid lines show the third order Fourier regression lines adapted by the least square method and dotted lines show standard deviations.

locomotor related mechanical feedback such as muscle pumping action. On the other hand, it was demonstrated in paralyzed decerebrate animals that centrally generated locomotor rhythm modulates cardiac rhythm probably by interacting with the efference feedforward signals from the locomotor centers[4]. Such central interactions may contribute in some part to coupling.

The fraction of coupling periods to total exercise duration for one trial was not large enough during both HG and KE exercise, as well as in previous walking and running studies[5,8,9]. This leads us to suspect that coupling may not have important role during exercise. However, we observed that the frequency of the occurrence of coupling was increased while breathing in time with muscle contraction. Further studies will be needed to clarify the functional meaning of the existence of coupling of locomotor and cardiac cycles, especially regarding its relationship with respiratory rhythm.

ACKNOWLEDGMENT

This work was partly supported by a grant from the Murata Science Foundation of Japan.

REFERENCES

1. Donville, J.E., R.L. Kirby, T.J. Doherty, S.K. Gupta, B.J. Eastwood, and D.A. MacLeod. Effect of cardiac-locomotor coupling on the metabolic efficiency of pedalling. Can. J. Appl. Physiol. 18:379–391, 1993.
2. Franz, M.R., R. Cima, D. Wang, D.Profitt, and R. Kurz. Electrophysiological effects of myocardial stretch and mechanical determinants of stretch-activated arrhythmias. Circulation 86:968–978, 1992.
3. Jalife, J. and C. Antzelevitch. Phase resetting and annihilation of pacemaker activity in cardiac tissue. Science Wash. DC 206:695–697, 1980.
4. Kawahara, K., T. Yoshioka, Y. Yamauchi, and K. Niizeki. Heart beat fluctuation during fictive locomotion in decerebrate cats: locomotor-cardiac coupling of central origin. Neurosci. Lett. 150:200–202, 1993.
5. Kirby, R.L., S.T. Nugent, R.W. Marlow, D.A. Macleod, and A.E. Marble. Coupling of cardiac and locomotor rhythms. J. Appl. Physiol. 66:323–329, 1989.
6. Kirby, R.L., D.A. Macleod, and A.E. Marble. Coupling between cardiac and locomotor rhythms: The phase lag between heart beats and pedal thrusts. Angiology 40:620–625, 1989.
7. Kirby, R.L., S.E., Carr, and D.A. Macleod. Cardiac-locomotor coupling while finger tapping. Percept. Mot. Skills 71:1099–1104, 1990.
8. McMahon, S.E. and P.N. McWilliam. Changes in R-R interval at the start of muscle contraction in the decerebrate cat. J. Physiol.(Lond.) 447:549–562, 1992.
9. Niizeki, K., K. Kawahara, and Y. Miyamoto. Interaction among cardiac, respiratory, and locomotor rhythms during cardiolocomotor synchronization. J. Appl. Physiol. 75:1815–1821, 1993.
10. Niizeki, K., K. Kawahara, and Y. Miyamoto. Cardiac, respiratory, and locomotor coordination during walking in humans. Folia Primatol. 66: 226–239, 1996.
11. Pavlidis, T. Biological Oscillations: Their Mathematical Analysis. New York: Academic, 1973.
12. Pickar, J.G., J.M. Hill, and N.P. Kaufman. Dynamic exercise stimulates group III muscle afferents. J. Neurophysiol. 71:753–760, 1994.
13. Sejersted, O.M., A.R. Hargens, K.R. Kardel, P.B.O. Jensen, and L. Hermansen. Intramuscular fluid pressure during isometric contraction of human skeletal muscle. J. Appl. Physiol. 56:287–295, 1984.
14. Walloe, L and J. Wesche. Time course and magnitude of blood flow changes in the human quadriceps muscles during and following rhythmic exercise. J. Physiol.(Lond.) 405:257–273, 1987.

$\dot{V}E$-$\dot{V}CO_2$ RELATIONSHIP IN TRANSIENT RESPONSES TO STEP-LOAD EXERCISE FROM REST TO RECOVERY

Tatsuhisa Takahashi, Kyuichi Niizeki, and Yoshimi Miyamoto

Department of Electrical and Information Engineering
Faculty of Engineering
Yamagata University
4-3-16 Joh-Nan, Yonezawa, Yamagata 992, Japan

1. INTRODUCTION

The ventilatory responses to dynamic leg exercise did not differ between voluntary and electrically induced exercise, and a significant correlation between pulmonary ventilation ($\dot{V}E$) and carbon dioxide output ($\dot{V}CO_2$) was observed consistently during the two types of exercise (1). It was also reported that a close $\dot{V}E$-$\dot{V}CO_2$ relationship was maintained under the nonsteady-state conditions, i.e. during exercise with sinusoidal variations in limb movement frequency (3) and during recovery with and without limb movement from exercise (14). However, during incremental-load cycling exercise below the point of respiratory compensation, the slope of the $\dot{V}E$-$\dot{V}CO_2$ relationship at a pedaling rate of 60 rpm was steeper than that at 30 rpm (15). These results suggest that in addition to humoral stimuli, neural stimuli originating from the cortical and subcortical regions and/or from contracting muscles are involved in the control of exercise hyperpnea in the non-steady-state. It has been proposed that both the centrally-generated commands and the afferent signals from the contracting muscles to the respiratory control center are eliminated and/or reduced during stationary rest. In an earlier study (14), we examined the hypothesis that if both the central command and muscle mechanoreflex neural drives play an important role in ventilatory responses to exercise, these responses should be exaggerated out of proportion to metabolism through the associated CO_2 production during active recovery compared to during passive recovery. Consequently, we failed to ascertain the importance of the neural drives contributing to the ventilatory responses, since the tight coupling of $\dot{V}E$-$\dot{V}CO_2$ relationship was almost identical between the two recoveries. However, the lack of difference between the $\dot{V}E$-$\dot{V}CO_2$ relationships may be explained in part by a significant reduction of the neural drive arising in the central motor commands (4) before the end of

Advances in Modeling and Control of Ventilation, edited by Hughson *et al.*
Plenum Press, New York, 1998.

moderate exercise followed by light exercise (unloaded pedaling) in the active recovery. Therefore, in this study to reexamine whether the neural drives of the central command and muscle mechanoreceptor afferents are responsible for the control of the ventilatory response to exercise, we compared both the $\dot{V}E$-$\dot{V}CO_2$ relationship and breathing pattern in the on-transition from rest to moderate exercise with those in the subsequent off-transition to rest.

2. METHODS

Five healthy young male subjects, aged 22.2 ± 0.8 years (meanSD), with weight of 59.2 ± 1.6 kg, height of 1.69 ± 0.05 m, and maximal oxygen uptake of 2.41 ± 0.25 l/min, volunteered for this study. Their informed consent was obtained prior to the study.

Measurements of minute expiratory ventilation ($\dot{V}E$), tidal volume (VT), respiratory frequency (f), end-tidal pressures of O_2 and CO_2, O_2 uptake ($\dot{V}O_2$), and CO_2 output ($\dot{V}CO_2$) were obtained on a breath-by-breath basis using an on-line automated system (9). Ventilatory airflow was monitored with a hot-wire-type pneumotachograph (RF-2, Minato). The composition of expired gas was continuously analyzed with a medical mass spectrometer (WSMR-1400, Westron). The mass spectrometer was calibrated with a standard reference gas mixture before each study. All data for respiratory variables were stored on diskettes for subsequent analysis by a personal computer (PC-98, NEC).

Prior to the experiments, each subject performed a 15 W/min incremental ramp-exercise test until exhaustion on an electromagnetically-braked cycle ergometer (Corival 300, Lode) for determination of maximal oxygen uptake ($\dot{V}O_2$max). The $\dot{V}O_2$max for each subject was determined by the criteria described by Hughson et al. (7).

Each subject completed two repetitions of the experiment. After 5 min of rest in an upright, seated position on the cycle ergometer, they performed the exercise at an exercise intensity of 170 W for 10 min and then recovered for 7 min in the seated position. As there were no large differences in the physical characteristics or $\dot{V}O_2$max among the subjects, the same absolute intensity was applied for the exercise in all subjects. The subjects pedaled at a constant rate of 60 rpm paced by a metronome. The subjects placed their feet on the footrest near the flywheel at rest before and after exercise.

All data were rearranged with a 5-s interval time base using a Lagrange interpolation (11). To characterize the kinetic behavior of VT, $\dot{V}CO_2$, and $\dot{V}E$ in the on- and off-transitions of exercise, the averaged response data for each subject were fitted by a first-order exponential function with a time constant and no pure time delay. Curve fittings were performed using the least-squares method. Group mean values were obtained in consecutive 10-s averages from the five individual data sets. All values are expressed as the meanSD. Differences in average values were examined using one-way analysis of variance. When a significant F ratio was observed, the post-hoc Scheffe's test was used to identify significant differences. For all statistical analyses, differences were considered significant at $p < 0.05$.

3. RESULTS

The time courses of $\dot{V}CO_2$ and $\dot{V}E$ responses to the transition from rest to exercise at 170 W (on-response) were exponential increases, whereas those to the reverse transition from exercise to rest (off-response) were exponential decreases. The time constants of these exponential changes of $\dot{V}CO_2$ and $\dot{V}_E$ were 61.3 ± 9.8 s and 69.5 ± 25.3 s for the on-response and 47.4 ± 9.8 s and 62.8 ± 20.1 s for the off-response, respectively. Both the on-response and the off-response were significantly faster for $\dot{V}CO_2$ than for $\dot{V}_E$. The temporal change in V_T was similar to that in $\dot{V}_E$, but the transient kinetics of V_T were slightly slower than those of $\dot{V}_E$ (the time constants of on- and off-response of V_T, being 77.1 ± 26.4 s and 85.0 ± 15.8 s, respectively). However, after the start of exercise f increased rapidly to a relatively constant level, and vice-versa after the offset of exercise. Thus, the exponential changes of both $\dot{V}_E$ on-response and off-response were achieved primarily by the changes in V_T rather than in f.

To further investigate the relationships between V_T and $\dot{V}CO_2$, between f and $\dot{V}CO_2$, and between $\dot{V}_E$ and $\dot{V}CO_2$, each of these variables was plotted against $\dot{V}CO_2$, as shown in Fig. 1, in which the on-response and off-response data were separately indicated. The V_T for the on-response was directly proportional to the $\dot{V}CO_2$, whereas V_T against $\dot{V}CO_2$ for the off-response exhibited anticlockwise looping. At the matched average $\dot{V}CO_2$ of 0.85 l/min, the averaged V_T obtained from the first minute after the end of exercise was significantly higher ($p < 0.05$) than those from the first minute after the start of exercise. At the same $\dot{V}CO_2$ level, f was lower for the off-response than for the on-response, but the difference was not significant. However, for the two responses, each of the similar levels of $\dot{V}_E$ at the matched $\dot{V}CO_2$ was achieved by the combination of high respiratory frequency and low tidal volume for the on-response, and of low respiratory frequency and high tidal volume for the off-response. Although each of f and V_T against $\dot{V}CO_2$ was different between the on- and the off-response, the change in $\dot{V}_E$ was proportional to that in $\dot{V}CO_2$ at both transitions with a resultant indication of similar $\dot{V}_E$-$\dot{V}CO_2$ regression lines.

4. DISCUSSION

We confirmed in the incremental-load exercise test that the linearly-increasing $\dot{V}_E$ with the increase in $\dot{V}CO_2$ for each subject was maintained up to the exercise intensity of 170 W. For all subjects, the constant-load exercise at 170 W, corresponding to 684% of $\dot{V}O_2$max, may be the upper limit of the linear relationship, since such parallel increases in $\dot{V}_E$ and $\dot{V}CO_2$ usually continue up to approximately 70–90% of $\dot{V}O_2$max (8, 17).

The result that the change in $\dot{V}_E$ highly correlates to that in $\dot{V}CO_2$ during the transitions of exercise from and to rest is in agreement with those from steady-state and non-steady-state exercises (1–3, 10, 14–16). The high correlation between $\dot{V}CO_2$ and $\dot{V}_E$ during exercise has been taken as evidence of a major role of carbon dioxide in determining $\dot{V}_E$ (1–3, 14, 16). Since the elimination of CO_2 from pulmonary capillary blood to alveolar gas was achieved by ventilation, this situation would inevitably result in a tight coupling between $\dot{V}_E$ and $\dot{V}CO_2$. However, there have been many studies which have shown that the transient kinetics of $\dot{V}_E$ in response to various exercises usually lagged behind those of $\dot{V}CO_2$ (3, 13, 14, 16, 17). We also confirmed that the responses of $\dot{V}_E$ in the transitions to and from exercise were slower than those of $\dot{V}CO_2$. These results suggest that the change in $\dot{V}_E$ is not cause of the change in $\dot{V}CO_2$. Moreover, it has been demonstrated in animal experiments that the change in the rate of alveolar ventilation is proportional to the change in the rate of CO_2 delivery to the lungs at rest by venous CO_2 loading or unloading, with little alteration in arterial levels of PCO_2, PO_2, and pH from rest values (12). In human studies, the slope of $\dot{V}_E$-$\dot{V}CO_2$ relationship in normal human subjects during voluntary exercise was almost identical to that in paraplegic subjects during electrically-induced exercise without either the central command or the neural afferent pathways from the

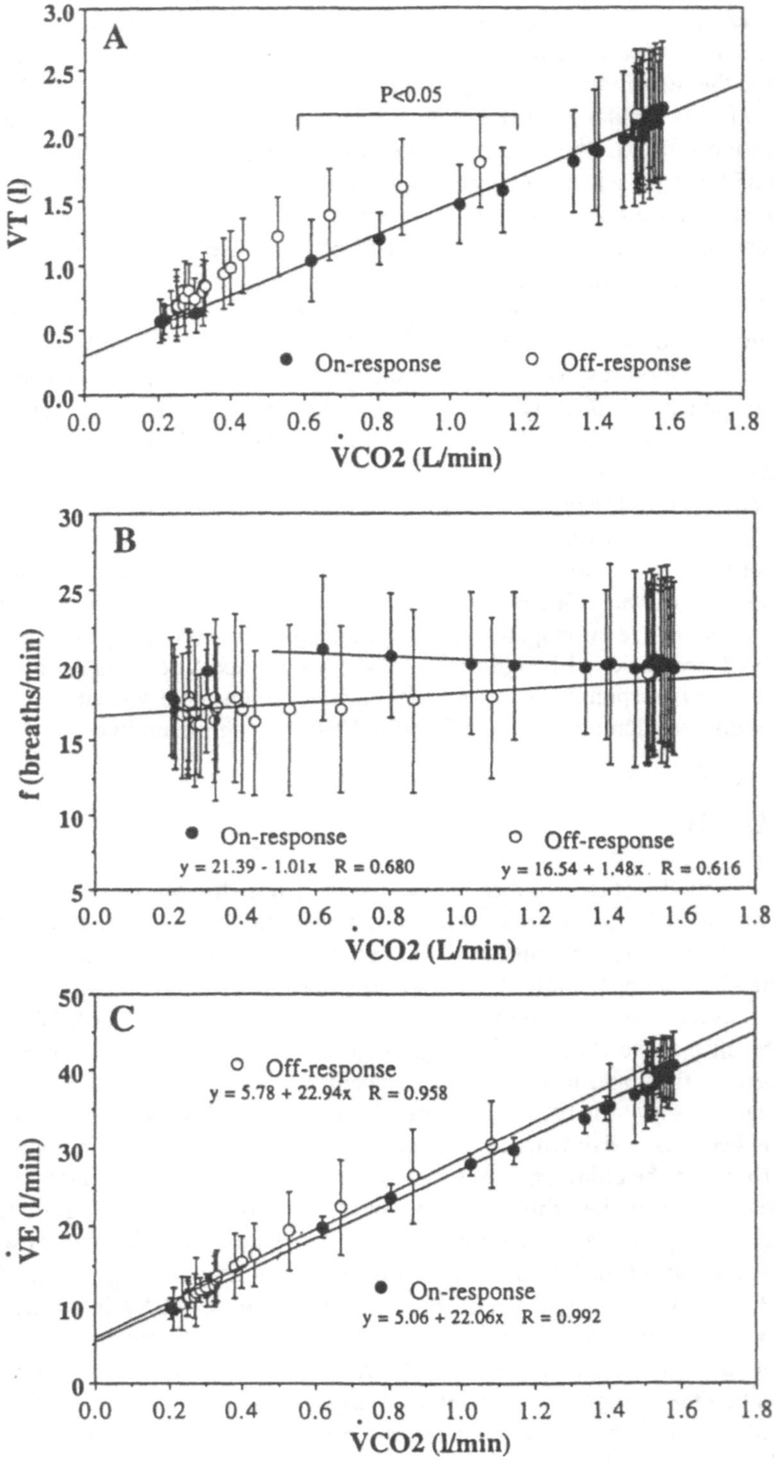

Figure 1. Group mean values for tidal volume (VT), respiratory frequency (f), and expiratory minute ventilation ($\dot{V}E$) as a function of CO_2 output ($\dot{V}CO_2$) for on-response (rest-exercise, ●) and off-response (exercise-rest, ○). (Mean and SD, n = 5).

contracting muscles (2). It was also reported that such a close $\dot{V}E$-$\dot{V}CO_2$ relationship was maintained under non-steady-state conditions, i.e., during exercise at sinusoidally altering rates of limb movement frequency (3) and during recovery with and without limb movement from exercise (14). Thus, on the basis of these many lines of circumstantial evidence, the delivery of CO_2 to the lungs seems to be a dominant determinant of ventilatory drive during exercise and recovery, regardless of the presence of central command and muscle mechanoreflex neural mechanisms.

It has been also demonstrated that the neural components of central command and muscular mechanoreflex mechanisms are critical in regulating the respiratory control system (6, 8, 15, 17). The contributions of these neural stimuli to exercise hyperpnea were studied in an incremental-load cycle exercise with high pedaling frequency (15) and with a neuromuscular blockade (6). The increases in $\dot{V}E$ at the matched metabolic rate of $\dot{V}CO_2$ or $\dot{V}O_2$ were greater during exercise with high pedaling frequency or with the curarization of contracting muscles as compared to the controls. The dissociation between $\dot{V}E$ and $\dot{V}CO_2$ or $\dot{V}O_2$ may be induced by additional neural stimuli arising in the muscle mechanoreceptors, depending on the frequency of limb movement, or in the motor cortex attaining exercise with more effort.

In the present study, it was found that the increase in VT for the on-response was directly proportional to that in $\dot{V}CO_2$, whereas the decrease in VT for the off-response was no longer proportional to that in $\dot{V}CO_2$. This loose relationship between VT and $\dot{V}CO_2$ for the off-response was evident from the large difference between the time constants ($p < 0.05$, 85.0 vs. 47.4 s, respectively). The slower reduction of VT after exercise may be ascribed to the afterdischarge mechanism (5). In contrast, the early change in f at the end of exercise was smaller than that at the start of exercise, as observed in this study. This is in accordance with the difference of early ventilatory changes between the start and end of exercise observed by Duffin (4), who explained the cause of the smaller reduction after exercise as the result of a decline of central neural drives during exercise. Although both the VT-$\dot{V}CO_2$ and the f-$\dot{V}CO_2$ relationship differed significantly between the on- and the off-response, the same $\dot{V}E$-$\dot{V}CO_2$ relationship between the two was achieved by the combination of VT and f. These results may support in part the concept that respiratory rhythm is critically dependent on stimuli related to metabolic CO_2 production, with a secondary influence of other afferent stimuli on the respiratory rhythm generation (12). However, on the basis of evidence that there was no difference between the $\dot{V}E$-$\dot{V}CO_2$ relationship under the different conditions, several investigators (1–3, 14) might mistake the significance of the neural components and might thus have been led to the incorrect conclusion that neither central command nor muscular afferent neural mechanisms play a dominant role in the control of ventilation to match the pulmonary gas exchange to the tissue metabolism.

In conclusion, although the appropriate mechanisms responsible for the tight coupling of $\dot{V}E$-$\dot{V}CO_2$ dynamics remain unclear, the breathing pattern was modulated redundantly by the central command and/or neural afferents from contracting muscles, in which minute ventilation could closely parallel the rate of elimination of CO_2 to the lungs.

ACKNOWLEDGMENTS

This research was supported in part by grants from the Ministry of Welfare and grants from the Ministry of Education, Science, Sports, and Culture of Japan (No. 0778076 to Dr. Takahashi and No. 07680932 to Dr. Miyamoto).

REFERENCES

1. Brice, A.G., H.V. Forster, L.G. Pan, A. Funahashi, T.F. Lowry, C.L. Murrhy, and M.D. Hoffman. Ventilation and $PaCO_2$ responses to voluntary and electrically induced leg exercise. J. Appl. Physiol. 64: 218–225, 1988.
2. Brice, A.G., H. Forster, L.G. Pan, A. Funahashi, M.D. Hoffman, C.L. Murrhy, and T.F. Lowry. Is the hyperpnea of muscular contractions critically dependent on spinal afferents? J. Appl. Physiol. 64: 226–233, 1988.
3. Casaburi, R., B.J. Whipp, K. Wasserman, and S.N. Koyal. Ventilatory and gas exchange responses to cycling with sinusoidally varying pedal rate. J. Appl. Physiol. 44: 97–103, 1978.
4. Duffin, J. Neural drives to breathing during exercise. Can. J. Appl. Physiol. 19: 289–304, 1994.
5. Eldridge, F.L., and T.G. Waldrop. Neural control of breathing during exercise. In: B.J. Whipp and K. Wasserman (Eds.), Exercise, Pulmonary Physiology and Pathophysiology, pp. 309–370. New York: Marcel Dekker.
6. Galbo, H., M. Kjaer, and N.H. Secher. Cardiovascular, ventilatory and catecholamine responses to maximal dynamic exercise in partially curarized man. J. Physiol. (Lond.) 389: 557–568, 1987.
7. Hughson, R.L., H.C. Xing, C. Borkhoff, and G.C. Butler. Kinetics of ventilation and gas exchange during supine and upright cycle exercise. Eur. J. Appl. Physiol. 63: 300–307, 1991.
8. Mateika, J.H., and J. Duffin. A review of the control of breathing during exercise. Eur. J. Appl. Physiol. 71: 1–27, 1995.
9. Miyamoto, Y., T. Hiura, T. Tamura, T. Nakamura, J. Higuchi, and T. Mikami. Dynamics of cardiac, respiratory, and metabolic function in men in response to step work load. J. Appl. Physiol. 52: 1198–1208, 1982.
10. Newstead, C.G., G.C. Donaldson, and J.R. Sneyd. Potassium as a respiratory signal in humans. J. Appl. Physiol. 69: 1799–1803, 1990.
11. Niizeki, K., K. Kawahara, and Y. Miyamoto. Interaction among cardiac, respiratory, and locomotor rhythms during cardiolocomotor synchronization. J. Appl. Physiol. 75: 1815–1821, 1993.
12. Phillipson, E.A., J. Duffin, and J.D. Cooper. Critical dependence of respiratory rhythmicity on metabolic CO_2 load. J. Appl. Physiol. 50: 45–54, 1981.
13. Takahashi, T., K. Niizeki, and Y. Miyamoto. Effects of base line changes in work rate on cardiorespiratory dynamics in incremental and decremental ramp exercise. Adv. Exp. Med. Biol. 393: 159–164, 1995.
14. Takahashi, T., K. Niizeki, and Y. Miyamoto. Respiratory responses to passive and active recovery from exercise. Jpn. J. Physiol. 47: 59–65, 1997.
15. Takano, N. Effects of pedal rate on respiratory responses to incremental bicycle work. J. Physiol. 396: 389–397, 1988.
16. Wasserman, K., B.J. Whipp, R. Casaburi, and W.L. Beaver. Carbon dioxide flow and exercise hyperpnea. Am. Rev. Respir. Dis. 115(Suppl.): 225–237, 1977.
17. Wasserman, k., B.J. Whipp, S.A. Ward, and R. Casaburi. Respiratory control during exercise. In: Handbook of Physiology. The respiratory System. Control of Breathing, edited by Cherniack NS and Widdicombe JG. Bethesda, MD: Am. Physiol. Soc., 1986, sect. 3, vol. II, pt. 2, chapt. 17, p. 595–619.

THE INFLUENCE OF HYPERCAPNIC HYPERPNEA ON THE INTERACTION BETWEEN BREATHING AND FINGER TRACKING MOVEMENTS IN HUMANS

Beate Raßler, Ingo Nietzold, and Siegfried Waurick

Carl Ludwig Institute of Physiology
University of Leipzig
D-04103 Leipzig, F.R.G.

1. INTRODUCTION

Numerous studies on entrainment of breathing to simultaneous limb movements, for instance walking, running, cycling, or finger tapping, exclusively referred to effects of the limb movements on breathing[1,2,5,6,10,17]. None of them considered effects of breathing on the limb movement which are much less obvious than movement-induced changes in breathing rhythm. In a previous study on finger tracking movements we could show that both influences of finger movements on the respiratory rhythm and breathing influences on the finger movements do exist[13]. The effect of breathing on the tracking movement consisted in more precise tracking in mid-inspiration and mid-expiration and greater tracking errors at the respiratory phase-transitions. Tracking movements influenced the respiratory rhythm by shortening the coinciding inspiration or expiration, particularly when they were started during inspiration.

The interactions between breathing and additional movements depend on various conditions such as movement rate[13,15], work load[3,15], number of limbs involved in the movement[13], or hypoxia[11]. Hence, we supposed that an increased respiratory drive, for instance during hypercapnia, can modify the interaction between breathing and finger tracking movements. Ebert et al.[4] reported that the incidence of coupling phenomena (i.e. coordination) between breathing and rhythmical forearm movements rose in hypercapnic conditions.

The purpose of the present study was to investigate whether hypercapnia intensifies the influences in both directions (respiratory effects on movement and movement effects on breathing) to the same extent. The results should elucidate the mechanisms of enhanced coordination in hypercapnia.

Advances in Modeling and Control of Ventilation, edited by Hughson *et al.*
Plenum Press, New York, 1998.

2. METHODS

We investigated 18 healthy volunteers in a finger tracking procedure under normocapnic and hypercapnic (3.5% CO_2 in air) conditions. They had to track a visually presented step-function as fast and as precisely as possible by flexion or extension of their right index finger. The evaluation included only finger flexions corresponding with the upward flank of the step-function (pre-set signal, S). The normocapnic and the hypercapnic experiments consisted of 6 series each with S given at a particular phase-relation with breathing: 0i, at start of inspiration; 30i, after 30% of inspiration time (TI); 60i, after 60% of TI; 0e, at start of expiration; 40e, after 40% of expiration time (TE); 70e, after 70% of TE.

Finger movements were transduced by a goniometer, the respiratory flow was recorded by a Fleisch pneumotachograph. Parameters were measured as follows: *Movement*: **tL:** latency, from pre-set signal until 10% of the pre-set amplitude was exceeded; **tF:** flexion time, from end of latency until 80% of the pre-set amplitude was exceeded; **E:** tracking error, area between pre-set and tracking curves, calculated over 2 min; **RS**tF+E: rank sum RtF + RE. Ranks (RtF and RE) from 1 to 6 were assigned to the individual mean values of tF and E calculated from the 6 series. *Breathing*: **T**I: inspiration time; **T**E: expiration time; ΔTI, ΔTE: TI/E[S−1] – TI/E[S]. [S] denotes the breath during which the pre-set signal was given, [S−1] the immediately preceding breath.

3. RESULTS

3.1. Effects of Breathing on Movement Parameters

The movement parameters, tL, tF, and E, varied with their phase-relation to the respiratory cycle (Fig. 1, left side). The tracking movements were accomplished faster and more precisely in the middle of both inspiration and expiration, but less fast and precisely when coinciding with respiratory phase-switching (inspiration to expiration, I/E and expiration to inspiration, E/I).

Under the hypercapnic condition, we found that—independently of phase-relation—E was smaller and tL was longer than during air breathing (Wilcoxon matched pairs signed ranks test: E: $p < 0.01$, tL: $p < 0.001$). The tracking overshoot (from 80% of required magnitude to peak value) amounted 25.1% of the required magnitude. It was smaller than in normocapnia (27.5%) and correlated positively with the tracking error ($r = 0.76$).

The phase-dependent effects of breathing on tracking movement were only slightly affected under hypercapnia (Fig. 1, right side). The movement parameters decreased in mid and late expiration and increased in mid-inspiration. Differences between hypercapnic and normocapnic results were more pronounced in movements performed during expiration.

This finding is also reflected by RStF+E (Fig. 2) even though the differences between the series did not reach significance. The lowest values expressing fast and precise tracking as well were achieved in mid-expiration. This profile was more pronounced under the hypercapnic condition.

3.2. Effects of Movement on Breathing

The finger movements were associated with a shortening of the current respiratory half-cycle (Table 1). Movements with S during inspiration reduced the related TI significantly more than movements with S during expiration reduced the current TE. The immedi-

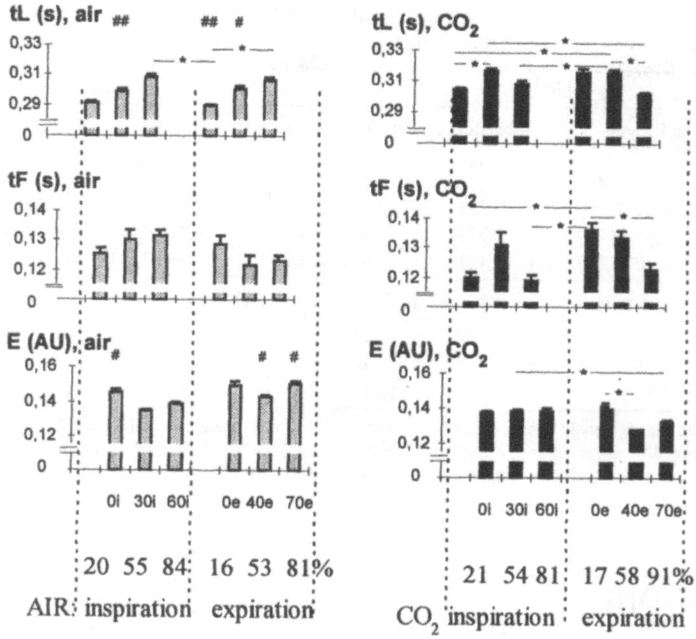

Figure 1. Mean values + variance of tL, tF and E in the 6 test series. Left side: normocapnic, right side: hypercapnic condition. Percentaged values at the bottom of the diagram indicate the relative phase-relation of movement onset to inspiration or expiration. Significance marks: —*—: significant differences between the series, #: significant differences between normocapnic and hypercapnic values (# $p < 0.05$, ## $p < 0.01$).

ately subsequent half-cycles (TE[S] and TI[S+1], respectively) were shortened, too. Under hypercapnic conditions the reduction of TI in 0i, 30i and 60i experiments was more pronounced than under normocapnic conditions. The related expiration was significantly shortened, too ($p < 0.01$). On the contrary, in 0e, 40e and 70e experiments the effects on the current TE as well as on the following TI[S+1] did not differ from those in normocapnia.

The comparison of ΔTI and ΔTE (Figure 3) between the 6 series revealed significant differences under hypercapnic (ANOVA: ΔTI: $p = 0.02$, ΔTE: $p = 0.03$) but not under normocapnic conditions.

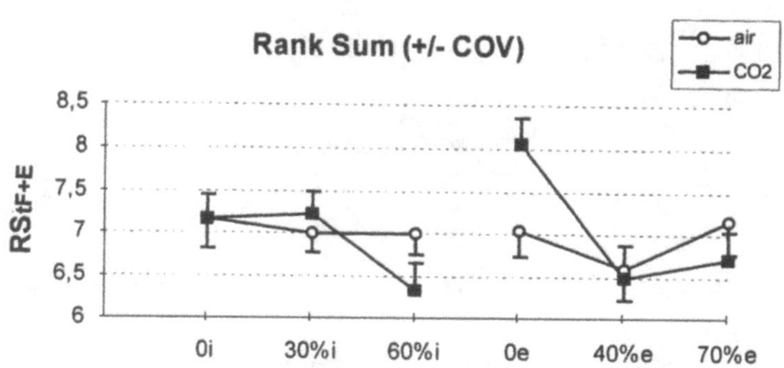

Figure 2. Rank sum RStF+E calculated from individual values of tF and E ranked across the series (mean values ± coefficient of variation). O: normocapnic, ■: hypercapnic condition.

Table 1. Mean values (in brackets: variances) of TI and TE coinciding with tracking movements (current half-cycle) as well as of the immediately following TE and TI (subsequent half-cycle)[a]

	Air				CO_2			
	Current half-cycle		Subseqent half-cycle		Current half-cycle		Subseqent half-cycle	
	TI[S−1]	TI[S]	TE[S−1]	TE[S]	TI[S−1]	TI[S]	TE[S−1]	TE[S]
0i	1,60 (0,15)^b	1,51 (0,12)	2,27 (0,23)	2,24 (0,31)	1,64 (0,11)^c	1,52 (0,09)	2,04 (0,11)^c	1,93 (0,11)
30i	1,60 (0,27)^b	1,51 (0,24)	2,33 (0,37)	2,29 (0,54)	1,65 (0,14)^c	1,52 (0,10)	2,01 (0,08)^b	1,89 (0,07)
60i	1,60 (0,17)^a	1,50 (0,13)	2,41 (0,42)	2,33 (0,48)	1,68 (0,12)^a	1,59 (0,12)	2,02 (0,14)^b	1,92 (0,15)
	TE[S−1]	TE[S]	TI[S]	TI[S+1]	TE[S−1]	TE[S]	TI[S]	TI[S+1]
0e	2,23 (0,35)	2,18 (0,35)	1,53 (0,11)^a	1,56 (0,11)	2,01 (0,11)	1,95 (0,16)	1,64 (0,07)	1,64 (0,09)
40e	2,37 (0,30)	2,36 (0,31)	1,53 (0,13)^a	1,54 (0,10)	2,06 (0,13)	2,05 (0,12)	1,69 (0,15)	1,68 (0,13)
70e	2,41 (0,37)	2,39 (0,30)	1,61 (0,12)	1,60 (0,11)	2,02 (0,15)	1,98 (0,13)	1,69 (0,14)	1,67 (0,16)

[a]Breaths preceding the trigger signal (labeled with [S−1]) are considered to be unaltered by the tracking movement. Superscript letters mark significant differences between TI/E[S] and TI/E[S−1] ($^a p < 0.05$, $^b p < 0.01$, $^c p < 0.001$). In 0e, 40e and 70e experiments, the subsequent half-cycle is TI[S+1], the reference inspiration is TI[S].

4. DISCUSSION

4.1. Effects of Breathing on Movement Parameters

Very few studies on coordination between breathing and rhythmic movements give hints on effects of breathing on the simultaneous movement[12,13,15]. The results of the normocapnic experiment prove the mutuality of interrelations between breathing and limb movements. They agree with findings described previously[14]. The phase-related differences in movement parameters, in particular the higher values at the respiratory phase-transitions, result from stronger respiratory influences on the tracking movement. This may account for the preferred phase-relationships found between breathing and other rhythmical movements, which are coincidence of the onset of movement and the respiratory phase-switching[5,8,9,13,17].

Under hypercapnic conditions, we observed significantly longer latencies but smaller tracking errors than in normocapnia—an effect that was independent of the phase-

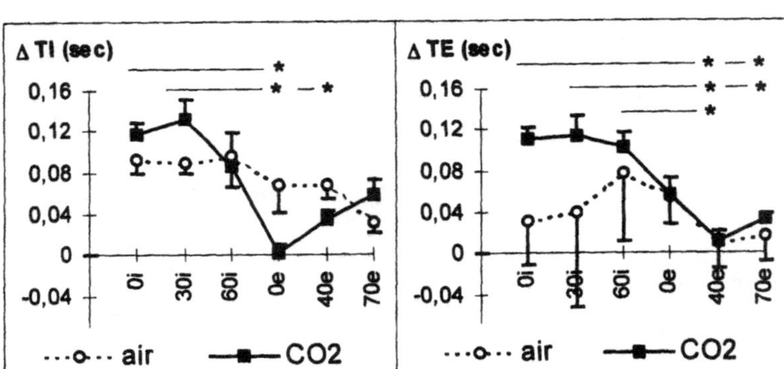

Figure 3. Left panel: ΔTI, right panel: ΔTE (mean values ± variances).O: normocapnic, ■: hypercapnic condition. Differences between the series were not significant in normocapnia but in hypercapnia (ΔTI: $p = 0.02$, ΔTE: $p = 0.03$; —*— between two columns mark a significant contrast).

relation with breathing. The smaller tracking errors result in part from a reduced movement magnitude, as indicated by the smaller tracking overshoot. This reflects a greater damping of the tracking movement accompanied by prolonged tL.

In contrast, phase-dependent differences between normocapnia and hypercapnia were considered to be consequences of the increased respiratory drive. The increase of tL and tF in the early expiration in hypercapnia suggested that breathing exerted a stronger influence on movement during this period. On the contrary, the respiratory influence on tracking parameters was reduced during mid and late expiration. We assume that the relatively low neural expiratory activity has no appreciable effect on simultaneous motor actions.

The tracking test is an optimization task requiring that the movement is performed both fast and precisely. Low RS_{tF+E} values (see Figure 2) mean good optimization and point toward a reduced impairment of the tracking movement by breathing. The flat profile during inspiration expresses that the subjects compensated for affections of velocity at the expense of precision and vice versa. Under both normocapnic and hypercapnic conditions, the optimization of velocity and precision was most successful in mid-expiration. Hypercapnia enhanced the respiratory effects on finger movements only slightly.

4.2. Effects of Movement on Breathing

Tracking movements were accompanied by a shortening of the coinciding breath. The extent of this response depended on the phase-relationship between movement and breathing and was larger when movements were performed during inspiration. This result confirms findings published in an earlier report[14]. In this study, we supposed that T_I and T_E of the current breath are determined in the first stage of inspiration. Disturbances in this period imply a stronger impairment of the breath than affections at other periods in a breath. This is clearly demonstrated by ΔT_I and ΔT_E expressing the respiratory response to the tracking movement.

It is well-known that exercise and hypercapnia additively act on the respiratory controller increasing the rate of rise of the central inspiratory activity[7]. In our experiment, only movements elicited during inspiration, particularly during the first half of inspiration, evoked a stronger respiratory response than in normocapnia. That means, that the interplay between chemical and neurogenic drives varies with the respiratory phase. We suggest that in the first half of inspiration the respiratory controller is sensitized by the increased chemical drive.

Studies on coordination between breathing and limb movements at different work loads[3,9,10,15] revealed that coordination improved with increasing work load. Since limb movements usually are the attractive process, one can assume that enhancement of the attractive force of the movement is the reason for closer coordination. On the other hand, driving the respiratory rhythm as the attracted process should reduce coordination as found by Paterson et al.[11] under hypoxic conditions. Therefore, we had expected that hypercapnia as a potent respiratory drive intensifies much more the respiratory effects on tracking movements and diminishes the movement effects on breathing. The results presented seem to disprove this hypothesis. We suggest that an increased respiratory drive does not simply increase the strength of the respiratory rhythm and thus counteracts influences from simultaneous motor activities. Modulations of attractivity and attractability rather seem to be phase-dependent with particular increase of attractability during the first half of inspiration. It also must be taken into consideration that external influences (in this case, hypercapnia) do not only modulate the properties of the oscillators (in this case, of the respiratory controller) but also those of the coupling mechanisms between the two processes as can be concluded from a model of Schöner and Kelso[16].

REFERENCES

1. Bechbache, R.R., J. Duffin. The entrainment of breathing frequency by exercise rhythm. *J. Physiol.* 272: 553–561, 1977.
2. Bernasconi, P. and J. Kohl. Analysis of co-ordination between breathing and exercise rhythms in man. *J. Physiol. (London)* 471: 693–706, 1993.
3. Bernasconi, P., P. Bürki, A. Bührer, E. A. Koller, and J. Kohl. Running training and co-ordination between breathing and running rhythms during aerobic and anaerobic conditions in humans. *Eur. J. Appl. Physiol.* 70: 387–393, 1995.
4. Ebert, D., B. Raßler, S. Waurick. Phase relations between rhythmical forearm movements and breathing under normocapnic and hypercapnic conditions. In: *Advances in Modeling and Control of Breathing*, edited by R. Hughson, D. A. Cunningham, and J. Duffin, New York: Plenum, 1998, (this volume).
5. Hill, A.R., J.M. Adams, B.E. Parker, D.F. Rochester. Short-term entrainment of ventilation to the walking cycle in humans. *J. Appl. Physiol.* 65: 570–578, 1988.
6. Jasinskas, C.L., B.A. Wilson, J. Hoare. Entrainment of breathing rate to movement frequency during work at two intensities. *Respir. Physiol.* 42: 199–209, 1981.
7. Kao, F.F., S.S. Mei, and M. Kalia. Interaction between neurogenic exercise drive and chemical drive. In: *Central Nervous Control Mechanisms in Breathing*, edited by C. v.Euler and H. Lagercrantz, Oxford, UK: Pergamon, Vol. 32 (Wenner-Gren Ctr. Int. Symp. Ser.), 1979, pp. 75–89
8. Kohl, J., E.A. Koller, M. Jäger. Relation Between Pedalling- and Breathing Rhythm. *Eur. J. Appl. Physiol.* 47: 223–237, 1981.
9. Lafortuna, C.L., E. Reinach, and F. Saibene. The effects of locomotor-respiratory coupling on the pattern of breathing in horses. *J.Physiol. (London)* 492: 587–596, 1996.
10. Loring, S.H., J. Mead and T.B. Waggener. Determinants of breathing frequency during walking. *Respir. Physiol.* 82: 177–188, 1990.
11. Paterson, D.J., G.A. Wood, R.N. Marshall, A.R. Morton and A.B.C. Harrison. Entrainment of respiratory frequency to exercise rhythm during hypoxia. *J. Appl. Physiol.* 62: 1767–1771, 1987.
12. Perségol, L., M. Jordan, D. Viala, and C. Fernandez. Evidence for central entrainment of the medullary respiratory pattern in the rabbit. *Exp. Brain Res.* 71: 153–162, 1988.
13. Raßler, B. and J. Kohl. Analysis of coordination between breathing and walking rhythms in humans. *Respir. Physiol.* 106: 317–327, 1996.
14. Raßler, B., D. Ebert, S. Waurick and R. Junghans. Coordination Between Breathing and Finger Tracking in Man. *J Motor Behavior* 28: 48–56, 1996.
15. Raßler, B., S. Waurick, D. Ebert. Einfluß zentralnervöser Koordination auf die Steuerung von Atem- und Extremitätenmotorik des Menschen. *Biol. Cybern.* 63: 457–462, 1990.
16. Schöner, G. and J.A.S. Kelso. A Synergetic Theory of Environmentally-Specified and Learned Patterns of Movement Coordination. I: Relative Phase Dynamics. *Biol. Cybern.* 58: 71–80, 1988.
17. Wilke, J.T., R.W. Lansing, C.A. Rogers. Entrainment of respiration to repetitive finger tapping. *Physiogical Psychology*, Vol. 3 (4): 345–349, 1975.

CHARACTERISTICS OF THE $\dot{V}O_2$ SLOW COMPONENT DURING HEAVY EXERCISE IN HUMANS AGED 30 TO 80 YEARS

C. Bell,[1] D. H. Paterson,[1] M. A. Babcock,[2] and D. A. Cunningham[1,3]

[1]Centre for Activity and Ageing
School of Kinesiology
The University of Western Ontario
London, Ontario, Canada, N6A 3K7
[2]Neurodiagnostics
John D. Dingell VA Medical Center
4646 John R, Detroit, Michigan 48202
[3]Department of Physiology
The University of Western Ontario

1. INTRODUCTION

At the onset of moderate intensity exercise oxygen uptake ($\dot{V}O_2$) increases exponentially from baseline to a new steady-state value. The steady-state $\dot{V}O_2$ is linearly related to work rate, such that: $\Delta\dot{V}O_2/\Delta WR \approx 10$ (ml·min^{-1})·W^{-1}. When the exercise intensity is heavy, inducing a sustained increase in blood lactate concentration, then the attainment of steady-state may be delayed, or even prevented, and a slow component of increasing $\dot{V}O_2$ is observed (10). Thus, during heavy exercise the $\Delta\dot{V}O_2/\Delta WR$ relationship is increased ($\Delta\dot{V}O_2/\Delta WR > 10$) and becomes non-linear.

The characteristics of the slow component have not been thoroughly studied, however: 1) With endurance training, the $\dot{V}O_2$ slow component was smaller when measured at the same pre-training absolute work rate (6), due to a lower relative exercise intensity post-training; 2) In children, the $\dot{V}O_2$ slow component has been absent, or difficult to observe, but the $\Delta\dot{V}O_2/\Delta WR$ relationship during heavy exercise was greater than observed for adults working at a similar relative intensity.

With ageing, relatively little is known about the $\dot{V}O_2$ slow component. Babcock *et al.* (1994) however, have shown that $\dot{V}O_2$ kinetics during the onset of moderate exercise were slower in older adults, but there was no age-related difference in the $\Delta\dot{V}O_2/\Delta WR$ relationship in the moderate intensity domain. The lack of information on the supra-threshold slow component in older adults, together with current debate concerning its mechanism provides

Advances in Modeling and Control of Ventilation, edited by Hughson *et al.*
Plenum Press, New York, 1998.

incentive for research. Specifically, one proposed mechanism of the slow component relates to the recruitment of fast twitch muscle fibres. The older adult provides a useful model to test this theory as ageing has been shown to reduce the proportion of fast twitch muscle fibres relative to slow twitch fibres (1). Hence, if older individuals have fewer fast twitch fibres to recruit, one might expect to observe a smaller $\dot{V}O_2$ slow component than that observed in younger adults. Thus, the purpose of our investigation was to examine the $\dot{V}O_2$ slow component, and the $\Delta\dot{V}O_2/\Delta WR$ relationship, during heavy exercise in a cross-sectional sample of men spanning a 50 year age range.

2. METHODS

Forty-seven adult males volunteered for the study and were assigned to groups depending on their age: Group 1 (G1) 30–44 years, n = 20; Group 2 (G2) 45–59 years, n = 18; and Group 3 (G3) 65–80 years, n = 9. Subjects recruited met the criteria that: 1) their habitual daily physical activity was characterised as sedentary, 2) none were actively engaged in a formal programme of exercise training, and 3) they were free of any medical condition which would contraindicate vigorous exercise.

Subjects were initially tested twice to determine maximal oxygen uptake ($\dot{V}O_{2max}$) and ventilatory threshold (T_{VE}). The test was a ramp function of continuously increasing work rate ranging from 15 to 30 $W{\cdot}min^{-1}$, according to subject age such that the test was completed within 8 and 12 minutes. Ventilatory threshold was determined by two independent investigators following the criteria outlined by Davis *et al.* (1979). Subjects then reported to the laboratory for the performance of square wave cycle exercise. This consisted of six minutes of loadless (0 W) cycling after which the work rate was abruptly increased to an intensity which would elicit a $\dot{V}O_2$ response approximately halfway between $\dot{V}O_2$ at T_{VE} and $\dot{V}O_{2max}$. This continued for six minutes when the work rate was abruptly decreased and the subjects exercised for a further six minutes of loadless cycling. Breath-by-breath data were collected following the methods outlined by Babcock *et al.* (1994). Respired gases were analysed continuously (1 $ml{\cdot}s^{-1}$) for concentrations of O_2, CO_2 and N_2 via mass spectrometry (Perkin Elmer MGA-1100). The mass spectrometer was calibrated prior to each test with precision analysed gas mixtures. Inspired and expired flow rates were measured using a low dead space (90 ml) bidirectional turbine (Alpha Technologies VMM 110) which was calibrated using a syringe of known volume (3.01 l). Alveolar $\dot{V}O_2$ data were calculated using the algorithms of Beaver *et al.* (1981).

The $\dot{V}O_2$ slow component was analysed by quantifying the change in $\dot{V}O_2$ between minutes 3 and 6 of heavy exercise. The $\Delta\dot{V}O_2/\Delta WR$ relationship was calculated as the change in $\dot{V}O_2$ between baseline and end-exercise at the end of the step change in work rate, assuming that loadless cycling was equivalent to 15 W (given the internal resistance of the pedals and the mass of the subjects' legs). One-way repeated measures analysis of variance techniques were used to examine differences between groups for $\dot{V}O_{2max}$, $\dot{V}O_2$ at T_{VE}, the $\dot{V}O_2$ slow component, and $\Delta\dot{V}O_2/\Delta WR$. Post-hoc pairwise comparisons were made using Student Newman-Keuls methods.

3. RESULTS

Data are presented in Table 1. $\dot{V}O_{2max}$ and $\dot{V}O_2$ at T_{VE} were greater in G1 than G3 and greater in G2 than G3. The $\dot{V}O_2$ slow component was lower in G3 (100 $ml{\cdot}min^{-1}$) and G2 (140 $ml{\cdot}min^{-1}$) than in G1 (210 $ml{\cdot}min^{-1}$), and was inversely related with age (r = −0.40,

Table 1. Cardiorespiratory data collected during ramp and square wave cycle exercise from G1, G2, and G3

Group	$\dot{V}O_{2max}$ (l·min^{-1})	$\dot{V}O_2$ @ T_{VE} (l·min^{-1})	$\dot{V}O_2$ @ $T_{VE}/\dot{V}O_{2max}$ (%)	$\Delta\dot{V}O_{2\,(3\text{–}6\,min)}$ (l·min^{-1})	$\Delta\dot{V}O_2/\Delta WR$ [(ml·min^{-1})·W^{-1}]	WR (W)
1	3.20 ± 0.57	1.66 ± 0.28	52 ± 5^{2,3}	0.21 ± 0.11	12.2 ± 0.5	166 ± 33
2	2.58 ± 0.42^{3}	1.57 ± 0.25^{3}	62 ± 7^{1,3}	0.14 ± 0.07^{1}	12.5 ± 0.5	138 ± 24^{1,3}
3	1.81 ± 0.21^{1,2}	1.23 ± 0.13^{1,2}	69 ± 4^{1,2}	0.10 ± 0.07^{1}	12.5 ± 0.3	84 ± 15^{1,2}

Data are presented as mean ± SD. Significant differences between groups are denoted by superscript group number ($p < 0.05$). WR = work rate required to elicit $\dot{V}O_2$ response halfway between $\dot{V}O_2$ at T_{VE} and $\dot{V}O_{2max}$.

$p = 0.006$). The absolute work rate however, was also greater in G1 (166 W) than in G2 (138 W) and G3 (84 W) and greater in G2 than in G3. The slow component was positively correlated with the work rate at which the response was elicited ($r = 0.49$, $p < 0.001$). Thus, the $\Delta\dot{V}O_2/\Delta WR$ was similar across age groups and, during this relatively heavy exercise, exceeded 12 (ml·min^{-1})·W^{-1} in all groups compared with the expected 10 (ml·min^{-1})·W^{-1} for moderate intensity exercise. The $\Delta\dot{V}O_2/\Delta WR$ was not different across age groups exercising at intensities designed to be similar relative to their $\dot{V}O_{2max}$ and T_{VE}.

4. DISCUSSION

The present study demonstrated that during heavy exercise a slow component of $\dot{V}O_2$ was observable in the elderly. The absolute magnitude of this slow component was lower in older than in younger adults; however, this appeared due to the lower absolute work rates for heavy exercise in the older group. When exercise was expressed in relative terms, as a proportion of the difference between T_{VE} and $\dot{V}O_{2max}$ in each individual, and across age groups, it was clear that the $\Delta\dot{V}O_2/\Delta WR$ relationship was not different across age. For all age groups the $\Delta\dot{V}O_2/\Delta WR$ in heavy exercise exceeded that expected for moderate exercise.

The mechanism of the $\dot{V}O_2$ slow component is of some debate. One hypothesis is concerned with muscle fibre type recruitment. During oxidative phosphorylation, pathways for transferring the NADH-linked reducing equivalents differ between muscle fibre types. Fast-twitch fibres utilise the α-glycerophosphate (αGP) shuttle (12), whereas slow twitch fibres favour the malate-aspartate (M-A) shuttle. The αGP shuttle bypasses one phosphorylation site, and thus to produce the same amount of ATP fast twitch fibres would require more oxygen than slow twitch fibres. During heavy exercise ($>T_{VE}$) a greater proportion of fast twitch fibres would be recruited than during moderate exercise ($<T_{VE}$). Thus, the $\dot{V}O_2$ slow component would represent an inefficiency of aerobic metabolism. The extra O_2 required by the fast twitch fibres to produce ATP may contribute to the slow component of $\dot{V}O_2$. The excess $\dot{V}O_2$ slow component has been shown to have a close relationship with the increase in blood lactate concentration measured over the same time period (11). Fast twitch fibre recruitment in exercise above $T\dot{V}_E$ would support this observation.

With respect to our study, the observation of a smaller $\dot{V}O_2$ slow component across age may relate to a decrease in muscle mass. Cunningham *et al.* (1997) reported a significant age-related decrease in body mass (0.45 kg·yr^{-1}) in a cross-sectional sample of men aged 55 to 85 yr. This loss of body mass was undoubtably related to a decrease in fat-free mass. Aniansson *et al.* (1986) have reported with increasing age a decrease in both the size and number of muscle fibres. In this respect the lower muscle mass in the old would relate to the smaller absolute work rate. With ageing there is also the possibility that the proportion of fast twitch fibres is reduced as some fast twitch fibres are lost and, through re-innervation, exhibit slow twitch properties (1, 9). In this regard one might expect a reduced $\dot{V}O_2$ slow component in the elderly as they have a relatively smaller pool of fast twitch fibres to recruit during heavy exercise. However, our finding was that for the same relative work intensity (as well as it could be equated) the $\Delta\dot{V}O_2/\Delta WR$ relationship was not different between age groups. It is quite possible that in the older subjects of the age in this report there may not be an extensive preferential loss of fast twitch fibres. Also muscle mass loss may represent a relative loss in slow and fast twitch muscle fibres. The $\Delta\dot{V}O_2/\Delta WR$ relationship was not different between age groups suggesting that oxidative metabolism during heavy exercise was not any more or less inefficient in older than in younger subjects. Babcock *et al.* (1994) had a similar finding for older subjects during moderate exercise, with $\Delta\dot{V}O_2/\Delta WR$ ranging between 10.3 and 11.2 $(ml \cdot min^{-1}) \cdot W^{-1}$, and not significantly changed across age. These findings support ideas proposed by Taylor *et al.* (1987) who suggested that cardiovascular and metabolic function are well matched such that changes in $\dot{V}O_2$ could not take place without concomitant changes in each of the other two systems. That is, in the older adult decreases in cardiovascular function are matched by decreases in metabolic function so when exercising at the same relative intensity as younger adults the older adults display the same degree of efficiency.

In summary, we were able to observe a $\dot{V}O_2$ slow component during heavy exercise in older adults. The magnitude of the slow component was reduced across age, in relation to the decreased absolute work rate required to elicit the same relative intensity of exercise. $\Delta\dot{V}O_2/\Delta WR$ in heavy exercise was not different across age.

REFERENCES

1. Aniansson, A., G. Grimby, I. Krotkiewska, M. Krotkiewski, and A. Rundgren. Muscle strength and endurance in elderly people, with special reference to muscle morphology. In A. Asmussen and A. Jorgensen (Eds.), *International Series on Biomechanics Vol. 2A, Biomechanics VI*-(pp. 100–110). Amsterdam: Elsevier/North Holland Biomedical Press.
2. Aniansson, A., M. Hedberg, G.-B. Henning, and G. Grimby. Muscle morphology, enzymatic activity, and muscle strength in elderly men: A follow up study. *Muscle Nerve* 9: 585–591, 1986.
3. Armon, Y., D.M. Cooper, R. Flores, S. Zanconato, and T.J. Barstow. Oxygen uptake dynamics during high-intensity exercise in children and adults. *J. Appl. Physiol.* 70: 841–848, 1991.
4. Babcock, M.A., D.H. Paterson, D.A. Cunningham, and J.R. Dickinson. Exercise on-transient gas exchange kinetics are slowed as a function of age. *Med. Sci. Sports Exerc.* 26: 440–446, 1994.
5. Beaver, W.L., N. Lamarra, and K. Wasserman. Breath-by-breath measurement of true alveolar gas exchange. *J. Appl. Physiol.* 51: 1662–1675, 1981.
6. Casaburi, R., T.W. Storer, I. Ben-Dov, and K. Wasserman. Effect of endurance training on possible determinants of$\dot{V}O_2$ during heavy exercise. *J. Appl. Physiol.* 62: 199–207, 1987.
7. Cunningham, D.A., D.H. Paterson, J.J. Koval, and C.M. St. Croix. A model of oxygen transport capacity changes for independently living older men and women. *Can. J. Appl. Physiol.* 22: 439–453, 1997.
8. Davis, J.A., M.H. Franks, B.J. Whipp, and K. Wasserman. Anaerobic threshold alterations caused by endurance training in middle-aged men. *J. Appl. Physiol.* 46: 1039–1046, 1979.
9. Keh-Evans, L., C.L. Rice, E.G. Noble, D.H. Paterson, D.A. Cunningham, and A.W. Taylor. Comparison of histochemical, biochemical and contractile properties of triceps surae of trained aged subjects. *Can. J. Ageing* 11: 4, 412–425, 1992.
10. Paterson, D.H., and B.J. Whipp. Asymmetries of oxygen uptake at the on- and offset of heavy exercise in humans. *Journal of Physiology* 443: 575–586, 1991.
11. Poole, D.C., S.A. Ward, G.W. Gardner, and B.J. Whipp. Metabolic and respiratory profile of the upper limit for prolonged exercise in man. *Ergonomics* 31: 1265–1279, 1988.
12. Schantz, P.G., and J. Henriksson. Enzyme levels of NADH shuttle systems: measurements in isolated muscle fibres from humans of differing physical activity. *Acta Physiol Scand* 129: 505–515, 1987.
13. Taylor, C.R., R.H. Karas, E.R. Weibel, and H. Hoppeler. Adaptive variation in the mammalian respiratory system in relation to energetic demand. *Resp. Physiol.* 69: 1–127, 1987.

VOICE, BREATHING, AND THE CONTROL OF EXERCISE INTENSITY

R. C. Goode, R. Mertens, S. Shaiman, and J. Mertens

Exercise Science Unit, Faculty of Physical Education and Health
Departments of Physiology and Speech Pathology, Faculty of Medicine
University of Toronto and the Toronto Rehabilitation Centre

1. INTRODUCTION

Exercise duration and its effect on performance have been much studied[8,2]. Our laboratory[5] has demonstrated that six minutes on a daily basis with heart rate at 170 beats per min, over a three month period would result in a training effect, a reduction in heart rate for a given amount of work[12].

Our present interest is to develop simple techniques for the public such that they can determine an appropriate exercise intensity for themselves without resorting to laboratory procedures.

In 1957 Karvonen[12] employed heart rate as a means of measuring exercise intensity in a series of experiments and was able to demonstrate that there is a minimum exercise intensity required before a training effect occurs. The results of this study demonstrated that physical activity which resulted in a heart rate greater than 60% of the range from rest to maximum would result in a training effect.

While the public can be taught to measure their pulse, the skill is often not readily acquired. Comparison of an individual's ability to count heart rate pulsations in a ten second interval to the actual rate resulted in errors from a minimum of 12 to a maximum of 30 beats per min[4]. These errors were not corrected in some subjects who subsequently had received additional instruction and participated in "pulse-taking practice" sessions. Recommendations such as the use of electrocardiograms, sports testers, while correct in that the error of measurement would be eliminated, have the problem of cost and equipment. Pollock, Wilmore and Fox[10] concluded, regardless of how some people measure heart rate, such as by palpitation of the radial artery in the wrist or the carotid artery, they will not be able to determine it.

Karvonen's method[12] necessitates determination or estimation of maximum heart rate. Determination of maximum heart rate for the public is generally not recommended as it is expensive, impractical and possibly unsafe[1]. A further restriction of the use of heart

Advances in Modeling and Control of Ventilation, edited by Hughson *et al.*
Plenum Press, New York, 1998.

rate as means to determine the exercise intensity at which to exercise is based on the knowledge that the heart rate does not correlate well with anaerobic threshold and which in our definition is analogous to the first increase in blood lactate concentration with increasing exercise intensity and the Ventilatory Threshold.

The first series of experiments to be described were designed to investigate the hypothesis that if one can "hear their breathing" while exercising the subject will have reached at the minimum intensity of exercise for a training effect and is analogous to 60–90% of maximum heart rate[1].

2. METHODS

The experiments involved 19 male subjects, mean age 22 yrs (Table 1), who volunteered to complete seven experiments. The initial experiment was a familiarization experience during which subjects pedalled an electrically driven ergometer. (Ergomed 920).

This experiment was repeated on a subsequent day and the measurements recorded. This was followed by three test experiments on separate days during which the subject was asked to raise the hand when they could "hear your breathing", as the workload was increased by 25 watts at 60 sec intervals. Once the subject was aware of their breathing they were asked to continue pedalling, with no change in load, until 5 min was completed (Fig. 1). Heart rate was continually monitored by an electrocardiogram throughout the experiment.

In the final experiment subjects were asked to jog on an indoor 200 metre track. They were asked to jog at a pace such that they could "hear your breathing". Once the subject was aware of their breathing, they were asked to continue jogging until 10 minutes was completed. They were again asked to maintain the same sound of breathing throughout the exercise period. Heart rate was continually monitored by telemetry and recorded throughout the experiment. (Fig. 2)

Table 1. Subject characteristics

Subject #.	VO2max (ml/kg/min)	Sex:	Age:	Height (in).	Ht (m): (in*0.0254)	Weight (lbs):	Wt (kg). (lbs/2.205)	BMI (kg/m2).
1	57.46	M	18	68.25	1.73	130.20	59.05	19.65
2	47.4	M	18	67.50	1.71	167.50	75.96	25.84
3	43.97	M	22	66.50	1.69	135.00	61.22	21.46
4	55.57	M	30	74.02	1.88	185.22	84.00	23.76
5	45.72	M	20	70.50	1.79	168.00	76.19	23.76
6	55.62	M	22	68.90	1.75	147.85	67.05	21.89
7	50.41	M	19	63.00	1.60	137.00	62.13	24.26
8	50.99	M	22	65.00	1.65	111.00	50.34	18.47
9	46.53	M	23	71.00	1.80	161.00	73.02	22.45
10	71.18	M	22	67.00	1.70	148.00	67.12	23.18
11	50.38	M	22	63.50	1.61	118.00	53.51	20.57
12	53.69	M	20	72.50	1.84	164.00	74.38	21.93
13	42.15	M	20	74.00	1.88	220.00	99.77	28.24
14	31.06	M	21	66.00	1.68	195.00	88.44	31.47
15	54.59	M	20	64.50	1.64	137.00	62.13	23.15
16	49.65	M	24	70.50	1.79	160.00	72.56	22.63
17	49.26	M	23	69.75	1.77	175.50	79.59	25.36
18	42.33	M	24	67.00	1.70	142.00	64.40	22.24
19	33.3	M	24	70.00	1.78	195.50	88.66	28.05

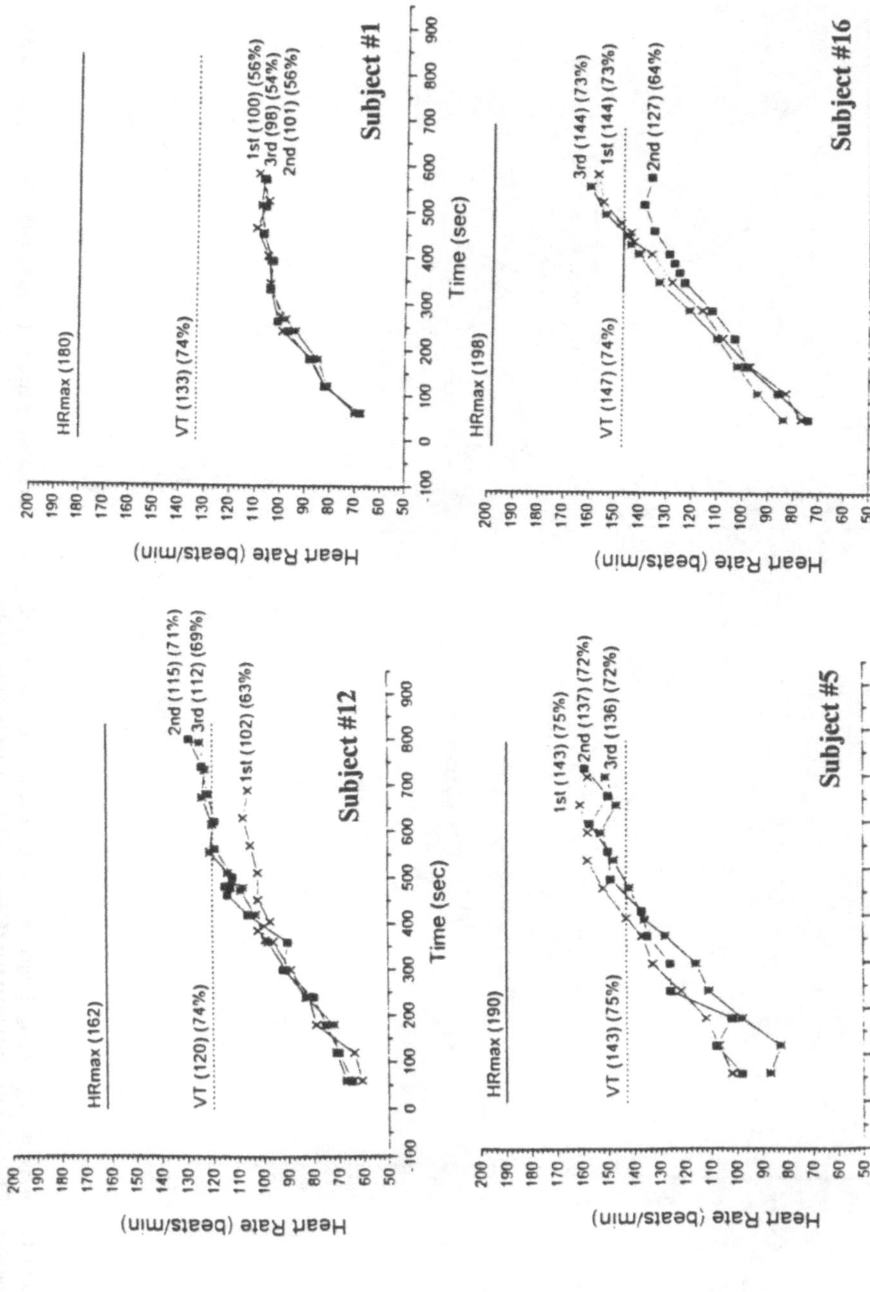

Figure 1. Heart rate response during cycle ergometry experiment. (): Heart rate when subject raised his or her hand; —×—: 1st submaximal cycle ergometer; —■—: 2nd submaximal cycle ergometer; *: 3rd submaximal cycle ergometer; ·······: ventilatory threshold (VT); ——: maximum heart rate (HRmax).

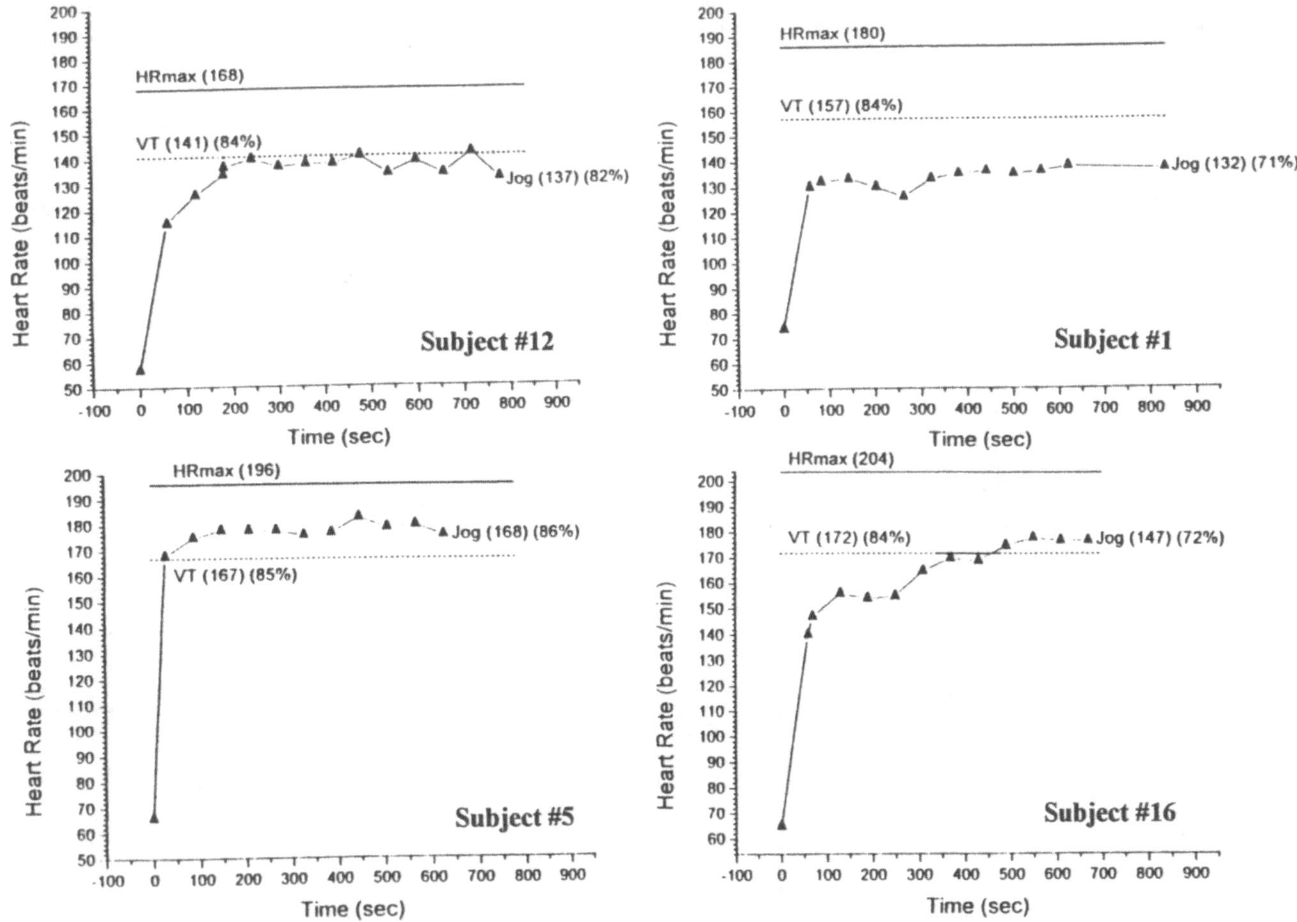

Figure 2. Heart rate response during jogging. Subject selected pace at his "hear your breathing" point (indicated by arrow). (): Heart rate when subject raised his or her hand; ········: ventilatory threshold (VT); ——: maximum heart rate (HRmax); —▲—: field trial (jogging).

VO_2 consumption values (Table 2) while pedalling the ergometer during the "hear" your breathing experiments were determined by matching the heart rates determined in the breathing sound experiments with the heart rates and oxygen consumption values obtained during VO_2 max determinations.

3. RESULTS

The results of eight experiments on the ergometer and while jogging are displayed in Figs. 1 and 2.

The mean heart rate while pedalling the ergometer for minute two to five was 127 b/min or 70 ± 0.03% of maximum heart rate. The mean heart rate for jogging, determined from the last five minutes of the 10 minute exercise period was 158 beats per min. Table 3, estimated to be 70% of maximum heart rate. A maximum heart rate, while running was not determined for the joggers. The maximum heart rate determined from the ergometer maximal test was adjusted upward by six beats[3].

The oxygen consumption data indicates the subjects were at 50% of their VO_2 max, on the cycle ergometer during the in level two to five minutes.

VO_2 max scores, maximum heart rates and Ventilatory Thresholds determined on a treadmill are reported to be significantly higher than when obtained from a maximal cycle ergometer test[3]. Buckfuhren et al reported the mean Ventilatory Thresholds to be some 13% higher for 12 subjects. If one calculates a similar increase in Ventilatory Threshold for the (jogging) subjects described in this experiment it could raise the obsered Ventilatory Thresholds while pedalling the cycle ergometer from ~136 b/min to ~154 b/min (Fig. 1). While jogging it is possible that our subjects at a "hear your breathing" pace were at or close to their Ventilatory Threshold" (~158 b/min).

4. DISCUSSION

These results suggest that when a subject can "hear your breathing" while cycling or jogging they are at or near their Ventilatory Threshold and their heart rate is above the minimum and below the maximum for a training effect (60 to 90% of maximum heart rate) as recommended by the ACSM (1990).

An earlier study from our laboratory[6] was designed to investigate if the phrase "just capable of talking" approximated the respiratory compensation threshold, (RCT),[14]. Thirty male subjects age 20–30 yrs cycled for two minutes at 50 watts, the load was increased by 25 watts each minute to a power output greater than that required to reach the RCT as defined by Wasserman. Subjects were instructed to read three sentences from a pre-rehearsed cue card at three steady states; rest, half the distance from the RCT and at the RCT. The voice was recorded and later played back in a random order. Eight listners were asked to identify when the subject was "just capable of talking". The correlation between the phrase "just capable of talking" and the RCT was 0.91.

Another interest of our laboratory was the establishment of a simple feedback mechanism that could ensure that physical activity, especially for beginners, was well within their aerobic capacity (below the Respiratory Compensation Threshold). Grayson (personal communication) in 1974, when presented with the problem suggested a phrase that had been employed by himself and other climbers in Scotland as early as 1937, "climb no faster than you can talk". We applied this phrase to any activity, that is, one should be able to "talk" while ex-

Table 2. VO_2 consumption values during cycle ergometry

Subject #:	1	2	3	4	5	6	8	9	10	12	13	14	15	16	17	18	19	Group Average:
HRmax:	180	162	171	182	190	196	184	193	178	162	182	193	187	198	178	194	180	
Time:																		
indicated	100	107	128	119	139	117	130	106	82	110	110	113	128	138	94	154	131	
1min	104	112	129	120	148	122	136	112	85	114	118	116	139	148	97	159	145	
2min	104	114	135	125	152	125	142	115	85	114	117	118	143	152	98	164	154	
3min	108	113	138	127	156	127	144	118	79	117	119	120	151	136	97	160	156	
4min	106	113	139	129	153	126	146	119	87	118	119	120	148		98	163	153	
5min	107	113	137	130	156	126	147	121	89	120	120	120	138		99	162		
Mean(2-5min):	106	113	137.3	128	154	126	145	118	85	117	119	119.5	145	136	98	162.3	154	127
%of Hrmax:	0.59	0.70	0.80	0.70	0.81	0.64	0.79	0.61	0.48	0.72	0.65	0.62	0.78	0.69	0.55	0.84	0.86	0.70
VO2:	1275	1683	1607	2174	2104	1360	1396	1523	684	2410	1718	1363	2270	1937	1321	1901	2292	
maxVO2:	3396	3603	2682	4668	3474	3729	2550	3397	4769	4027	4215	2764	3384	3625	3930	2752	2959	
%VO2	0.38	0.47	0.60	0.47	0.61	0.36	0.55	0.45	0.14	0.60	0.41	0.49	0.67	0.53	0.34	0.69	0.77	0.50

Table 3. Heart rate response during field trial (jogging)

Heart Rate for each subject during Field Trial

Subject #:	**1**	**2**	**3**	**4**	**5**	**6**	**7**	**8**	**9**	**10**	**11**	**12**	**13**	**14**	**15**	**16**	**17**	**18**	**19**
Rest HR:	74	82	77	65	66	57	69	71	78	63	64	57	69	99	74	65	63	76	66
Hear:	132	136	156	137	168	141	123	153	157	114	163	137	149	182	176	147	137	167	146
1min	133	139	158	141	175	137	139	160	164	118	172	140	159	190	179	156	148	175	151
2min	130	139	158	131	178	135	138	154	166	122	174	137	160	190	182	154	150	178	152
3min	126	141	156	143	178	137	142	146	170	124	181	138	160	195	184	155	151	175	157
4min	133	141	160	144	178	140	143	149	173	128	177	138	165	196	179	165	155	180	160
5min	135	142	160	142	176	148	144	148	172	122	174	141	166		180	170	153	180	160
6min	136	141	163	145	177	142	146	151	171	127	172	134	166		176	169	154	175	162
7min	135	142	164	150	183	145	146	152	175	134	183	139	168		176	175	158	176	163
8min	136	140	165	145	179	144	151	151	179	130	176	134	167		176	178	157	175	165
9min	138	142	161	145	180	143	149	148	177	138	177	142	166		177	177	157	177	164
10min	137	146	164	151	176	144	148	153	179	134	184	132	167		174	177	161	182	160

ercising in order to ensure the activity is below the respiratory compensation threshold. This was introduced to the public in a series of publications[7,8,9].

REFERENCES

1. American College of Sports Medicine. The Recommended Quantity and Quality of Exercise for Developing and Maintaining Cardiorespiratory and Muscular Fitness in Healthy Adults. Med. Sci. Sports Exercise 265–274, 1990.
2. Astrand, P.O. and K. Rodahl. Textbook of Work Physiology, 3rd Ed. New York: McGraw Hill, 1986.
3. Buckfuhren, M.J., J.E. Hansen, T.E. Robinson, D.Y. Sue, K. Wasserman and B.J. Whipp. Optimizing the exercise protocol for cardiopulmonary assessment. J. Appl. Physiol. 55(5):1558–1564, 1983.
4. Cunningham, G.R. and J. Glenn, Evaluation of the Canadian Home Fitness Test in middle-aged men. Can. Med. Assin. J. 117:346–349, 1977.
5. Goode, R.C. Active Living. Queen' Printer, Prov. of Ontario ISBN 0–7729–8120–5, 1994. Goode, R. C., A. Virgin, T.T. Romet, P. Crawford, J. Duffin, T. Pallandi, and Z. Woch. Effects of a short period of physical activity in adolescent boys and girls. Can. J. Appl. Sports Sci. 1:241–250, 1976.
6. Goode, R.C., A. Sharp and S. Shaiman. Speech, Breothlessness, and monitoring Exercise Intensity. Can. J. Physio. Pharma Col. 72:AVI, 1993.
7. Goode, R.C. Physical Fitness in Your School. The Lung Associateion, Toronto, Ontario, 1976.
8. Goode, R.C. A Guide to Personal Fitness. Queen's Printer, Prov. of Ontario, 1978.
9. Goode, R.C. Active living. Ministry of Tourism and Recreation, Toronto, Ontario. ISBN 0–7729–8120–5, 1994.
10. Gordon, N.F. and C.B. Scott. Exercise intensity prescription in cardiovascular disease: theoretical basis for anaerobic threshold determination. J. at Cardiopul. Rehab. 15:193–196, 1995.
11. Hartung, G.H., M.H. Smolensky, R.B. Harrist, R. Rangel, and C. Skrovan. Effects of varied durations of training on improvement in cardiorespiratory endurance. J. Human Ergol. 6:61–68, 1977.
12. Karvonen, M., K. Kentala, and O. Mustala. The effects of training heart rate: a longitudinal study. Ann. Med. Exp. Biol. Fenn 35:307–315, 1957.
13. Pollock, M.L., J.H. Wilmore, and S.M. Fox. Exercise in Health and Disease: Evaluation and Prescription for Prevention and Rehabilitation. Philedelphia, PA: Saunders, 1974.
14. Wasserman, K., G.G. Burton, A.L. Van Kessel. The physiological significance of the "anaerobic threshold" Physiologist, 7:297.

PULMONARY TRAINING MAY ALTER EXERTIONAL DYSPNEA AND FATIGUE VIA AN EXERCISE-LIKE TRAINING EFFECT OF A LOWERED HEART RATE

George D. Swanson

Department of Physical Education and Exercise Science
California State University
Chico, California

1. INTRODUCTION

Recent small studies utilizing respiratory muscle training have incorporated a voluntary hyperventilation maneuver via an open chamber gas mixing system (to maintain isocapnic conditions). These studies suggest that the time to exercise exhaustion may be increased at exercise levels near but above the so-called anaerobic threshold after several weeks of intense pulmonary training (1,2).

Two mechanisms have been proposed to explain such findings (15). First, pulmonary training might reduce exertional dyspnea and, therefore, improve exercise fatigue tolerance (9). Furthermore, respiratory muscle training results in increased ventilatory endurance. This improved ability to maintain high levels of ventilation might help to maintain arterial blood gasses and pH homeostasis during prolonged intense exercise. Enhanced homeostasis may improve exercise fatigue tolerance.

Respiratory muscle fatigue can occur during heavy exercise (3,8,10) and respiratory muscle fatigue (i.e. induced by prolonged isocapnic hyperpnea) impairs subsequent exercise performance (11). However, some investigators argue that this respiratory muscle fatigue does not limit exercise tolerance because pulmonary training does not improve fatigue time at exercise near maximal oxygen consumption (4,6,12). One interpretation is that respiratory muscle endurance is more important during prolonged, moderate exercise as compared to short-term high intensity exercise (15).

We shall offer an additional explanation. Suppose pulmonary training induces a small exercise-like training effect of a lowered heart rate for a given exercise load. Using a power duration curve (14) analysis, the consequence for exercise duration time will be

Advances in Modeling and Control of Ventilation, edited by Hughson *et al.*
Plenum Press, New York, 1998.

most obvious near but above the critical power point. Furthermore, an effect on endurance time will not be observable near maximal oxygen consumption.

From this point of view, we can proceed to motivate the pulmonary training, heart rate connection, by analysis of available data from the literature. We shall demonstrate that the effects of pulmonary training should be tested at exercise levels near the critical power point so as to maximize effect sensitivity. Furthermore, we shall report the results of a pilot study, which utilize a re-breathing type pulmonary training, to demonstrate the pulmonary training heart rate connection.

2. PRELIMINARY ANALYSIS

Available data from the literature can be used to motivate the concept that pulmonary training may induce an exercise-like training effect. In 1991, Boutellier and Piwko published a small study with four sedentary subjects designed to investigate the respiratory system (breathing) as an exercise limiting factor (2). The four subjects were pre-tested for endurance time at an exercise cycle work rate level near their so-called anaerobic threshold. Subsequently, they underwent pulmonary training for four weeks by breathing daily at approximately 90 l/min for 30 min. Isocapnic conditions were maintained via an open circuit gas-mixing chamber. Their results indicate that cycle endurance was improved on average from 26.8 min to 40.2 min after the four-week pulmonary training period.

The corresponding heart rate data are shown in Fig. 1. Note the initial data point for each subject represents the heart rate at exhaustion before the pulmonary training period. After pulmonary, subjects 2,3, and 4 follow a characteristic pattern of a decreased heart rate at the given exercise workload when the pre-pulmonary training endurance time ends. At the endurance time after pulmonary training, the heart rate returns to near pre-pulmonary training levels. Thus, for three out of four subjects, pulmonary training lowers heart rate at the given exercise level until exhaustion. Its interesting to note that subject number 1, who did not show this pattern, had the lowest exercise capacity of the four sedentary subjects.

If in fact, four weeks of pulmonary training induces a small exercise-like training effect, then the test exercise level can be optimized to maximize effect sensitivity. A power duration curve can be utilized in the analysis (14). We shall assume that the exercise-like training effect of a lowered heart rate raises the critical power, defined as the asymptote that separates the aerobic and so-called anaerobic regions. That is, in the anaerobic region,

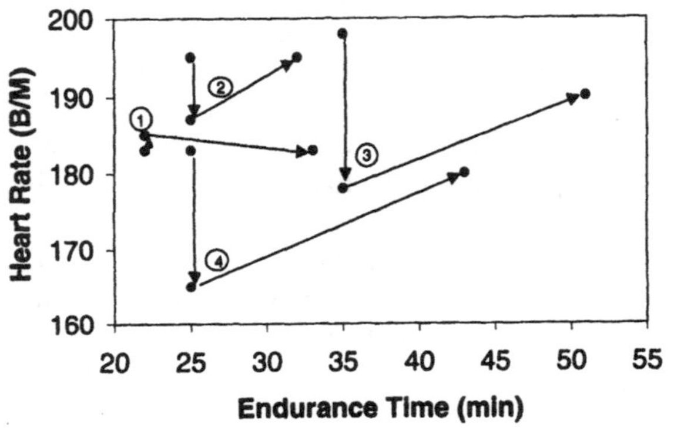

Figure 1. Heart rate data showing pattern of heart rate decrease after pulmonary training for three out of four subjects (2).

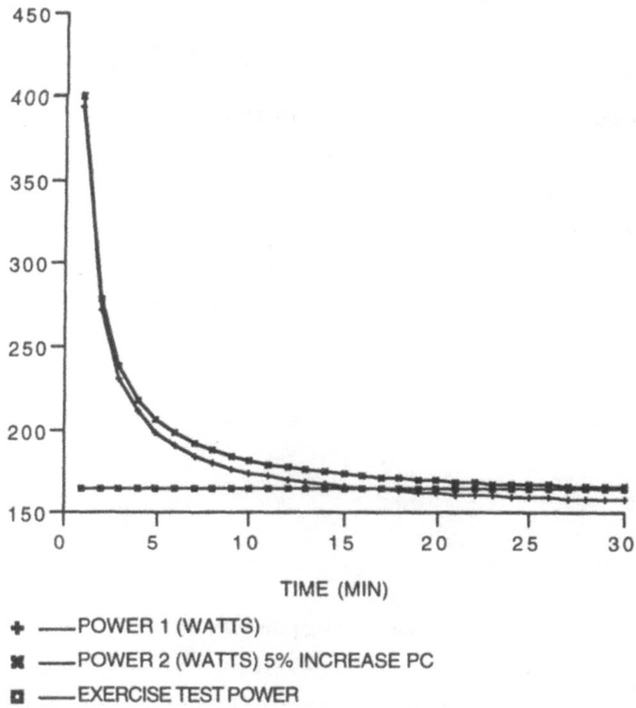

Figure 2. Power duration curve (14).

a rectangular parabola characterizes the exercise endurance time whose asymptote (parallel to the time axis) is the critical power point.

A typical power duration curve is shown in Fig. 2. Note two curves are shown with slightly different (5 %) critical power points. As shown, a given exercise test near, but above the critical power point maximizes the sensitivity of the hypothetical pulmonary training effect. Furthermore, an exercise test near maximum work capacity would mask the hypothetical pulmonary training effect.

3. PILOT STUDY

A new re-breathing pulmonary trainer has been developed at our laboratory. As shown in Fig. 3, the pulmonary trainer consists of two "PEPSI" bottles interconnected back-to-back so as to slide back and forth for an adjustable re-breathing volume. A medical facemask has been attached to one end and an outlet hole created in the other end so as to create a non-resistance respiratory exercise. This device yields near isocapnic conditions under a hyperventilation maneuver.

In our pilot-study, eight male cyclists volunteered with four dropping out for various reasons. The remaining four subjects maintained their regular exercise training schedule while adding pulmonary training at twenty minutes a day, five days a week, for six weeks. These subjects were tested for time to exercise exhaustion before and after six weeks of pulmonary training. The results for average heart beat data are listed in Table 1.

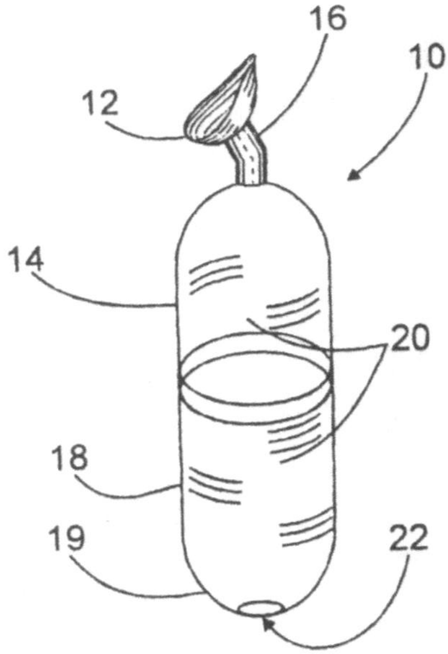

Figure 3. Re-breathing pulmonary trainer.

Our previous studies have indicated that isocapnic voluntary hyperventilation elevates heart rate (17). For the present study, seven male subjects volunteered to use the pulmonary re-breathing trainer for the twenty-min isocapnic hyperventilation maneuver. Heart rate was measured during a five-phase test (Fig. 4). Phase 1 consisted of 10 min of resting breathing while the subject was seated. Phase 2 consisted of 10 min of resting data while breathing through the re-breathing pulmonary trainer. Phase 3 and phase 4 consisted of 20 min of near isocapnic hyperventilation and phase 5 consisted of 10 min of recovery breathing through the device.

The corresponding heart rate response is shown in Fig. 4. Note that breathing through the device elevates the heart rate by about 10 beats/min on average. The hyperventilation maneuver elevates the heart rate by another 10 beats/min with a gradual decline over the twenty-minute hyperventilation period.

4. DISCUSSION

Endurance exercise training facilitates a peripheral muscular/cellular adaptation as well as a central circulatory adaptation. The result is a lower heart rate and ventilation for

Table 1

	Pre-pulmonary training		Post-pulmonary training	
Power (W)	Time (min)	Heart rate	Time (min)	Heart rate
226	48.0	179	68.0	169
226	65.1	157	70.4	152
240	53.1	167	87.2	150
240	24.2	180	31.0	167

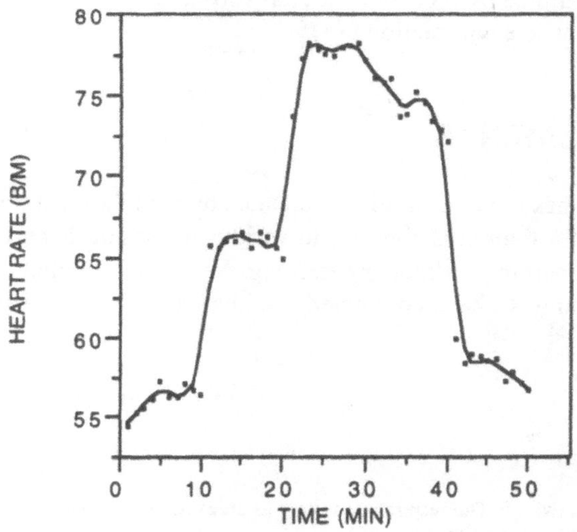

Figure 4. Heart rate response for pulmonary training maneuver. —— : Smoothing spline fit, lambda = 1.

a given workload, especially above the lactate threshold. However, does exercise endurance training yield a respiratory muscular/cellular adaptation?

To date, there are no published reports regarding the effects of endurance exercise training on cellular alterations in human respiratory muscle (15). Alternatively, endurance exercise training does appear to result in improved ventilatory muscle endurance as evidenced by elevated maximal sustained ventilation and an increased maximal voluntary ventilation (13,16). However, subjects developing a more efficient breathing strategy and, therefore, a reduced energy requirement of ventilation could also explain these observations (15).

In fact, there may be an extensive pulmonary muscle reserve. If this is the case, then endurance exercise training (and pulmonary training via voluntary isocapnic hyperventilation) may not produce a training adaptation in respiratory muscle per se. That is, breathing practice may lead to a more efficient breathing pattern, which may in fact alter the associated cardiac response to breathing.

The effects of the parasympathetic nervous system on the sinus node are mediated solely by efferent vagal neural stimulation (5). During the inspiratory phase of resting breathing, the activity of cardiac vagal efferent nerve fibers is greatly reduced. Thus cardiac vagal efferent discharge occurs predominately during expiration. The periodic effects of cardiac vagal efferent discharge in the sinus node are mediated and "smoothed" by acetylcholine release. The result is the characteristic "respiratory sinus arrhythmia."

During isocapnic hyperventilation, the elevated heart rate is most likely due to parasympathetic withdrawal. If pulmonary training leads to a more efficient breathing pattern during endurance exercise, then there may be less parasympathetic withdrawal and a lower heat rate with a corresponding increase in stroke volume ---an exercise-like training effect. In addition, a more efficient breathing pattern may lead to a reduced pulmonary blood flow requirement, leaving potentially a higher blood flow available for peripheral muscle and cooling needs as the point of exhaustion is approached (7).

Furthermore, essentially all of the spiritual traditions associate an increase of breathing with an increase of "life force" (prana, chi etc.). Therefore, pulmonary training may enhance exercise tolerance by increasing a vital substance necessary for prolonged peripheral and pulmonary muscle contraction (18).

ACKNOWLEDGMENTS

The author wishes to acknowledge graduate students Paul Daniels, Chuck McKenna and Richard Valdez for their contributions to various phases of this project. Chuck was co-inventor of the re-breathing pulmonary training device. Paul conducted the exercise endurance pilot study and Richard conducted the heart rate, pulmonary training maneuver, study.

REFERENCES

1. Boutellier U. and Piwko P. The respiratory system as an exercise limiting factor in normal sedentary subjects. Eur J Appl Physiol. 64:145–152, 1992.
2. Boutellier U., R. Buchel, and A. Kindest, A. Kundert and S. Spengleu. The respiratory system as an exercise limiting factor in normal trained subjects. Eur J Appl Physiol. 65: 347–353, 1992.
3. Coast R J., P. S. Clifford, T. W. Henrich, J. Stray-Gunderson and R. L. Johnson. Maximal inspiratory pressure following maximal exercise in trained and untrained subjects. Med Sci Sports Exerc. 22:811–815, 1990.
4. Fairbarn M. S., K. C. Coutts, R. L. Pardy, and D. C. McKensic. Improved respiratory muscle endurance of highly trained cyclists and the effects on maximal exercise performance. Int J Sports Med. 12: 66–70, 1991.
5. Goldberger, J.J, and A. H. Kadish. Influence of sympathetic and parasympathetic maneuvers on heart rate variability. In: Noninvasive Electrocardiology, edited by A.J. Moss and S. Stem. Philadelphia, PA: Saunders, 1996, p. 210.
6. Hanel B. and N. Secher. Maximal oxygen uptake and work capacity after inspiratory muscle training: a controlled study. J Sports Sci. 43–52, 1990.
7. Harms, C. A., M. A. Babcock, S. R. McClarin, D. F. Pegelow, G. A. Nickele, W. B. Nelson and J. A. Denpsey. Respiratory muscle work compromises leg blood flow during maximal exercise. J. Appl. Physiol. 82: 1573–1583, 1997.
8. Johnson, B., M. Babcock, and J. Dempsey. Exercise-induced diaphragmatic fatigue in healthy humans. J Physiol (London). 460: 385–405, 1993.
9. Killian K., N. Jones. Respiratory muscles and dyspnea. Clin. Chest Med. 9:237–248, 1988.
10. Mador M. J., U. J. Magalang, A. Rodis and T. J. Kufel. Diaphragmatic fatigue after exercise in healthy subjects. Am Rev Respir Dis. 148:1571–1575, 1993.
11. Martin, B, M. Heintzelman and H. Chen. Exercise performance after ventilatory work. J Appl Physiol 52:1581–1585, 1982.
12. Morgan D. W., W. M. Kohrt,B. J. Bates and J. S. Skinner. Effects of respiratory muscle endurance training on ventilatory and endurance performance of moderately trained cyclists. Int J Sports Med. 8:88–93, 1987.
13. O'Kroy J., and J Coast. Effects of flow and resistive training on respiratory muscle strength and endurance. Respiration 60:279–283, 1993.
14. Poole D. C., S. A. Ward, and B. J. Whipp. The effects of training on the metabolic and respiratory profile of high intensity cycle-ergometer exercise. Eur J Appl Physiol 59:421–429, 1990
15. Powers, S. K., J. Coombes and H. Demirel. Exercise training-induced changes in respiratory muscles. Sports Med. 24:120–131, 1997
16. Robertson E., J. Kjeldgaard. Improvement in ventilatory muscle function with running. J Appl Physiol 52:1400–1406, 1983.
17. Swanson, G. D., D. S. Ward, and J. W. Bellville. Posyhyperventilation isocapnic hypernea. J Appl Physiol 40:592–596, 1976.
18. Swanson, G. D. Redundancy structure in respiratory control. In: Control of Breathing and Its Modeling Perspective. Edited by Y. Honda, Y. Miyamoto, K. Konno and J.G. Widdecombe. New York: Plenum, 1992, p.171.

INDEX

FSC
www.fsc.org
MIX
Papier | Fördert
gute Waldnutzung
FSC® C083411

Kolibri 360 GmbH
Ferdinand-Jühlke-Straße 7
99095 Erfurt, Deutschland
produktsicherheit@kolibri360.de